THE MIGRATIONS OF HAWKS

THE MIGRATIONS OF
H A W K S

BY

Donald S. Heintzelman

INDIANA UNIVERSITY PRESS
Bloomington and Indianapolis

Library of Congress Cataloging-in-Publication Data

Heintzelman, Donald S.
    The migrations of hawks.

    Rev. and enl. ed. of: Autumn hawk flights. 1975.
    Bibliography: p.
    Includes index.
    1. Hawks—Migration.   2. Bird watching—North America.
3. Birds—Migration.   4. Birds—North America—
Migration.   1. Heintzelman, Donald S.   Autumn hawk
flights.   II. Title.
QL696.F3H44   1986   598'.916   85-45412
ISBN 0-253-33821-2

1 2 3 4 5 91 90 89 88 87

To the memory of my parents,
who always enjoyed hearing about each day's
hawk flight and helped to make this book possible.

CONTENTS

vii

PART THREE
Hawk Migrations and Weather Conditions

PART FOUR
Migration Routes, Geography, and Hawk Counts

FIGURES AND MAPS

Figures

Maps

PREFACE

It is more than ten years since the first edition of this book appeared under the title *Autumn Hawk Flights*. During that decade substantial growth and interest in hawk migrations and hawk watching developed in North America and elsewhere (Peterson, 1985). The volume of published hawk migration literature also increased significantly so that this new information, combined with earlier publications, produced this book's long literature cited section. In itself, that compilation is a significant and useful reference tool. Thus the time arrived to update and revise the book, now titled *The Migrations of Hawks* because autumn and spring flights are included.

My objective in this book is to organize and summarize the most important New World hawk migration literature and evaluate the current (through 1984) status of our knowledge of these migrations. I also include some previously unpublished analyses of certain hawk migration data from Bake Oven Knob, Pennsylvania, where I have observed hawk flights since 1956, and call attention to certain specific problems or field study opportunities at various other locations although my intent is not to provide a detailed catalog ·of them.

Appreciation is extended to the editors of the following journals for permitting me to reproduce illustrations from material originally appearing in their publications: *Auk, Chat, Journal of Field Ornithology, Raptor Research, Scientific American, Search, Western Birds,* and *Wilson Bulletin.* The authors in whose papers the illustrations appeared also are extended thanks, identified in the legends used with the illustrations, and more fully in the literature cited section.

The Pennsylvania State University Press also kindly granted permission to reprint the hawks per hour and hawks per day rating scales I presented in *A Guide to Hawk Watching in North America.*

A special note of appreciation is due Mark R. Fuller, Lynda Garrett, and Chandler S. Robbins for invaluable assistance in the offices and library at Patuxent Wildlife Research Center, Laurel, Maryland.

Other persons who provided particularly helpful assistance include D. Amadon, A. H. Bergey, A. Bihun, Jr., I. Black, J. Bond, A. Brady, F. Brock, E. A. Choate, G. B. David II, H. E. Douglas, H. Drinkwater, E. Eisenmann, T. W. Finucane, F. T. Fitzpatrick, C. L. Garner, E. W. Graham,

A. Grout, P. Grout, R. L. Haines, G. Hanisek, R. R. Hendrick, B. C. Hiatt, C. M. Hoff, P. B. Hofslund, J. B. Holt, Jr., D. Hopkins, J. A. Jacobs, J. W. Key, D. L. Knohr, B. Lake, C. L. Leck, W. Lent, G. H. Lowery, Jr., R. and A. MacClay, E. R. Manners, F. Mears, E. O. Mills, L. and N. Moon, I. Morrin, R. Moser, B. G. Murray, Jr., A. Nagy, O. S. Pettingill, Jr., E. L. Poole, R. H. Pough, B. Reed, C. J. Robertson, J. L. Ruos, F. Scott, A. A. Sexauer, R. W. Smart, V. Smiley, R. W. Smith, D. Steffey, G. Steffey, S. Thomas, F. Tilly, J. Tobias, W. C. Townsend, A. Webster, C. Wellman, C. E. Wonderly, E. Wonderly, and K. Zindle.

Maurice Broun introduced me to the wonder of hawk migrations at Hawk Mountain, Pennsylvania, and many years later still retained his wonder of hawk flights at Bake Oven Knob and elsewhere along the Kittatinny Ridge. Thus, in many respects, he directed me into a lifetime of study and exploration of the migrations of hawks.

Portions of my field studies, reports on which already have been published, were conducted while I served as Associate Curator of Natural Science and Curator of Ornithology at the William Penn Memorial Museum and the New Jersey State Museum.

ALLENTOWN, PA.   DONALD S. HEINTZELMAN

PART
ONE

*History and
Field Study Methods*

C H A P T E R   1

Introduction

The migrations of hawks in the Americas have been observed for almost five centuries although detailed field studies of hawk flights, especially in North America, began only during the past 90 years. Despite the brief period of time devoted to modern hawk migration studies, a rich literature developed as is demonstrated by the number of entries in the Literature Cited section of this book. In order to understand more fully how this body of knowledge developed it is useful to consider a brief history of hawk watching and migration study in North America.

HISTORICAL BACKGROUND

Considered from a broad point of view, hawk watching and migration study in North America exhibits a trend in which human use of birds of prey began as objects of curiosity, later changed to conservation campaigns to stop the slaughter of hawks and other raptors, and now promotes nonconsumptive educational, recreational, and scientific uses of hawks—a trend not unlike that for whale watching and migration study (Heintzelman, 1983b). When examined more closely, however, Robbins (1975) identified five more or less distinct time periods that divide the history of hawk watching and migration study in North America.

Curiosity Period (1500–1895)

The first person in the Americas to record information on hawk migrations was the naturalist Oviedo. During the period from 1526 to 1535, he noted hawks migrating across the West Indies and presented his data in *Historia general y natural de las Indias, islas y tierre-firme del mar oceano* (Baughman, 1947). Two centuries passed before another reference to hawk migrations was published in the New World. Then, in 1756, an account of what was probably a very large flight of Broad–winged Hawks over what is now Brattleboro, Vermont, on 20 or 21 September, appeared in the *New Hampshire Gazette* for 14 October 1756 (Goldman, 1970). A century later, on 17 September 1858 in Mercer County, Ohio, another large flight of migrating hawks (probably Broad-wings) was observed for 90 minutes as the birds followed a north-northeast to south-southwest direction (Deane, 1905), and a few years later spring and autumn hawk flights were observed in Iowa and Massachusetts (Allen, 1871). On 29 September 1886 at Germantown, Pennsylvania (part of Philadelphia) a flight of 250 hawks also was observed by Stone (1887), and at Lake Forest, Illinois, during the autumns of 1893–1895 additional observations were made of hawk migrations (Ferry, 1896). Townsend (1892), too, briefly described annual autumn hawk flights at Bay Ridge, Long Island, New York, as did Dunn (1895) for coastal Rhode Island. Cary (1899) also observed large migrations of Swainson's Hawks on 29 August 1899, in Nebraska, and the observations of hawk flights in New York (Barbour, 1908) also were isolated and not subject to a detailed analysis. Unusual, however, were Howland's (1893) comments on spring hawk migrations at Montclair, New Jersey.

Robbins (1975) collectively considered these early, scattered hawk migration observations as the curiosity period in the history of hawk migration study because they were neither consolidated into a body of knowledge nor studied to determine their significance.

Initial Study Period (1895–1930)

Within a few years of the observation of the hawk flight at Germantown, Pennsylvania, however, the curiosity period changed into the initial study period (Robbins, 1975). Although Ferry (1896) made a modest attempt to analyze his Lake Forest, Illinois, hawk migration data, the initial study period really began when Trowbridge (1895) published his careful analytical study of major hawk flights along the coastline of Connecticut. This was followed by the work of Burns (1911) whose classic monograph of the Broad-winged Hawk included 16 pages devoted to a summary of the migratory status of that species.

A decade later, Ferguson and Ferguson (1922) published the results of their five-year study of autumn hawk migrations at Fishers Island, New York. During that same year the spring hawk migrations at Whitefish Point, Michigan, received some study by Magee (1922). Each of these studies advanced our knowledge of hawk migrations and their relations to weather conditions and/or geography.

Conservation Period (1930–1950)

Robbins (1975) recognized 1930 to 1950 as the third time period in the history of North American hawk migration studies. This was a time of seriously conflicting human uses of birds of prey, such as hawk shooting versus hawk watching. In many respects, however, it also was the Golden

Age of Hawk Watching during which much of the foundation of current hawk watching and hawk migration study was constructed. At Cape May Point, New Jersey, for example, the pioneering hawk migration studies made by Allen and Peterson (1936) prepared the foundation for current work there whereas Stone (1937) made comprehensive ornithological studies at that site.

In Pennsylvania, in 1934, Hawk Mountain Sanctuary was established, hawk shooting was stopped at that site, and basic studies of hawk flights along the Kittatinny Ridge were begun (Broun, 1935, 1939, 1949a; Poole, 1934).

Above all, however, the period from 1930 to 1950 was a time devoted to raptor conservation—especially to stopping the widespread shooting of hawks and other birds of prey along migration routes and at concentration areas. Various people and organizations played key roles in these conservation efforts. In New Jersey in the 1930s, for example, the efforts of Robert Porter Allen, Roger Tory Peterson, William J. Rusling, and others working in cooperation with the National Audubon Society resulted in legislative protection for some hawks and a halt to hawk shooting at Cape May Point and vicinity (Allen and Peterson, 1936; Stone, 1937). At the Montclair Hawk Lookout Sanctuary in northern New Jersey, hawk shooting also was stopped when the town prohibited the discharge of firearms (Breck, 1960b).

It was along the Kittatinny Ridge in eastern Pennsylvania, however, during the period 1932 through 1956, that classic hawk conservation efforts occurred. Because of several important articles exposing the slaughter of hawks at what is now Hawk Mountain Sanctuary near Drehersville (Collins, 1933; Pough, 1932) various conservationists including Rosalie Edge and the Emergency Conservation Committee established Hawk Mountain Sanctuary and hired Maurice Broun as curator. He was assisted by his wife Irma. The Brouns achieved now legendary success in stopping hawk shooting at Hawk Mountain (Broun, 1949a), turning a hawk slaughter ground into an internationally celebrated sanctuary. Elsewhere along the Kittatinny Ridge, however, hawk shooting continued unabated through 1956, after which it was stopped by a new hawk protection law in Pennsylvania (Broun, 1956a, 1956b; Heintzelman, 1975, 1983d). An informal organization of conservationists known as the Pennsylvania Hawk Committee played a key role in securing the new 1957 hawk protection law (Senner, 1984a: 31–32).

During the conservation period the National Audubon Society and various regional conservation organizations and bird study societies also worked to secure hawk protection laws including those for the Bald Eagle, the national bird of the United States of America. In addition, a stream of educational materials was issued including classic books such as Maurice

Broun's (1949a) *Hawks Aloft*. Thus the conservation period witnessed scientific studies of hawk migrations as well as intensive efforts designed to stop hawk shooting everywhere but particularly at migration concentration_ areas.

Data-Gathering Period (1950–1970)

Although the necessary work of the conservation period was not completed, and actually continues still, Robbins (1975) correctly identified the time period 1950 to 1970 as a major hawk migration data-gathering period in North America. In particular, large-scale, cooperative, regional, or statewide hawk migration study projects were organized in a number of eastern states. These were the forerunners of the Hawk Migration Association of North America established in 1974.

In 1949, for example, the Maryland Ornithological Society conducted statewide hawk counts in September at 20 locations and in 1950 at 21 locations (Robbins, 1950). In 1950, in Tennessee, another regional autumn hawk migration project was launched (Behrend, 1950b, 1951), and later similar projects were conducted in West Virginia (DeGarmo, 1953).

From 1951 through 1955, the Bureau of Sport Fisheries and Wildlife (now the U.S. Fish and Wildlife Service) followed a recommendation made by Aaron M. Bagg (1947) and coordinated short-term September hawk migration studies in 29 states and Canadian provinces with particular emphasis on the migratory movements of Broad-winged Hawks (Robbins, 1975).

During the autumns of 1958 and 1959, George W. Breck (1959, 1960a, 1960b, 1960c) organized annual across-the-state hawk watches in New Jersey and parts of New York and Pennsylvania which provided a considerable quantity of short-term hawk migration data. These tri-state projects continued annually for a number of years.

At the 1964 meeting of the Wilson Ornithological Society a symposium on hawk migrations in the Greak Lakes area produced eight papers on the subject and further stimulated interest in hawk migration studies in that area (Robbins, 1975).

In addition to all of these efforts, a growing number of hawk migration lookouts in the eastern half of North America were identified and used annually for studies of hawk migrations (Robbins, 1975). Details of these projects, and the lookouts from which information was collected, are provided later in this book under the various site accounts. A growing number of hawk trapping and banding stations also began operating in various parts of the eastern United States and Canada during this data-gathering period (Heintzelman, 1975; Robbins, 1975).

Analytical Period (1970 onward)

The final time period in the history of North American hawk migration studies began in 1970 and continues to the present (Robbins, 1975). It is a period of explosive growth in both recreational hawk watching activities and in scientific hawk migration studies. This period also witnessed growth in the educational aspects of hawk watching in schools (Brett, 1981; Czajkowski, no date; Heintzelman, 1982a, 1983d; Keller, 1983a, 1983b, 1983c, 1984a, 1984b; Lee, 1977; Lee and Sykes, 1975) and in the publication of special hawk watching field guides (Brett and Nagy, 1973; Clark and Pramstaller, 1980; Heintzelman, 1972d, 1976a, 1979a).

Several factors played significant roles in this expanding public interest in hawk migrations and birds of prey generally. These include the conservation, education, and research activities of the Vermont Institute of Natural Science in Vermont, the hawk watch committee of the Connecticut Audubon Council in Connecticut, the Cape May Bird Observatory in New Jersey, the Bake Oven Knob hawk watch and the Hawk Mountain Sanctuary Association in Pennsylvania, the Hawk Ridge Nature Reserve in Minnesota, the Eagle Valley Environmentalists, Inc., in Illinois and Wisconsin, the Southwest Hawk Watch in Arizona, the Hawk Migration Association of North America, the Raptor Information Center of the National Wildlife Federation, the Raptor Research Foundation active throughout North America, and various other local and regional organizations and hawk watching projects.

Several books also played fundamental roles in helping to expand public interest in hawk migrations. They include Maurice Broun's (1949a) classic *Hawks Aloft*, James Brett and Alexander C. Nagy's (1973) *Feathers in the Wind*, Michael Harwood's (1973) *The View from Hawk Mountain*, and my own *Autumn Hawk Flights* (1975) and *A Guide to Hawk Watching in North America* (1979a and earlier editions). Thus, as a result of all of these factors, hawk watching and migration study has come of age (Peterson, 1985: 83–92).

TYPES OF HAWK WATCHERS

Hawk watching has evolved into a major branch of bird watching and ornithology with participants grouped into four basic categories (Heintzelman, 1983b).

Local Citizens

First are large numbers of local citizens who visit hawk lookouts for a few hours every autumn to watch migrating hawks, look at colorful autumn leaves, and enjoy spectacular panoramic views of the landscape. This category of people includes scouts, families on outdoor ventures, senior citizens, and many others. Most of these people know relatively little about hawk migrations and perhaps have limited interest in the subject. Nevertheless, they appreciate and enjoy the experience of seeing a hawk, eagle, or other raptor at close range. Some admit they were former hawk shooters who now derive great pleasure from watching the birds migrate past lookouts. It is from this category of people that some dedicated hawk watchers come.

Out-Of-The-Area Visitors

The second category of people consists of out-of-the-area visitors who come to hawk migration lookouts, or other raptor concentration areas, in spring or autumn. A few Bald Eagle concentration areas also attract out-of-the-area visitors in winter (Heintzelman, 1979a). At major sites such as Hawk Mountain in Pennsylvania, it is not uncommon to find visitors on the lookout representing a dozen or more states, and sometimes a few foreign countries, at any given time during the autumn migration season (Broun, 1949a). Many other important hawk migration lookouts also experience a similar representation of out-of-the-area visitors. Most of these people are active bird watchers, hawk watchers, ornithologists, or naturalists. They visit hawk migration lookouts because of the exceptional raptor viewing opportunities existing at these sites.

Student Hawk Watchers

The third category of people consists of school children, and other student hawk watchers of all ages, who visit hawk migration lookouts in autumn (or spring, at a few locations) as part of field trips conducted by local or more distant public or private schools (Brett, 1981; Broun, 1949a; Heintzelman, 1982a, 1983b, 1983d). Some schools have very well prepared and organized raptor teaching units designed for their students, such as programs in Maryland (Lee, 1977; Lee and Sykes, 1975), New Hampshire (Czajkowski, no date), New Jersey (Keller, 1983a, 1983b, 1983c, 1984a, 1984b), and Pennsylvania (Brett, 1981; Heintzelman, 1982a, 1983b, 1983d).

For convenience, I also include in this category student hawk watching field trips taken by undergraduate or graduate students from colleges or universities as well as teacher in in-service graduate level programs, such as those offered in some locations in Pennsylvania (Heintzelman, 1980a, 1983d).

Hawk Watching Enthusiasts

The final category of people who engage in hawk watching is composed of enthusiasts with varying degrees of interest. Some are general bird watchers who visit hawk lookouts for a few hours several times each spring or autumn in order to add new birds to their life lists. Others are hawk watching fanatics, some of whom spend hundreds of hours every autumn or spring at a lookout. Still others are advanced amateur or professional ornithologists engaged in serious hawk migration research also requiring many hundreds of hours of effort from a lookout. Thus hawk watching enthusiasts may participate in hawk watching merely for recreational purposes or do so in order to advance ornithological knowledge of these bird migrations. While the time an enthusiast spends on a lookout may vary greatly, the intensity of interest in hawk migrations is very great in all cases.

FIELD STUDY METHODS

There are many methods applicable to the study of hawk migrations. Some of these have been standard ornithological techniques for decades. These are described by DeGarmo (1967), Heintzelman (1972d, 1975, 1976a, 1979a), Haugh (1972), and Moore (1982). Other methods employ expensive and/or complex electronic equipment not readily available to most hawk watchers. Only the traditional methods are used at most North American hawk migration lookouts, but radar, radio telemetry, or other sophisticated techniques are beginning to come into use at some lookouts.

Field Guides

A fundamental requirement of all hawk migration studies is the ability of hawk watchers to identify the birds they observe correctly. Sometimes that can be a difficult task, even for very experienced observers, and some birds of prey cannot always be identified correctly to species level due to a variety of circumstances (Heintzelman, 1979a: 41–44). Those birds are then listed as unidentified individuals. Nevertheless, by referring to the text and illustrations in field guides, most hawks and other birds of prey generally can be identified correctly.

To assist hawk watchers in the identification task, two different types of field guides are available—general guides to bird identification, and advanced or special hawk watching field guides.

Complementing and supplementing the general bird identification guides are several advanced or special hawk watching field guides which provide more detailed and sophisticated criteria for the identification of hawks and other birds of prey.

In some respects *A Field Guide for Locating Bald Eagles at Cassville, Wisconsin* (Ingram, 1965) was a forerunner of today's advanced hawk and owl field guides, although it devoted attention to the Bald Eagle at only a few locations. Thus the first actual advanced hawk watching field guide for North America was my own *A Guide to Northeastern Hawk Watching* (Heintzelman, 1972d), later replaced by *A Guide to Eastern Hawk Watching* (Heintzelman, 1976a), and in turn replaced by *A Guide to Hawk Watching in North America* (Heintzelman, 1979a). Other hawk watching guides that followed include *Feathers in the Wind* (Brett and Nagy, 1973), *Field Studies of the Falconiformes of British Columbia* (Beebe, 1974), *Field I.D. Pamphlet for North American Raptors* (Clark and Pramstaller, 1980), *The Audubon Society Master Guide to Birding*, volume one (Farrand, 1983), and *A Guide for Hawks Seen in the North East* (Carrier, no date). Available, too, as an introduction to hawk watching and migration study are *A Beginner's Guide to Hawkwatching* (Moore, 1982) and *Hawk Watch/A Guide for Beginners* (Dunne, Keller, and Kochenberger, 1984). *Autumn Hawk Flights* (Heintzelman, 1975) also provided a detailed and comprehensive summary of our knowledge through 1973 of eastern North American autumn hawk migrations.

Hawk Lookout Rating Scales

One of the difficult problems encountered when trying to compare yearly seasonal hawk counts made at a hawk watching lookout is the lack of uniformity of coverage from year to year as well as enormous variation in the amount of coverage from lookout to lookout. To resolve that problem, and develop a simple quantitative method suitable for intra- and inter-lookout comparisons, I invented and developed two standard hawk lookout rating scales—one based upon the number of hawks observed per hour of site coverage, the other based upon the number of hawks observed per day of site coverage (Heintzelman, 1979a).

The rating scale using hawks counted per hour of site coverage is the most sensitive of the two and the scale whose use is recommended if the number of hours of observation per season at a lookout are known. There is a considerable amount of historic hawk count information preserved in the hawk watching and migration study literature, however, as in *Autumn Hawk Flights* (Heintzelman, 1975), that contains only the number of days of observation at a site. To use that information, therefore, the rating scale

HAWKS PER HOUR RATING SCALE

Lookout Rating	Average Number of Hawks Observed Per Hour
Poor	0 to 11
Fair	12 to 22
Good	23 to 33
Excellent	34 or more

HAWKS PER DAY RATING SCALE

Lookout Rating	Average Number of Hawks Observed Per Day
Poor	0 to 46
Fair	47 to 92
Good	93 to 138
Excellent	139 or more

Source: *A Guide to Hawk Watching in North America* by Donald S. Heintzelman (1979a, Pennsylvania State University Press). Reprinted with permission of the publisher.

using the number of hawks counted per day of observation is reasonably satisfactory and usually produces results remarkably similar to the more sensitive hawks per hour rating scale.

To illustrate the use of these rating scales one can select seasonal Broad-winged Hawk counts made during the autumn of 1983 at Bake Oven Knob, Pennsylvania, where 3,000 Broad-wings were counted during 441 hours of observation (Heintzelman, 1983c). Using the hawks counted per hour rating scale, these raw data convert into a seasonal passage rate of 6.8 hawks counted per hour at that site. Similar analysis of data from other lookouts are possible, thus making the information from those sites comparable with Bake Oven Knob. The only important restriction in the use of these rating scales is that a reasonable number of hours or days of observation is necessary and that coverage at a hawk watching lookout should be distributed reasonably uniformly throughout the migration season (August or September through November in autumn and March through May or early June in spring).

Benz (1982b) expanded the use of the hawks counted per hour rating scale by applying its use to data collected at most of the hawk watching lookouts in the Northern Appalachian region of the Hawk Migration Association of North America. He then produced what he described as a regional hawks per hour seasonal passage figure. That application of the rating scale, however, is a distorted and inappropriate use of the scale because some of the raptors counted at one hawk watching lookout on a flyway doubtless also are recounted at other lookouts thus producing a distortion in the data the magnitude of which is unknown for most species.

Nevertheless, a few studies of variations in raptor use of flight-lines are available, some of which were summarized earlier (Heintzelman, 1975). Frey (1940), for example, determined that as much as 70 percent of the Hawk Mountain hawk flight drifted from the ridge before it reached Sterrett's Gap far downridge from Hawk Mountain. He further determined that the number of hawks counted at Hawk Mountain (using only the North Lookout without the psychological distortion of radio use) was 35 percent greater than the number of raptors counted at Sterrett's Gap. He cautioned, however, that great variations occur and the situation is complex.

My own analysis of the variations in use of the Kittatinny Ridge in eastern Pennsylvania in autumn by migrating Golden Eagles and Bald Eagles also helps to define the degree of overlap or lack thereof of eagle movements in the area (Heintzelman, 1975, 1982g). While great yearly variation exists in the number of eagles observed at Bake Oven Knob and downridge at Hawk Mountain, generally from 32 to 53 percent of the eagles observed at Bake Oven Knob are not observed and counted at Hawk

Mountain (Heintzelman, 1975, 1982g), suggesting that many more eagles use parts of the Kittatinny Ridge during autumn than are counted at any single lookout. On the other hand, enough birds pass two or more sites to produce considerable duplication of counts which thus negates the validity of the manner in which Benz (1982b) applied the rating scale to regional collections of hawk counts.

Visual Observations

Visual observation is the standard method used for studying hawk migrations. One or more observers are stationed on a favorable hawk lookout on a flyway and scan the sky and horizon for approaching hawks. The manner in which various observers conduct a scan doubtless varies from person to person and from site to site, but 9,240 feet may be the maximum range at which ground observers using an optical aid can detect a Broad-winged Hawk (Welch, 1980) and 2,700 feet may be the maximum range for detection of a Sharp-shinned Hawk moving against a blue sky (Kerlinger and Gauthreaux, 1983). A mathematical model designed to define the limits of the hawk counting effort was developed by Welch (1981).

During the Broad-wing season in eastern North America, in September, observations at many lookouts begin at 0700 or even earlier. However, on days immediately following a major flight, or when a major flight is expected, observations often begin as early as 0630 at Bake Oven Knob, Pennsylvania, and other lookouts. This is shortly before the first thermals of the day develop.

September hawk flights in eastern North America are the most prolonged of the season in respect to hours of duration per day. Broad-winged Hawks usually stop flying only when thermal activity is severely reduced in intensity. Then we often see hawks approaching head-on only to watch them plunge into the surrounding forest for the night. Thus it is sometimes necessary to continue observations until 1800 hours or later. On the other hand, Bald Eagles and Ospreys, which are often less dependent upon thermals, usually remain aloft long after Broad-winged Hawks land. These late afternoon flights are majestic. The birds approach in mellow light, often close to the lookout. The extra time spent watching eagles and Ospreys gliding by on set winds is not only very enjoyable but also a necessary part of making accurate hawk counts.

October hawk flights in the East differ from September flights in several respects. To begin, the bulk of the Broad-wings have passed and are far along on their passage to Central and South America. A far greater variety of raptors comprise the October flights. Sharp-shinned Hawks account for the bulk of the flights during early October at most hawk lookouts. During the latter part of the month, however, Red-tailed Hawks, Golden Eagles, and other large raptors appear in greater numbers. In addition, October hawk flights last for fewer hours per day because of the shorter days. At

Bake Oven Knob our observations usually begin at about 0800 hours and continue until about 1730.

November hawk flights in the East are still more time-compressed. Early in the month large Red-tailed Hawk flights occur. Northern Goshawks, Red-shouldered Hawks, Rough-legged Hawks, and Golden Eagles are some of the species counted. However, by the latter half of the month and into early December, only a trickle of hawks and a few eagles are still migrating southward. Observations during November usually begin at about 0900 hours and continue to about 1600.

In western North America, where relatively limited systematic hawk watching and migration study is conducted thus far, observations during autumn follow a schedule somewhat similar to that used in the East. In the West from September to mid-October (when most hawk migrations occur) observations tend to begin before mid-morning and continue until later into the afternoon (Hoffman, 1981a, 1982b).

Traditionally in North America spring hawk migrations received much less study than was devoted to autumn hawk migrations (Heintzelman, 1975), although that situation is beginning to change. At the relatively few major spring hawk migration concentration areas in the East where de-tailed hawk counts are made from late February through mid-June, how-ever, observations tend to begin at about 0800 or 0900 hours and end at about 1700 hours, although daily variations occur in hawk watching sched-ules (Haugh and Cade, 1966; Moon, 1977a–1978e; Moon and Moon, 1979a–1983c). In the West, where even fewer spring hawk migration stud-ies are made, daily hawk watching schedules are not yet established but probably those used in the East are adequate.

Binoculars and Telescopes
A good pair of binoculars and a telescope are essential pieces of field equipment for studying hawk migrations. Generally 10X center-focus bin-oculars are ideal. Sometimes even 10X binoculars are inadequate, however. This is particularly true when hawks are flying at very high altitudes, at a distance over valleys, or under dim light conditions. Then a 20X telescope is necessary to make positive identifications. For maneuverability, it is best mounted on a gunstock.

In the Isthmus of Panama, where migrating Broad-winged Hawks and Swainson's Hawks often fly at altitudes far beyond the limits of vision of observers on the ground, telescopes with focal lengths up to 2,000 mm are used to detect and photograph high altitude hawk flights (N. G. Smith, 1985: 277–279).

The correct use of these instruments (binoculars and telescopes) also is important. Experienced observers usually scan the skies and the horizon with binoculars at periodic intervals to detect distant flying hawks which otherwise might pass unseen and uncounted. But during the Broad-wing season in North America, when kettles of these birds rise to high altitudes and are difficult to detect against blue skies, a scan of the sky and horizon with a telescope sometimes produces additional birds.

Field Data Forms

Most hawk watchers engaged in hawk watching and migration study now keep hourly counts of the various species passing their lookouts. Printed data forms greatly facilitate this record keeping. In addition, it is also desirable to record the exact times (Standard Time) when rare birds such as Golden Eagles, Bald Eagles, Peregrine Falcons, and perhaps other species appear. For species such as the Rough-legged Hawk, the color phase of the bird also should be recorded. In other instances the sex of the bird can be documented. Such detailed records frequently are helpful in tracking movements of individual birds over limited geographic areas. Determination of the ages of endangered species is particularly important and should be carefully noted. During recent years at Bake Oven Knob, Pennsylvania, I also documented the ages (adult, immature, or undetermined) of as many hawks as possible (Heintzelman, 1982i, 1983c, 1984a), and many other hawk watchers stationed on other lookouts also are starting to record this type of information, too.

During years past a variety of field data forms have been designed and used by various hawk watchers (Heintzelman, 1975). During recent years, however, many hawk watchers selected the field data form distributed by the Hawk Migration Association of North America in order to maintain some degree of uniformity as an aid to entering data into computer systems for further analysis (Fuller and Robins, 1979).

In addition to the standard field data forms used for routine recording of hawk count data on an hourly basis, various other special data forms have been developed for specific research projects. At Bake Oven Knob, for example, I designed a special form to facilitate the collection of age and sex data on selected species; it is used as a second field data form along with the standard form in use at this site for many years. Doubtless other hawk watchers also designed special field data forms for use on special projects when standard field data forms were discovered to be inadequate for certain specific purposes.

The use of field data forms on hawk watching lookouts helps to organize the collection of essential data and also serves as a reminder that certain types of information are needed. Their use is strongly encouraged.

Estimating Wind Velocity

To make accurate measurements of wind velocity requires expensive and complex equipment. Haugh (1972), for example, used a Windscope to determine wind velocity in his hawk migration study. At most lookouts such equipment is not available. Nevertheless, reasonably accurate hourly measurements of wind velocity can be estimated. Some observers allow a cloth to flutter in the breeze and estimate velocity from this. With a little experience, reasonable approximations can be made. Plastic wind speed gauges are more accurate. They may be purchased from camping equipment supply stores. Many hawk watchers now use these instruments. Use of the Beaufort Scale of Wind Force is still another method of estimating wind velocity and the Hawk Migration Association of North America recommends its use to its cooperators.

Recording Wind Direction

Wind direction is an extremely important factor affecting hawk migrations. It is important that hourly records be maintained. Haugh (1972) used his Windscope to determine surface wind directions. At Bake Oven Knob, Pennsylvania, and most other lookouts, a cloth is used as a wind sock, and its movement is compared against known compass directions. Either method determines surface wind direction only.

It is more difficult to determine wind directions at higher altitudes. Edwards (1939) devised a technique to solve this problem by filling small balloons with gas (helium) and releasing them into the atmosphere. By watching their movements as they rose, an approximation of higher altitude wind directions was secured. In 1967, I experimented with this technique at Bake Oven Knob. Helium-filled balloons were released hourly. In most cases surface winds were roughly equal to higher altitude wind directions. In 1968, hawk watchers at Hawk Mountain Sanctuary, Pennsylvania, used this technique. Haugh (1972) also used a similar technique.

Recording Air Temperatures

Air temperatures also are components of the weather data which hawk watchers should develop for their lookouts. Any accurate thermometer can be used. At Bake Oven Knob I use an instrument calibrated in degrees Celsius (Centrigrade) or Fahrenheit and convert the data from one scale to another as necessary. The thermometer is suspended from a branch or limb which permits a free flow of air to circulate around it, but care is taken not to expose the thermometer directly to sunlight. Hourly air temperatures are entered onto a field data form. Robinson (1979) describes a variety of types of thermometers and other weather recording devices, some of which hawk watchers engaged in hawk migration study may find helpful.

Estimating Visibility

The degree of visibility at a lookout unquestionably affects the numbers of hawks counted. On clear days one can expect to see most hawks passing a lookout. But when extreme haze or air pollution occurs, many birds can pass uncounted. Hence estimates of visibility are desirable. They, too, are recorded hourly on data forms.

At Bake Oven Knob, Pennsylvania, the maximum visibility (expressed in miles) one can see from the lookout is estimated based upon the known distance from the lookout of certain landmarks. This information is recorded each hour on the day's field data form. Hawk watchers using the methodology developed by the Hawk Migration Association of North America record similar data expressed in kilometers (HMANA, no date).

Estimating Percentage of Cloud Cover

Hourly estimates of the percentage of cloud cover over a lookout can help to reveal important aspects of the mechanics of hawk migrations. My studies of migrating Broad-winged Hawks at Bake Oven Knob, Pennsylvania, for example, reveal that estimates of the percentage of cloud cover are essential to understanding the migrations of these birds. Clouds are important factors which help to regulate thermal activity, and Broad-winged Hawks are highly dependent upon thermals. Although Haugh (1972: 3) considered cloud cover relatively unimportant to migrating hawks, in point of fact relatively little is known still about the movements and navigational abilities of birds in opaque clouds (Griffin, 1969, 1973). Hawk watchers observing hawks flying above, or through clouds should make very detailed notes on the behavior of the birds. Such information may provide important insights leading toward a better understanding of this aspect of bird migration and navigation.

The Hawk Migration Association of North America (HMANA, no date) developed a sky code chart for use with their data forms in which 10 different sky conditions are listed. Many hawk watchers use this code as an expression of the percentage of cloud cover noted every hour at hawk watching lookouts. Code zero indicates a clear sky or a sky with up to 15 percent cloud cover, code one indicates 16 to 50 percent cloud cover, code two indicates mostly cloudly with 51 to 75 percent cloud cover, and code three indicates an overcast sky with 76 to 100 percent cloud cover. Other numbers in the sky code indicate other weather situations such as rain, snow, or other conditions.

Reporting Altitude of Flight

The altitude at which a day's or hour's hawk flight passes a hawk watching lookout also is of importance and often is noted on field data forms at some hawk watching lookouts. The Hawk Migration Association of North America, however, developed a formal code containing eight categories to indicate the altitude at which hawk flights occur (HMANA, no date). For example, code zero indicates that hawks passed below an observer's eye level, code three indicates the birds passed at the limit of unaided vision (but does not address variations in limit among observers), and code seven indicates that the hawks passed at no predominant height.

Duration of Daily Observations

It is becoming increasingly obvious that the amount of time that hawk watchers spend counting hawks migrating past a lookout is a critical factor in evaluating hawk count data from a site. Thus it is important that the daily starting and ending times be noted on each day's field data form. This vital information will permit the determination at the end of the season of the total number of hours spent hawk watching at the site and thus will permit the application of the rating scales discussed earlier in this chapter.

Hawk-Counting Techniques

Although numerous techniques are used to detect and count raptors during various seasons of the year, under various circumstances, and at various locations (Fuller and Mosher, 1981), in most instances hawks migrating past a lookout are relatively easy to count because they appear individually or in small numbers. When hundreds of Broad-winged Hawks or Swainson's Hawks form kettles in thermals, however, it is difficult or impossible to make accurate counts. Experienced observers then wait until the hawks glide overhead in bomberlike formations, after the thermals start to dissipate, before counting them. Since thousands of hawks sometimes pass within relatively brief periods of time, a hand tally facilitates the counting process. The observer rapidly clicks off each bird as it passes overhead. If a flight continues into a new hour period, another tally counter is used for the hawks appearing in the new time period. After the flight is completed, data for the previous hour (recorded on the first tally counter) are recorded on field data forms. Hand tally counters are used only for a single species (the most abundant migrant at the time).

On rare occasions in North America, very large numbers of hawks pass a lookout during such brief periods of time that it is impossible to count them. One then is forced to estimate the number of birds seen. Broun (1949a: 185–186) described such a flight on 16 September 1948 at Hawk Mountain, Pennsylvania. More than 11,392 hawks (mainly Broad-wings) were reported. The extraordinary American Kestrel flight on 16 October 1970 at Cape May Point, New Jersey, is another example. Choate (1972) estimated about 25,000 individuals passed the Point. Estimating the number of hawks passing a hawk watching lookout should be done only when it is impossible to make actual counts.

At a few migration bottlenecks in the world such enormous concentrations of hawks occur that even estimating the numbers of birds passing overhead becomes difficult. Smith (1973) described such a situation in the Canal Zone in Panama where spectacular numbers of migrating Broad-winged Hawks, Swainson's Hawks, Turkey Vultures, and Black Vultures stream overhead in a restricted flight line about a mile wide as they head for South America. He, therefore, photographed much of the visible migration passing overhead using a non-overlapping method and later counted the birds in his photographs with an electronic scanning counter to make reasonably accurate hawk counts at this site (Anonymous, 1974: 1–2; Smith, 1980). More recently, however, he expressed reservations about the accuracy of the counts because some birds fly at altitudes up to at least

18,700 feet under certain weather conditions—altitudes far too high to allow visual or photographic detection.

The number of hawks counted migrating past a lookout represents only the visible migration passing a site. Numerous factors, however, also can affect the actual number of hawks migrating past a lookout, some of which remain uncounted (Heintzelman, 1975; Fuller and Mosher, 1981). Thus use of the elaborate mathematical model suggested by Welch (1981) to calculate raptor population trends seems inappropriate.

Weather Maps

Weather systems play important roles in autumn hawk migrations either by acting as releasers, thereby causing hawks to begin migrating (Robbins, 1956), or by affecting their migrations once the movements have begun (Broun, 1949a; Haugh, 1972; Heintzelman, 1975; Mueller and Berger, 1961). Thus many hawk watchers have found weather maps useful aids in helping to predict good hawk flights (Broun, 1951b, 1963a). Weather maps also are valuable when correlating a season's data with general weather conditions. Such correlations may lead to a better understanding of hawk migrations.

Several sources of weather maps are available. Television stations and newspapers prepare reasonably good maps which contain sufficient detail on the positions of cold fronts and high- and low-pressure areas to make fairly accurate predictions of hawk flights. Weekly booklets of weather maps also are available on a subscription basis from the federal government. These are now issued at the end of each week rather than daily.

Decoys

The value of decoys in luring hawks to a shooting stand was well known to the gunners who frequented Bake Oven Knob, Hawk Mountain, and other shooting stands in Pennsylvania earlier in this century (Broun, 1949a, 1956b). The most effective decoys were live pigeons placed in cloth harnesses and suspended from long poles with wire or string. Occasionally live or artificial Great Horned Owls also were used as decoys as were wounded hawks. Sharp-shinned Hawks were especially attracted—particularly when a pigeon was lifted from its perch and flapped its wings.

At Fishers Island, New York, a decoy owl also was used as a hawk lure. Ospreys, Northern Harriers, Sharp-shinned Hawks, American Kestrels, Merlins, and Peregrine Falcons were among the species that attacked it (Ferguson and Ferguson, 1922).

After organized hawk shooting was banned in Pennsylvania most hawk watchers failed to continue the use of decoys at hawk watching lookouts until I began using a large papier-mâché Great Horned Owl at Bake Oven Knob, Pennsylvania, and reintroduced the widespread use of these decoys at hawk watching lookouts (Heintzelman, 1975, 1979a). The artificial owl is placed on a long pole in an upright position. Size and color of the decoy are unimportant. Indeed, some Sharp-shinned Hawks are readily attracted to decoys of various sizes and colors. Other species, too, occasionally can be lured closer to a lookout. Thus decoys are an aid to photography and also help to assure more complete hawk counts by bringing some birds into closer view than might otherwise occur.

Despite the current widespread use of Great Horned Owl decoys at hawk watching lookouts, Kerlinger and Lehrer (1982) completed the only detailed study of anti-predator responses of migrating Sharp-shinned Hawks directed toward owl decoys. They determined that hawks attracted to the decoy used a four-step sequence—recognition, reorientation, approach, and resumption of migration. At Bake Oven Knob, Pennsylvania, however, I sometimes noted a fifth step—a second or even third return to the decoy after the first behavior sequence was completed. Kerlinger and Lehrer (1982: 34) determined that of 131 Sharp-shinned Hawks studied 90 (68.7 percent) exhibited some form of anti-predator behavior when migrating past an owl decoy positioned on a hawk watching and migration lookout. When considered based upon age, 71.4 percent of adult and 68.0 percent of immature Sharp-shins reacted to the decoy. The significance of these anti-predator behavior responses by Sharp-shinned Hawks is unknown, but it was speculated that owls prey upon Sharp-shinned Hawks more frequently than is recognized (Kerlinger and Lehrer, 1982).

No other detailed field studies concerning anti-predator behavior of other raptors are available, but at Fishers Island, New York, Ferguson and Ferguson (1922) reported various anti-predator reactions to a decoy owl. Ospreys squealed, repeatedly attacked, and struck the decoy, Northern Harriers made a low squeal and attacked once with their feet extended outward, Sharp-shinned Hawks changed their course and flew close to the decoy once, American Kestrels darted toward the decoy but never struck it, Merlins squealed and repeatedly dashed toward the decoy but did not strike it, and Peregrine Falcons generally attacked the decoy only once before continuing their migration. At Bake Oven Knob, Pennsylvania, I also observed a variety of hawks react to decoy owls. Included were Northern Harriers, Sharp-shinned Hawks, Cooper's Hawks, Northern Goshawks, Red-shouldered Hawks, Broad-winged Hawks, Red-tailed Hawks, and American Kestrels.

Radio Communication

For years hawk watchers considered the value of radio communications between lookouts. In 1967, Hawk Mountain Sanctuary and I cooperated in an experimental radio use project to study the migration routes used by hawks crossing a small section of eastern Pennsylvania. A 25-watt transmitter was established at the Sanctuary headquarters building, and Motorola radiophones were used at nearby lookouts. The radios were operated on a Forestry Conservation Band frequency. One 10-watt unit was positioned on the North Lookout at Hawk Mountain. I operated the second unit at Bake Oven Knob, sixteen miles northeast of Hawk Mountain and also on the Kittatinny Ridge. Reception between the stations was improved by putting a small antenna in a tree. This communications network enabled observers at the upridge site (Bake Oven Knob) to notify observers at Hawk Mountain that eagles or other hawks had been seen there and might be approaching Hawk Mountain. When easterly and southerly winds occurred, evidence suggested that many birds seen at Bake Oven Knob failed to reach Hawk Mountain. Some data also were gathered on hawk flight speeds along this section of the Kittatinny Ridge. Unmarked birds limited the accuracy and validity of the project's data, however, although in the case of eagles careful records of the ages and times the birds were seen helped to compare sightings from the two stations.

In retrospect, I consider use of radios between sites such as Bake Oven Knob and Hawk Mountain a bad idea. Observer-expectancy bias always is a serious problem in ornithological and many other scientific studies (Balph and Balph, 1983), and the use of a radio at an upridge site to alert hawk watchers downridge that spectacular birds such as eagles might be approaching the downridge lookout merely adds to an already high degree of observer-expectancy bias existing at many lookouts as a result of unfortunate competition between hawk watchers and hawk lookouts. Under those circumstances it is likely that some observers might record a positive identification for an eagle or other rare raptor if it is seen at an extreme distance from the lookout as a result of observer-expectancy bias induced by the radio message from the upridge site.

Radio communication between lookouts was a novel experiment but failed to provide much not already available from examination of hourly records maintained at the two sites. Radios are more useful to hawk watchers using several lookouts within Hawk Mountain Sanctuary by permitting rapid coordination of field studies. They also may be useful as part of certain radar or telemetry studies dealing with radio-tagged hawks. Haugh (1972) also used walkie-talkies in his migration studies.

Short-Term Photoidentification Techniques
The use of photographs to identify otherwise unmarked individual animals is growing in acceptance in some branches of field biology such as cetology (Bigg, MacAskie, and Ellis, 1983; Bryant and Lafferty, 1983; Glockner-Ferrari, 1982; Katona, et al., 1980; Katona, Beard, and Balcomb, 1982; Kelly, 1983).

The technique also was used to identify unmarked California Condors (Snyder and Johnson, 1985), and a short-term photoidentification technique could be used by hawk watchers if its use is restricted to certain specific circumstances and species. Individual birds to be photographed would have to display a distinctive or unusual feature such as an albino feather, broken or missing feathers, or some other odd or unique trait so that the bird could not be confused with any other individual of the same species in the area. In addition, individual birds would have to remain in the vicinity of a particular hawk watching lookout for several days or longer to allow repeated recognition and identification. Selected Black Vultures and/or Turkey Vultures might be ideal study birds for use of the short-term photoidentification technique because they tend to remain in areas for periods of time longer than most other diurnal raptors do during migration but nevertheless engage in a slow drift-like migration early in the migration season (Heintzelman and MacClay, 1972: 3–4). Thus one could photograph odd-looking vultures and later compare the catalog of photographs with each odd-looking vulture passing a lookout. Of course the catalog of photographs would become obsolete after the birds engaged in their next molt if distinctive feather loss was the identification feature being used to recognize a particular bird. Despite its limitations, however, use of short-term photoidentification techniques could be a valid and helpful hawk migration research tool.

Telemetry

Although telemetry now is used widely in studying wildlife movements in relatively restricted geographic areas, especially in respect to the movements of game animals, its use as a tool for studying long-distance movements of migrating hawks remains mostly ignored. Dunstan (1972, 1975) provided a general summary of the techniques and possibilities that telemetry offers to hawk migration researchers. In addition, a few other reports are published dealing with specific uses of telemetry in hawk migration studies.

Unquestionably the projects conducted by Cochran (1972, 1975) are the most fascinating and informative in respect to long distance migration tracking. In his first effort (Cochran, 1972), an adult, female Sharp--shinned Hawk was captured at Cedar Grove, Wisconsin, on 9 October 1971. The hawk was banded and a 3-gram radio transmitter also was attached. For the next 11 days, and for 750 miles, the migration path of this bird was radio-tracked until it reached a position about 15 miles east of Huntsville, Alabama, where tracking efforts were terminated because of transmitter and logistical problems.

During the autumn of 1974, Cochran (1975) again placed a 1.4 gram radio transmitter on a raptor—this time an immature, male Peregrine Falcon and tracked the long-distance migration of the bird for the next 16 days from its capture point at Green Bay, Wisconsin, on 12 October, through Louisiana and Texas into Mexico.

In addition to tracking the autumn migration routes of a Sharp-shinned Hawk and a Peregrine Falcon via the use of telemetry, a great deal of additional important information also was secured using this technique.

Included were data on daily time budgets, altitude of migration, speed of migration, use of thermals, distance covered per day, hunting behavior, reaction to rivers or other diversion-lines, reaction to wind drift and orientation to opaque clouds, and other data for either or both of the species (Cochran, 1972, 1975). Some of these data are discussed as appropriate later in this book.

Beske (1982) also radio-tagged juvenile Northern Harriers and tracked the local and migratory movements of seven birds fledged from nests in Buena Vista Marsh in central Wisconsin. The harriers remained in the vicinity of their nests for approximately three weeks after fledging, then migrated southward. Another harrier, however, remained in the nest vicinity for 50–51 days before migrating. Telemetry tracking of these birds, via ground vehicle and fixed-wing aircraft, demonstrated that the juvenile harriers started their autumn migration alone, and established temporary home ranges at various points along the migration route. Generally the telemetry tracking of these migrating radio-tagged harriers demonstrated a slow, southeasterly movement often using low-level flapping or high soaring or gliding flight. The longest distance one of the radio-tagged harriers was tracked was about 102.5 miles (164 kilometers). Hunting and roosting behavior also was monitored as part of Beske's telemetry tracking effort.

In comparison to the long-distance migration telemetry studies just discussed, telemetry studies also can be used at migration traps to determine the movements of migrating hawks at such sites, the number of birds delaying their migrations after being counted, the rate of hawk count duplication, and other aspects of migration. Holthuijzen and Oosterhuis (1982), for example, used the technique to study the migration movements of female Sharp-shinned Hawks at Cape May Point, New Jersey. The radio transmitters selected each weighed about 4 grams and tracking was done from mobile vehicles, a stationary unit on top of the lighthouse at the Point, and an additional unit placed at Cape Henlopen, Delaware, across Delaware Bay from Cape May Point. Analysis of the telemetry tracking data demonstrated that hawks remaining in the area of the Point for a day or two were counted once, but birds were counted twice if they remained in the area for three days, and a hawk remaining for four days may have been counted six times.

Clearly the use of telemetry to track movements of migrating hawks, either locally or for long distances, can add significantly to our knowledge of hawk migrations. It should be used more frequently by qualified persons engaged in carefully designed field studies.

Satellite Tracking

One of the most sophisticated uses of electronic technology to aid in the study of hawk migrations currently remains unused but under active development and doubtless will receive limited experimental use on eagles in the future—satellite tracking of radio-tagged birds of prey (Buechner, Craighead, and Craighead, 1971; Craighead and Dunstan, 1976; Heintzelman and MacClay, 1976a:16). The Nimbus-3 satellite already was used to track and monitor free-ranging Elk (F. C. Craighead, Jr., et al., 1972) and a Black Bear (J. J. Craighead, et al., 1971). An ARGOS satellite system positioned on NOAA 7 and NOAA 8 satellites also was used to monitor signals emitted by a radio tag attached to a whale (Anonymous, 1983a: 1; Beamish and Carroll, 1983: 2; Mate, 1983: 2). A variety of additional environmental data also were gathered by the satellite system and correlated with the cetacean data.

Use of a similar system to monitor the migratory movements of birds of prey, such as those that Cochran (1972, 1975) and Beske (1982) radio-tagged, would allow scientists to secure more data at considerably less effort. Satellite tracking, for example, would eliminate the need for the use of aircraft and ground vehicles to travel along and monitor the movements of the radio-tagged raptors. It also would greatly reduce the number of people needed to monitor such birds, or at least the number of hours of effort required to do so. In addition, it would eliminate complicated political and logistical requirements imposed upon scientists when tracking radio-tagged birds crossing international borders. Finally the amount of data on the movements of radio-tagged birds, and related environmental factors including cloud cover, that such a satellite tracking system would produce would be much greater than are produced by manual methods now in use. Clearly, satellite tracking of radio-tagged raptors is a technique that should be used at the first available opportunity. It is likely to produce major new information on hawk migrations in North America and elsewhere.

Radar
Radar now is used with increasing regularity in various types of bird migration research because it allows ornithologists to detect and study various bird movements that could not be detected or studied in other ways. Eastwood (1967) provided an excellent summary of the techniques used in *Radar Ornithology.*

In general, there are at least two basic types of radar used in bird migration studies—broad-beam surveillance radars and narrow-beam tracking radars (Richardson, 1975; Kerlinger, 1980). Both types have received limited use in hawk migration studies.

Richardson (1975), for example, used broad-beam air traffic control surveillance radar in southern Ontario near London and Toronto to track autumn movements of what were presumed to be Broad-winged Hawks migrating near or along the northern shorelines of Lakes Erie and Ontario. He determined that hawks mostly formed concentrations near Lake Ontario on days with prevailing northerly winds. In addition, the radar studies suggested that many more hawks were migrating after a cold front passed through the area than during other times. Finally the radar studies detected flight-lines of hawks that previously were undetected by visual observation, and radar also allowed study of low-volume hawk migration movements over a broad area. Richardson recognized various limitations involved with the use of hawk migration radar studies, but nevertheless concluded that visual observations of hawk flights passing known concentration points produced biased impressions of raptor movements.

In mid-September of 1974, in Connecticut, limited use of weather surveillance radar also produced detection of various isolated kettles of migrating Broad-winged Hawks the identification of some of which were confirmed by visual observation (Mersereau, 1975: 10–11). As in the radar study in Ontario, the work in Connecticut also detected some previously unknown flight-lines of hawks.

In addition to the use of surveillance radar in hawk migration studies, some limited use of tracking radar also is reported in the literature. Kerlinger (1980), for example, used an automatic tracking radar in the central Helderberg Mountains near Albany, New York, during the autumn of 1978 to study altitude, speed, flight direction, and other aspects of migrating Sharp-shinned, Red-tailed, and Broad-winged Hawks. He secured tracks for 289 birds but positive identification was possible for only 171 of those birds. He further limited his use of data to tracks of birds for which data were collected for no less than 30 seconds. He concluded that tracking radar was well suited to the study of hawk migrations and that much important information could be secured by its use in respect to hawk flight behavior and dynamics.

During the autumn of 1982 at Cape May Point and Woodbine, New Jersey, tracking radar again was used to determine altitude and direction of migrating Sharp-shinned Hawks. The effort produced important new information discussed later in this book. Similar tracking radar studies of spring hawk migrations in southern Texas also were completed (Kerlinger, 1983; Kerlinger and Gauthreaux, 1983).

Although various types of radar thus far received only limited use in the study of North American hawk migrations, it is obvious from the data produced by the work already performed that radar offers great opportunities to hawk watchers to augment traditional visual observations of migrating hawks. Hopefully, much more use of radar will occur in the future in an effort to help to resolve various migration puzzles that visual observation apparently can't solve.

Observations from Aircraft

The potential value of aircraft as a base for studying hawk migrations received limited use but remains relatively little exploited. Nevertheless, use of various types of aircraft can provide observers with many insights into the routes which migrating hawks follow, the altitudes at which they fly, behavior aloft, and other related phenomena.

The earliest and most unusual use of aircraft in studying hawk migrations was suggested by E. I. Stearns (1948a, 1948b, 1949: 110) and conducted in New Jersey with the assistance of James L. Edwards and Alfred E. Eynon. A blimp was used and the technique was called "blawking."

The Urner Ornithological Club of Newark, New Jersey, selected 21 September 1948 to conduct the blawking experiment. The Tide Water Associated Oil Company supplied the blimp, and a second blimp also was sent aloft later in the day through the cooperation of the Tide Water and associated companies. The day was clear with an air temperature of 65 degrees Fahrenheit and a 10-mile-per-hour northwest wind. In addition to observers in the blimps, observers also were stationed on the ground at Upper Montclair, New Jersey, and elsewhere. Ground-to-blimp radio communications were not available, but visual signals were given from the ground stations via fluorescent panels (Stearns, 1948b).

Ground observers at Upper Montclair counted 2,150 Broad-winged Hawks, observers in the blimps only 290 hawks. This suggests that hawks are more difficult to locate from the air than from the ground. Hopkins and Mersereau (1975c: 23) also found it difficult to count hawks from a motor-glider.

Despite the difficulty of spotting hawks from the air, four separate kettles were found spiraling upward at 1,500, 2,000, 2,700, and 2,400–2,900 feet above sea level. The birds were flying above a valley with an elevation of 190 feet and rising over a ridge with an elevation of 590 feet. "It is not known that these were the highest kettles of the day, nor measured at their highest point except for the 2000 foot kettle. Although earlier ground estimates had placed the kettles at greater heights, the maximum height reached may well be only 3000 feet and many times the birds abandon their upward spiraling and 'peel off,' or enter their straight, downward glide, at only 2000 feet.

"The birds peeling off from the 2000 foot thermal, a rising column of air heated from a warmer ground area, were successfully followed until they roosted in trees 4 miles away at a ground elevation of about 450 feet. Thus the ratio of glide to fall was about 12 to 1. The air speed of the Broadwings in the glide was 32 mph in one measurement and 26 mph in another. Judged by the criterion that the hawks were not frightened if they continued their glide in an undeviating line, the birds did not seem to mind the airship provided it was more than 300 feet distant" (Stearns, 1949).

Ground observers overestimated the height which hawks reached in thermals and were unable to judge airspeeds and distance of glides accurately.

During recent years other types of aircraft also were used in the study of hawk flight and hawk migrations. Raspet (1960), for example, used a sailplane to study the flight of Black Vultures in the southern United States, and Pennycuick (1975) used a Schleicher ASK-14 motor-glider to study the soaring flight of vultures in Serengeti National Park in northern Tanzania. Pennycuick (1975) also discussed various aspects of the use of motor-gliders in the study of hawk migrations. In New England, a RF-5B Sperber motor-glider also was used to make inflight studies of hawk migrations—the first use of this technique in North America (Anonymous, 1975a: 13; Welch, 1975; Hopkins, Mersereau, and Welch, 1975; Hopkins, Mitchell, and Welch, 1978; Hopkins, et al., 1978, 1979). These New England motor-glider studies confirmed some earlier hawk migration observations and also produced much new information discussed later in this book (Hopkins and Mersereau, 1975c).

Trace-Element Analysis of Feathers

A fascinating new experimental technique of great potential value in hawk migration studies is the use of instrumental neutron-activation analysis (INAA) of samples of raptor secondary remiges to determine trace-element content as an indicator of the geographic origins (natal locales) of the birds sampled. Parrish, Rogers, and Ward (1983), for example, demonstrated a 100 percent correct predictability rate for the natal locales of a small sample of Peregrine Falcons based especially upon Mercury content, but also using Aluminum and Vanadium in conjunction with Mercury content.

If trace-element analysis of feathers proves to be a valid technique, it could be used by hawk trappers and banders to determine the geographic origin of raptors captured at various points along their migration routes. The information secured via INAA also might produce additional new insights into migration routes used by various raptors. Factors limiting the use of INAA, however, are the special types of equipment needed for analysis of feather samples and the cost of such equipment and analysis.

Trapping and Banding Migrating Hawks

Until about a decade ago, most hawk banding in North America was confined to nestlings. Several excellent summaries and analyses of recoveries of these birds have been published (Cross, 1927; Gillespie, 1960; Henny and Van Velzen, 1972; Kennedy, 1973; Lincoln, 1936; Melquist, Johnson, and Carrier, 1978; Stewart, 1977; Worth, 1936). Indeed, some of the work dealing with Osprey migrations is particularly fascinating.

In addition to banding raptor nestlings, a number of spring and/or autumn hawk trapping and banding stations also are operating now along hawk migration routes in North America. Thousands of migrating hawks and other diurnal birds of prey now wear leg bands after being trapped during their migrations. Along the Atlantic Coast, for example, a spring station operates at Sandy Hook, New Jersey (W. S. Clark, 1978a), an autumn station operates at Cape May Point, New Jersey (W. S. Clark, 1968, 1969, 1970, 1971, 1972, 1973, 1976, 1978b; Henny and Clark, 1982), and another autumn station is used at Assateague Island, Maryland (Berry, 1971; Ward and Berry, 1972). The latter concentrates mostly on migrating Peregrine Falcons. Another Peregrine trapping and banding station also is established along the Gulf Coast on Padre Island, Texas (Hunt, Rogers, and Slowe, 1975; Henny and Clark, 1982).

Still other stations are established along the Kittatinny Ridge in New Jersey and Pennsylvania (Heintzelman, 1975; Panzer, 1976; Soucy, 1976, 1983). Holt and Frock (1980) summarized the results of 20 years of these autumn raptor banding efforts along the Kittatinny Ridge in Pennsylvania.

Several stations also are established during autumn along the northern or western shorelines of the Great Lakes. Included in Ontario along Lake Erie are Hawk Cliff (Field and Field, 1979, 1980a, 1980b; Field and Rayner, 1974, 1976, 1977, 1978; Fowler and Fowler, 1981, 1982; Haugh, 1972) and Point Pelee (Gray, 1961). Cedar Grove in Wisconsin along the west shoreline of Lake Michigan (Mueller and Berger, 1961, 1966, 1967b, 1967c, 1968, 1973; Mueller, Berger, and Allez, 1977, 1981) and the Hawk Ridge Nature Reserve in Duluth, Minnesota, along the southwestern side of Lake Superior (Anonymous, 1973; Evans, 1974, 1975, 1976, 1978, 1979, 1980; Rosenfield and Evans, 1980), also are autumn banding stations. There are various newer stations being established elsewhere, too.

Various other spring banding stations also are operating now including those at Whitefish Point, Michigan (Baumgartner, 1979, 1980, 1981, 1982, 1983), and Braddock Bay, New York (Taylor, 1984).

STANDARDIZATION OF METHODOLOGY

An important requirement of every hawk migration study—especially a long-term study—is the need for careful record keeping (Heintzelman, 1972c) and the use of standardized and consistent methodology and species identification criteria for the duration of the study. One standardized methodology was just described. An alternative was developed by the Hawk Migration Association of North America (HMANA, no date) and issued as instructions for use of field data forms followed by more general instructions for beginning hawk watchers (Moore, 1982). Many North American hawk watchers now use the HMANA methodology, but numerous flaws appear in their data reporting process (Cooper, 1983). Indeed, numerous aspects of hawk watching methodology are subject to the need for more careful scientific study (Fuller, 1979; Fuller and Mosher, 1981).

The criteria used for making flight identifications of hawks and other diurnal birds of prey, for example, vary widely among hawk watchers and inject considerable variability into the hawk migration data gathering process. A listing of specialized hawk watching field guides, now used by various hawk watchers, was provided earlier in this chapter.

DEPTH AND DURATION OF MIGRATION STUDIES

The depth and duration of hawk migration studies are divisible into two basic types: (1) short-term field studies simultaneously conducted for a few days (often weekends) at numerous sites, and (2) long-term field studies conducted for many years or decades at a single site. Each type has certain advantages and disadvantages.

Short-Term Field Studies

Examples of short-term field studies are those used for a few weekends in spring and autumn at many sites in Vermont (Norse, 1979a, 1979b, 1980a, 1980b, 1981a, 1981b, 1982a, 1982b, 1983a, 1983b; Pistorius, 1977, 1978; Rowlett, 1977; Will, 1974, 1975, 1976, 1980) and southern New England (Currie, 1978, 1979; Currie and Cote, 1981, 1982, 1983; Currie and MacRae, 1979, 1980; Hopkins and Mersereau, 1973a, 1973b, 1974a, 1974b, 1975a, 1975b, 1976) to try to document the overall broad movements of migrating hawks passing across those states. At best, this approach provides a useful, but very brief, sample of a season's entire visible hawk migration picture. If unfavorable or atypical weather conditions prevail during the time the field studies are made, however, the data secured may be distorted or deceptive and the conclusions based upon them equally flawed. Sometimes little or no migration occurs at all which then negates the field study effort. In addition, a large number of hawk watchers of greatly varying ability and qualifications are used to conduct short-term field studies. Thus additional variables are introduced into the data gathering process the overall importance of which remain unknown.

Long-Term Field Studies

In comparison, a long-term annual hawk watching and migration study conducted daily or for much of a season every autumn and/or spring for many decades at one site produces a detailed cross-section of these visible migrations not obtainable in any other way. If the site selected for the field study is located at a bottleneck or concentration area along a major hawk flyway or migration route some sort of sample of the annual spring or autumn hawk migration results. For example, the long-term autumn hawk counts and other migration studies made at Bake Oven Knob, Pennsylvania (Heintzelman, 1963a through 1984a; Heintzelman and Armentano, 1964; Heintzelman and MacClay, 1971 through 1979; Heintzelman and Reed, 1982), are excellent examples of long-term migration studies conducted at a single site where standard methods (Heintzelman, 1975) and raptor identification criteria (Heintzelman, 1979a) are used. Long-term field studies also are conducted at Hawk Mountain Sanctuary, Pennsylvania (Anonymous, 1975b, 1976, 1977a, 1978, 1979c, 1981c; Benz, 1980a–1984b; Broun, 1935 through 1966; Edge, 1939b, 1940b, 1942, 1943; Nagy, 1967, 1968; Sharadin, 1972, 1973, 1974; Wetzel, 1969, 1970, 1971).

Some other long-term hawk migration field studies include those at Braddock Bay Park (Moon, 1978a through 1984; Moon and Moon, 1979a through 1984c; Dodge, 1981; Listman, Wolf, and Bieber, 1949; Wolf, 1984) and Derby Hill (Elkins, 1962; Haugh, 1966, 1972; Haugh and Cade, 1966; Anonymous, 1984; Baker, 1983; Muir, 1978, 1979, 1980, 1981, 1982; Peakall, 1962; Smith, 1973; Smith and Muir, 1978) in New York, Cape May Point (Allen and Peterson, 1936; Choate *in* Heintzelman, 1970e; Choate, 1972; Choate and Tilly, 1973; Clark, 1972, 1973, 1975b, 1976; Dunne, 1978b, 1979b, 1980b; Dunne, et al., 1981; Dunne and Clark, 1977; LeGrand, 1984; LeGrand, et al., 1983) and the Montclair Hawk Lookout Sanctuary (Bihun, 1967 through 1983; Breck, 1962, 1963; Breck and Breck, 1964, 1965, 1966; Redmond and Breck, 1961) in New Jersey, and the Hawk Ridge Nature Reserve in Minnesota (Hofslund, 1966; Kohlbry, 1980a; Peacock and Myers, 1982b; Sunquist, 1973).

The long-term raptor migration studies conducted at the Cedar Grove Ornithological Station near Cedar Grove, Wisconsin, where banding of birds of prey is emphasized, also are noteworthy because of the significant number of excellent analytical reports resulting from those field studies (Mueller and Berger, 1961, 1967a, 1967b, 1967c, 1968, 1973; Mueller, Berger, and Allez, 1977, 1981).

When conducted properly, long-term hawk migration studies may permit scientists to document a variety of difficult-to-detect trends and changes in populations in some species that short-term field studies could not reveal or might distort. Heintzelman (1975: 278–289, 1982e, 1982f, 1982h), Nagy (1977a), Snyder et al. (1973), and Spofford (1969) provide examples of efforts to detect trends and changes in long-term hawk migration data although the interpretation and validity of some of the conclusions based upon some of these studies may be subject to debate and further evaluation. A more detailed discussion of these and similar data is provided later in this book.

Long-term field studies also lessen the impact of brief but atypical weather conditions upon hawk migrations because one has the important advantage of using data from thousands of days and hours of observation collected over decades to evaluate conditions and data collected during any particular migration season. When possible, therefore, use of a long-term hawk migration study is the preferred method for studying these bird movements.

MIGRATION SEASONS

Seasonal hawk migrations in North America occur during spring and autumn with the latter receiving the bulk of the field study. Only in recent

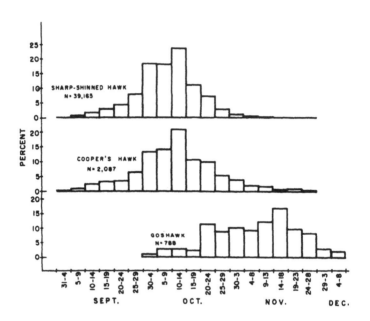

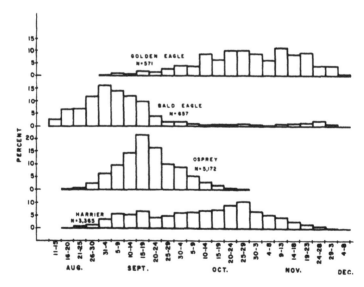

The charts on these two pages show the temporal migration periods of hawks observed at Hawk Mountain, Pa., from 1954 through 1968. Similar periods apply to hawk migrations at Bake Oven Knob, Pa. *Reprinted from Haugh (1972).*

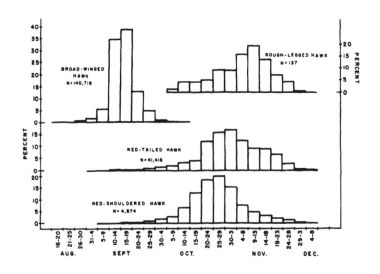

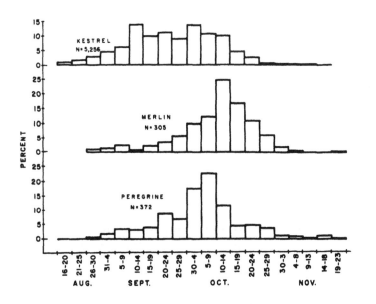

years have spring hawk migrations begun to receive widespread study. Nevertheless enough basic information is available for both seasonal migrations to plot their general months of movement.

Within a given season, however, each species uses a reasonably well-defined portion of the season to migrate. Thus, during autumn, Broad-winged Hawks typically migrate southward in eastern North America in largest numbers during September in the northern parts of the range of the species and October farther south, Sharp-shinned Hawks occur in largest numbers during October, and Red-tailed Hawks reach their peak numbers during late October and November (Broun, 1949a; Haugh, 1972; Heinzelman, 1975, 1979a).

Spring Migrations

Although some early studies of spring hawk migrations exist (Magee, 1922; Tyrell, 1934), generally more recent studies such as those of Haugh and Cade (1966), Haugh (1972), Kranick (1982, 1984), Moon (1977a–1978e), Moon and Moon (1979a–1984c), Morton (1982, 1983), C. C. Sutton (1981), Sutton and Sutton (1984), and others define the spring migratory periods for diurnal raptors—at least in eastern North America. Collectively these studies document hawk flights from February through June or even into early July with the bulk of the flights occurring during April and May. In western North America, as in southern Texas, abundant species such as Broad-winged Hawks appear from late March to early May (Oberholser, 1974).

Autumn Migrations

As a result of the widespread and extensive autumn hawk migration studies conducted in eastern North America, and to a much lesser extent recently in western North America, the seasonality of these migrations is relatively well known. In general, they occur from early August to mid-December, especially during September, October, and November.

Numerous studies define the autumn seasonality of North American hawk migrations in the eastern half of the continent including those of Allen and Peterson (1936), Bagg (1947–1950), Broun (1935–1966), Haugh (1972), Heintzelman (1975), Hofslund (1954a, 1954b, 1962, 1966), and Laurie, McCord, and Jenkins (1981). In western North America the reports from Binford (1979), Hoffman (1981a, 1982b), and Tilly (1980–1984) are noteworthy.

SPEED OF MIGRATION

The speed at which hawks and other birds of prey migrate from their breeding grounds to their wintering grounds is relatively little studied (Brown and Amadon, 1968: 65). Nevertheless, some speed-of-migration data are available based upon recoveries of leg-banded birds, telemetry tracking of radio-tagged birds, observations of color-marked birds or birds with individually distinctive features, and flight speed measurements over short distances. A summary of this information is instructive, therefore, in providing a brief view of a rarely considered aspect of North American hawk migrations.

Vultures

Very limited data document the speed-of-migration of vultures. Parmalee and Parmalee (1967) studied returns and recoveries of 2,382 Black Vultures banded in eastern North America (especially Avery Island, Louisiana) and concluded that an undetermined number of individuals are sedentary in a restricted local area for indefinite time periods, but that more data from color-marked birds are needed to determine accurately the local or migratory behavior of Black Vulture populations in eastern North America. Nevertheless, most recoveries of these vultures from Avery Island were within a 200-mile radius and within a two-year period. Six vultures from the island, however, were recovered more than 300 miles from the point of banding—one 720 and another 600 miles away. Thus the Avery Island data suggest a modest degree of wandering by some segments of southeastern North American Black Vulture populations but no distinctive seasonal migrations occur. In Texas, however, Parmalee's (1954) field studies suggest that Black Vultures wander to a considerable extent, tend to concentrate at locations where a large food supply exists, then disperse after the food supply is depleted.

On the other hand, the number of Black Vultures appearing during the autumn migration season in eastern Pennsylvania are very limited and appeared only recently. This is a cooler and more temperate part of the range of the species. At Bake Oven Knob, Pennsylvania, for example, where Black Vultures never were observed prior to 1974 (Heintzelman and MacClay, 1975), and only with some limited degree of regularity since 1982 (Heintzelman, 1982i, 1983c) as a result of the northward range expansion of the species, the limited number of birds in the population provide unusual study opportunities to determine the speed of migration of non-color-marked individuals.

During 1982 and 1983, for example, sightings of unmarked Black Vultures at the Knob extended over periods no longer than two or three days before long gaps developed in sightings again (presumably of different individuals). It appears, therefore, that the newly expanding Black Vulture population in northeastern North America is migratory, which supports Leck's (1980) thesis that some newly expanded bird populations may adopt migration or other behavior strategies different from long-established sedentary populations. Stewart (1977) also demonstrated that

Turkey Vulture populations from more northerly parts of their range engage in much longer-distance migrations than do individuals from more southern parts of their range.

Turkey Vultures, indeed, are at least partly migratory in parts of North America. Stewart (1977) analyzed the distance of migration and mortality rates of these birds but did not mention the speed of migration. Smith (1980: 54), however, reported 28 days as the average speed-of-migration of some migrant Turkey Vultures moving between northern breeding grounds and nonbreeding grounds in Panama. He reported further that the Turkey Vulture passage extends longer than that for Broad-winged Hawks and Swainson's Hawks.

In comparison, at Bake Oven Knob, Pennsylvania, I observed a few color-marked or otherwise distinctive Turkey Vultures during several autumn migration seasons. In 1978 one bird with a greenish wing marker appeared on 2 September and was last seen on 23 September indicating the bird remained in the Bake Oven Knob area for 22 days before disappearing—presumably as a result of migrating southward. In 1982 another individual appeared and was distinguished by a peculiar white retrix. This Turkey Vulture remained in the vicinity of the Knob between 9 October and 21 October, for a 13-day period.

These observations help to confirm the suggestion of Heintzelman and MacClay (1972: 3–4) that Turkey Vultures observed early in the season drift slowly from the Bake Oven Knob area and thus are migratory rather than sedentary, contrary to popular opinion.

At Hawk Mountain, Pennsylvania, Broun and Goodwin (1943) measured the flight speed of one Turkey Vulture at 34 miles per hour as the bird crossed a short course.

Osprey

Osprey speed-of-migration data can be approximated from various sources. For example, the last week of August is the start of the autumn migration period for many juvenile Ospreys reared in Maryland and Virginia (Kennedy, 1973: 180) and 21 September is the approximate peak migration date for many Ospreys passing Bake Oven Knob and Hawk Mountain, Pennsylvania (Heintzelman, 1969b; Heintzelman and Armentano, 1964; Broun, 1949a). Moreover, almost all juvenile Ospreys fledged in Maryland and Virginia arrive in South America by 15 October (Kennedy, 1973: 183). In Panama, 11 October is an arrival date (Wetmore, 1965: 257), and in Surinam Ospreys were noted by 19 October (Haverschmidt, 1968: 69). Along the River Amazon in 1974 and 1975, Ospreys were seen at about 11 locations (Heintzelman, 1981) with first sightings on 7 October (1975) and 27 October (1974). By using 21 September as the departure date in North America, the birds arrived in Amazonia in approximately 17 and 37 days respectively, in Surinam in about 29 days, and in Panama in about 21 days.

Using the Osprey recovery data published by Kennedy (1973: 181), and 25 August as an arbitrary start-of-migration date, the speed-of-migration calculated for 22 recoveries ranged from 22.8 miles per day to 103.5 miles per day with a mean of 48.8 miles per day.

The flight speeds of 16 Ospreys measured passing Hawk Mountain, Pennsylvania, by Broun and Goodwin (1943) ranged between 20 and 80 miles per hour with a mean of 41.5 miles per hour.

Eagles

Bald Eagle migrations in eastern North America are reasonably well known as a result of the classic banding studies conducted by Broley (1947). He demonstrated that newly fledged eagles exhibited a rapid northward speed-of-migration, "within a few weeks of learning to fly." One eagle-of-the-year, for example, migrated from Florida to New Brunswick—approximately 1,600 miles—within 32 days for a speed-of-migration of 50 miles per day. Griffin, Southern, and Frenzel (1980: 163–164) also reported that an immature Bald Eagle radio-tagged in Missouri migrated to Ontario in 70 days. Another Bald Eagle, banded as a nestling in Minnesota on 10 June 1968, was found dead in Texas on 10 October 1968—123 days later and a linear distance of about 1,185 miles for a speed-of-migration of about 9.6 miles per day (Dunstan, 1969: 92). During the spring migration of Bald Eagles near Saskatoon, Saskatchewan, one bird exhibited a speed-of-migration of 25 kilometers per hour, two others moved at 27 kilometers per hour, and another at 33 kilometers per hour (Gerrard and Gerrard, 1982: 103).

The flight speeds of two Bald Eagles measured at Hawk Mountain, Pennsylvania, by Broun and Goodwin (1943) ranged between 36 and 44 miles per hour and averaged 40 miles per hour.

Recoveries of Golden Eagles banded in North America suitable for providing speed-of-migration data are limited. Kuyt (1967), however, reported an immature banded in the Northwest Territories, Canada, and recovered in Montana 32 days later. The eagle migrated approximately 1,000 miles for a speed-of-migration of about 31.2 miles per day.

At Hawk Mountain, Pennsylvania, flight speeds of two Golden Eagles ranged from 28 to 32 miles per hour and averaged 30 miles per hour (Broun and Goodwin, 1943).

Harrier

The most sophisticated method used to determine speed-of-migration of Northern Harriers was telemetry tracking of radio-tagged juveniles that migrated from a nest site in Wisconsin (Beske, 1982). He determined that the juvenile harriers he studied engaged in "a slow, generally southeasterly movement interrupted by pauses at temporary home ranges" at the onset of the autumn migration season. One male, for example, migrated about 102.5 miles from its nest site in six days (about 17 miles per day), a female migrated about 107 miles in 46 days (about 2.3 miles per day), and another female migrated about 44.3 miles in 10 days (about 4.4 miles per day) although these birds disrupted their migrations to establish temporary home ranges which distorts their speed-of-migration times.

Northern Harrier speed-of-migration data also are based on limited numbers of recoveries of leg-banded birds. Lincoln (1936), for example, reported one hawk banded in Alberta, Canada, and recovered in Washington 59 days later and the recovery of another which moved from Ohio to Oklahoma in 60 days. Houston (1968a) also reported recoveries of harriers banded in Saskatchewan, Canada, and noted one migrated from its banding location to Louisiana in 119 days.

At Hawk Mountain, Pennsylvania, Broun and Goodwin (1943) measured the average flight speeds of four Northern Harriers at 28.7 miles per hour with a range between 21 and 38 miles per hour.

Accipiters

Using telemetry to track the autumn migratory movements of a radio-tagged Sharp-shinned Hawk from Wisconsin to Alabama, Cochran (1972) determined the bird moved a distance of 750 miles in 11 days for a speed-of-migration of about 68.1 miles per day. Another Sharp-shinned Hawk banded at Hawk Cliff, Ontario, reached Point Pelee, Ontario, in six days (Field, 1970: 68). Holt and Frock (1980) also reported three recoveries of Sharp-shinned Hawks banded along the Kittatinny Ridge in eastern Pennsylvania. One hawk reached North Carolina in 49 days, another reached Maryland in 39 days, and a third reached Maryland in 5 days. At Braddock Bay Park, New York, however, one banded Sharp-shinned Hawk remained in the area where it was banded for ten or more days in May, whereas more than 750 additional birds of the species were observed migrating past the park during the same time period (McKinney, 1984).

A sample of 37 Sharp-shinned Hawks migrating past Hawk Mountain, Pennsylvania, was measured at flight speeds of 16–30 miles per hour with a mean of 30 miles per hour (Broun and Goodwin, 1943).

Some speed-of-migration data also are available for the Cooper's Hawk. Holt and Frock (1980), for example, reported a recovery of one bird only three miles from the banding station on the Kittatinny Ridge in eastern Pennsylvania 34 days after being banded, indicating no migration. At Hawk Mountain, Pennsylvania, however, a sample of 12 Cooper's Hawks was measured with an average flight speed of 29.3 miles per hour with a range between 21 and 55 miles per hour (Broun and Goodwin, 1943).

Evans and Sindelar (1974) reported a recovery of an adult female Northern Goshawk in Louisiana—some 1,160 miles from its capture and banding point at Duluth, Minnesota—62 days later for a speed-of-migration of about 18.7 miles per day. At Hawk Mountain, Pennsylvania, Broun and Goodwin (1943) measured the flight speed of a Northern Goshawk at 38 miles per hour.

Buteos

Speed-of-migration data for Red-shouldered Hawks apparently are not available except for flight speeds measured by Broun and Goodwin (1943) at Hawk Mountain, Pennsylvania. A sample of seven birds averaged 28.3 miles per hour and ranged between 18 and 34 miles per hour.

Broad-winged Hawks are among the most abundant seasonal raptor migrants. Unfortunately these hawks are almost impossible to capture for banding purposes during migration (Holt and Frock, 1980: 9) which seriously restricts the amount of speed-of-migration data available from banding returns. Holt and Frock (1980), for example, reported no band recoveries useful for determination of migration speed for this species. Lincoln (1936), however, reported one bird banded in Massachusetts and retrapped in Michigan—a remarkable retrap location far west (rather than south) of the banding location—54 days after banding. A second Broad-wing banded in Massachusetts was recovered in Virginia. This hawk probably did not begin its southward migration until mid-August, at the earliest, so its speed-of-migration could have been about 25 days to cover the distance, although 8 to 10 days may be more accurate. Smith (1980:54) stated that Broad-winged Hawks average 28 days engaged in spring and autumn migration.

At Hawk Mountain, Pennsylvania, Broun and Goodwin (1943) reported the mean flight speeds of eight Broad-winged Hawks as 31.7 miles per hour with a range between 20 and 40 miles per hour.

Lincoln (1936) reported a record of a Swainson's Hawk banded in Saskatchewan, Canada, and shot in Oklahoma 62 days later. Houston (1968b) also reported on recoveries of Swainson's Hawks banded in Saskatchewan and recovered elsewhere. One bird, for example, migrated to Montana in 59 days. Houston (1974) later reported two even more fascinating recoveries of Swainson's Hawks. One bird migrated from Saskatchewan to El Salvador within 106 days, and another reached the pampas of Argentina—about 7,200 miles away—in 126 days or less. This latter recovery represents a speed-of-migration of about 57.1 miles per day. Smith (1980: 54), however, stated that most Swainson's Hawks average 70 days engaged in spring or autumn migrations.

Four Red-tailed Hawks banded by Holt and Frock (1980) on the Kittatinny Ridge in eastern Pennsylvania were recovered elsewhere during the same season. One immature was retrapped 29 days after banding in Virginia, another 36 days later in Maryland, the third 10 days after banding at Cape May Point, New Jersey, and the last 45 days later in Virginia. Of several

Red-tails banded in Saskatchewan, Canada, and recovered during the year of banding, one migrated to Oklahoma in 148 days and another reached Louisiana in 178 days (Houston, 1967).

Red-tailed Hawks migrating past Hawk Mountain, Pennsylvania, were measured using an average flight speed of 29 miles per hour, with a range between 20 and 40 miles per hour, based upon a sample of 54 hawks (Broun and Goodwin, 1943).

Falcons

Limited speed-of-migration data are available for the American Kestrel in eastern North America. Layne (1982: 96–97, 99) reported band recovery records from Florida and determined the minimum speed-of-migration for four kestrels to be 80 to 120 kilometers (50 to 75 miles) per day. One falcon banded at Cape May Point, New Jersey, was recovered 13 days later in south-central Florida, another 16 days later in northeastern Florida, and a third 17 days later in southwestern Florida (Layne, 1982: 97). Two banding recoveries reported from southern Canada also are pertinent (Field, 1970: 68). One falcon banded at Hawk Cliff, Ontario, was recovered in Ohio 71 days later. Another bird banded and released at Hawk Cliff arrived in Louisiana 101 days later. Still another kestrel banded at Hawk Cliff was found dead in northwestern Florida 50 days later (Field, 1971: 75).

At Hawk Mountain, Pennsylvania, Broun and Goodwin (1943) reported a mean American Kestrel flight speed of 26.2 miles per hour for four birds with a range between 22 and 36 miles per hour.

Migration speed data derived from recoveries of banded Merlins in North America are unavailable in the ornithological literature, but at Hawk Mountain, Pennsylvania, the measured flight speed of one Merlin was 28 miles per hour (Broun and Goodwin, 1943).

The speed-of-migration for the Peregrine Falcon is documented as thoroughly as for any raptor. Cochran (1975), in particular, contributed greatly toward this data base via his use of telemetry tracking of radio-tagged falcons over very long distances. He tracked one Peregrine Falcon from Wisconsin to Mexico continuously for 20 days, of which 15 involved migration. During the 15 days the falcon flew 1,637 miles during 4,766 minutes—an average speed-of-migration of 21 miles per hour or 111 miles per day. Low migratory flight occurred during 7.5 percent of each day and circling and soaring migratory flight required 28.5 percent of each day's time budget.

Various banding data also provide Peregrine Falcon speed-of-migration data. Lincoln (1936: 43) reported one record of a falcon banded in New Jersey and recovered in Illinois 20 days later. The westward movement was noteworthy. Holt and Frock (1980: 23) also reported a fascinating recovery of a Peregrine banded on the Kittatinny Ridge in eastern Pennsylvania. The bird was retrapped three days later along coastal Virgina, and retrapped a day later farther south at another coastal Virginia site. Henny and Clark (1982), too, reported various Peregrine Falcon band recovery data. One bird migrated from western Greenland to southern New Jersey in 77 days, three birds flew about 62.5 miles from southern New Jersey to coastal Virginia in two days, two additional Peregrines reached coastal Texas from southern New Jersey in 24 and 25 days respectively, a falcon from southern New Jersey reached Venezuela in 18 days, another bird migrated between coastal Texas and El Salvador in 51 days, and a Peregrine from north-eastern Alberta migrated to coastal Texas in 79 days. Kuyt (1967) also reported a remarkable recovery of a Peregrine, banded as a nestling in the Northwest Territories, Canada, and killed in Argentina—about 9,000 miles away—174 days later for an approximate speed-of-migration of 51.7 miles per day.

At Hawk Mountain, Pennsylvania, Broun and Goodwin (1943) reported an average flight speed of 30 miles per hour for three Peregrine Falcons with a range between 28 and 32 miles per hour.

DISTANCE OF MIGRATION

Haugh (1972) pointed out that long-distance migrants (birds which winter in Central and South America and/or the West Indies), generally appear early in autumn whereas medium- and short-distance migrants move later in autumn. Immature Sharp-shinned Hawks migrate earlier than adults.

Haugh (1972: 4) grouped Broad-winged Hawks and Ospreys as long-distance migrants. Turkey Vultures, Sharp-shinned Hawks, Cooper's Hawks, Red-shouldered Hawks, Rough-legged Hawks, Bald Eagles, and Northern Harriers are medium-distance migrants. Northern Goshawks, Red-tailed Hawks, and Golden Eagles are short- to medium-distance migrants. The American Kestrel is a short- to long-distance migrant, and Merlins and Peregrine Falcons are medium- to long-distance migrants.

NOMADISM

Nomadism is a variation of annual seasonal migrations in which a raptor species moves in varying numbers in various directions at irregular intervals perhaps because of variations in food supply (Brown, 1977: 128). The nomadic movements of Gyrfalcons provide an example of a nomadic species. These falcons seemingly appear irregularly at various hawk lookouts. Nevertheless, Fingerhood and Lipschutz (1982) and Fingerhood (1984a, 1984b) demonstrated that a compilation of isolated Gyrfalcon records in Pennsylvania revealed a strong attraction to a migration corridor and wintering area in eastern Pennsylvania. Thus seemingly isolated nomadic raptor sightings may actually reveal a migration pattern when enough records can be gathered and studied.

NOCTURNAL HAWK ACTIVITY

Occasionally published statements or speculation suggest that some species of diurnal birds of prey are nocturnal as well as diurnal in behavior and/or that their migrations are partly nocturnal. Although these assertions lack proof and are scientifically unverified, a brief discussion of a few such accounts is appropriate.

Breninger (1897: 54–55), for example, observed what apparently were Turkey Vultures flying to and from a roost after dark in mid-May in Arizona but the impact of his nearby presence upon the roosting birds was undetermined. No similar observations for the species were reported by Wilbur and Jackson (1983), although Black Vultures and Turkey Vultures sometimes fly from roosts at first morning light (Rabenold, 1983).

Enderson (1965: 331) also speculated that some migrant Peregrine Falcons observed along coastal Texas during autumn may have arrived during the night although late evening or predawn arrival was not ruled out. Christie (1983b) also reported migrant Sharp-shinned Hawks in flight over Seal Island, Nova Scotia, at sunup during autumn and speculated that the hawks engaged in nocturnal flight earlier. Harwood (1980: 3), too, speculated that some diurnal birds of prey migrate at night, or in very dim light, perhaps to accompany flocks of avian prey species. No acceptable data were provided, however, to support his speculation.

At Vickery Bat Cave in western Major County, Oklahoma, inhabited between mid-April and mid-October by 700,000 to 800,000 Mexican Free-tailed Bats, Harden (1972) and Looney (1972) observed Northern Harriers, Swainson's Hawks, Red-tailed Hawks, and American Kestrels engaged in crepuscular hunting activity during evening twilight. Gerow (1943) also observed a Merlin at dawn in Washington darting into the headlight beams of an automobile to prey upon a Horned Lark. The falcon remained behind the car, kept out of sight of the lark until an opportunity developed to dart forward and seize the prey, then captured the lark. Was this hunting technique an example of tool-using behavior by a raptor?

Based upon what is known about North American Falconiform raptors, no satisfactory scientific data support statements that some species normally engage in nocturnal activity and/or migrations in addition to their diurnal behavior. Nevertheless, a few examples of crepuscular behavior during migration, and at other seasons, are documented. In addition, there is the important case of a radio-tagged Peregrine Falcon being tracked by telemetry by William Cochran as the falcon twice soared over the ocean off the Atlantic Coast for much of the night in comparison to usual Peregrine behavior which brings the birds ashore at various locations toward sunset after following over-the-water migration routes during daylight hours (Floyd, 1983: 145). These two Peregrine night flights add some evidence to Enderson's (1965: 331) speculation that some migrant Peregrines may arrive during the night along coastal Texas.

PART TWO

The Hawk Lookouts

CHAPTER 2

Eastern Canada Hawk Migrations

Eastern Canada, exclusive of Ontario where hawk migration studies are conducted along the northern shorelines of the Great Lakes, is a major breeding ground for birds of prey. Nevertheless, comparatively little detailed information is available regarding hawk migrations in this part of North America (Christie, 1980, 1981a, 1981b, 1982a, 1982b, 1983a, 1983b; Godfrey, 1979; Heintzelman, 1975).

LABRADOR

Todd (1963) mentioned scattered hawk migrations in various sections of the Labrador peninsula but did not identify any well defined flight-lines. Austin (1932) also did not provide any major information on Labrador hawk migrations. This province is so large, however, and portions so remote, that some regularly used flight-lines could easily be overlooked.

QUEBEC

Autumn and spring hawk migrations in Quebec tend to be scattered (Heintzelman, 1975: 51), but some consistent flight-lines exist at some locations. Christie's (1981a) compilation of many isolated records help to suggest important areas needing further field study.

Bromont
Spring flights of Rough-legged Hawks are reported at Bromont (Christie, 1981a), but detailed data are lacking.

Cap Tourmente
Christie (1981a) reported numerous autumn hawk flights at Cap Tourmente, east of Quebec City, suggesting it is a productive flight-line and

observation area. Sharp-shinned Hawks, Red-tailed Hawks, and Rough-legged Hawks occur in largest numbers.

Godbout

Napoleon A. Comeau reported large numbers of Rough-legged Hawks migrating over Godbout during October and November (Todd, 1963: 277).

Granby

Granby produces some spring hawk migrations, but data from the site are fragmentary (Christie, 1981a). Ospreys and Sharp-shinned Hawks were noted.

Île Perrot

Both autumn and spring hawk flights are reported at this location (Heintzelman, 1979a: 234). Sharp-shinned Hawks, Broad-winged Hawks, and Red-tailed Hawks, in particular, are seen (Copeland, 1977b: 20; David and Gosselin, 1976: 37, 1977: 153).

Indian House Lake

This site is located 350 miles north of Mingan on the St. Lawrence River. Northern Goshawks, Rough-legged Hawks, and Gyrfalcons are reported migrating across the area during autumn—the latter in regular autumn migrations (Todd, 1963).

James Bay

Mid-September flights of Rough-legged Hawks along the Quebec side of James Bay suggest a flight-line for these birds there (Carleton, 1963: 16).

Montmagny

An April flight of Rough-legged Hawks was reported at Montmagny but detailed data are lacking (Christie, 1981a).

Montreal

Thousands of migrating Broad-winged Hawks, and smaller numbers of other raptors, sometimes pass Montreal during autumn (David and Gosselin, 1982a: 156). In addition, both spring and autumn hawk migrations are reported at Valleyfield some 60 miles southwest of Montreal (Heintzelman, 1979a: 235).

The spring hawk flights are poor (8.7 hawks per hour) at Valleyfield with Ospreys, Sharp-shinned Hawks, Broad-winged Hawks, Red-tailed Hawks, Rough-legged Hawks, and American Kestrels forming the bulk of the migrations (Klabunde, 1981a, 1982a, 1984a).

Fair autumn hawk migrations (17.6 hawks per hour) are reported at Valleyfield. The most frequently observed species are Ospreys, Sharp-shinned Hawks, Broad-winged Hawks, Red-tailed Hawks, and American Kestrels (Kleiman, 1981, 1982).

Mount Saint Bruno

As many as 300 Broad-winged Hawks are reported at this site in autumn, but further details are unavailable (David and Gosselin, 1976: 37).

Pointe-au-Père

Spring hawk flights (19.1 hawks per day) occur at this site with Northern

Harriers, Sharp-shinned Hawks, Red-tailed Hawks, and Rough-legged Hawks noted in largest numbers (Christie, 1982a: 11).

Rimouski

Rimouski, on the south shore of the St. Lawrence River on the Gaspé Peninsula, produces some fine spring flights of Sharp-shinned Hawks, Red-tailed Hawks, Golden Eagles, and Gyrfalcons (David and Gosselin, 1982b: 832).

Saint Jean Cherbourg

A September flight of 21 American Kestrels might suggest a flight-line at this site on the northern side of the Gaspé Peninsula (Christie, 1981a).

Valleyfield

See Montreal.

NEW BRUNSWICK

Current knowledge of spring and autumn hawk migrations in New Brunswick is fragmentary although some Sharp-shinned Hawk flights pass along the Bay of Fundy, cross Grand Manan Island, and elsewhere (Squires, 1952, 1976; Christie, 1981a).

Campobello Island

Limited spring hawk flights, especially of Ospreys and Sharp-shinned Hawks, sometimes cross this island (Vickery, 1977: 974).

Chignecto Isthmus

An April flight of American Kestrels in the Chignecto Isthmus area between New Brunswick and Nova Scotia suggests a flight-line there (Christie, 1981a: 8).

Fundy National Park

Fundy National Park is located along Chignecto Bay at the northeastern end of the Bay of Fundy. Both spring and autumn hawk migrations occur here and not infrequently are observed from the park's headquarters.

The spring migrations (36.7 hawks per day) contain various species, with Ospreys, Sharp-shinned Hawks, and American Kestrels appearing in largest numbers (Christie, 1981a: 8). During autumn (7.1 hawks per hour), migrations of Sharp-shinned Hawks and Broad-winged Hawks seem to be the largest (Christie, 1981b: 11).

Grand Manan Island

This charming island is located east of Lubec, Maine, near the mouth of the Bay of Fundy. During the autumn migrations (38.7 hawks per day) Sharp-shinned Hawks, American Kestrels, and Merlins are reported most frequently (Christie, 1981a: 10; Finch, 1973: 25–26). The spring migrations often contain Sharp-shinned Hawks and American Kestrels (Christie, 1981a: 10).

Machias Seal Island

Occasionally flights of Merlins and other hawks are seen over this island located 10 miles south of Cutler, Maine, and southwest of Grand Manan Island (Christie, 1981a: 8)—part of the offshore hawk migrations across the Gulf of Maine and Bay of Fundy (Anderson and Powers, 1979; Kerlinger, Cherry, and Powers, 1983).

Miscou Island
Various spring and autumn hawk flights are seen at this island in the Gulf of St. Lawrence off the northeastern tip of New Brunswick (Christie, 1981a: 9). Northern Harriers, Sharp-shinned Hawks, and American Kestrels are reported most frequently.

Point Lepreau
According to Christie (1981a: 9), there is an autumn 1885 record of Sharp-shinned Hawks and Merlins migrating past this site.

NEWFOUNDLAND

Relatively little is known about hawk migrations in Newfoundland (Peters and Burleigh, 1951; Austin, 1932), but recently Burrows (1981) mentioned a number of locations where some raptor movements are seen.

Cape Ray
Field studies conducted by Stuart Tingley indicate autumn hawk flights of Sharp-shinned Hawks, American Kestrels, and Merlins at the Cape Ray headland at the southwestern tip of Newfoundland (Burrows, 1981: 52).

L'Anse-Aux-Meadows
This site is located on the northern tip of the Northern Peninsula and is noted for mid-October to early May migrations of Gyrfalcons (Burrows, 1981: 83; Vickery, 1978b: 978, 1979b: 752, 1980: 755, 1982: 153).

Long Point
Autumn migrations of Northern Harriers, Sharp-shinned Hawks, Rough-legged Hawks, American Kestrels, and Peregrine Falcons are reported at Long Point—a long peninsular funnel about 10 miles in length on the southwestern side of Newfoundland (Burrows, 1981: 58–59).

Otter Bay Provincial Park
Autumn flights of Sharp-shinned Hawks and American Kestrels are reported at this park (Burrows, 1981: 51).

Saint Pierre and Miquelon Islands
These islands are located south of Newfoundland, but actually are part of France rather than Canada. Spring migrations of American Kestrels and Merlins are reported (Burrows, 1981: 37) as are autumn hawk flights including some Peregrine Falcons (Burrows, 1981: 38).

Southern Long Range Mountains
Some October migrations of Sharp-shinned Hawks cross the Twin Hills area of these mountains (Burrows, 1981: 51).

Terra Nova National Park
During autumn migrations of Sharp-shinned Hawks, American Kestrels,

and Merlins are reported at this park as are sightings of a few other raptors (Burrows, 1981: 110).

NOVA SCOTIA

Both spring and autumn hawk migrations occur in Nova Scotia, but relatively little is known about the spring flights (Christie, 1981a: 5; Tufts, 1962, 1973). More, however, is known about autumn hawk flights in the province (Christie, 1981a).

Bon Portage Island
Occasionally small Sharp-shinned Hawk flights are reported at this small island near the southeastern tip of Nova Scotia (Christie, 1981a: 7; Tufts, 1962, 1973).

Brier Island
Located off the southwestern tip of the North Mountain which borders the Bay of Fundy for 120 miles, Brier Island probably is Nova Scotia's most important hawk migration concentration area (Tufts, 1962: 3; Heintzelman, 1979a: 230). Both spring and autumn hawk flights are reported on the island.

Broad-winged Hawks form the bulk of the spring flights (Vickery, 1977: 974, 1979b: 752; Nikula, 1982: 829, 1983: 846).

The autumn hawk flights at Brier Island are much larger than the spring migrations and sometimes contain a varied assortment of species. Thousands of Sharp-shinned Hawks and Broad-winged Hawks have been seen along with smaller numbers of American Kestrels, Merlins, and other species (Bagg and Emery, 1962: 9, 1966: 14; Christie, 1982b; Finch, 1969: 18, 1974: 113; Hawkes, 1958; Squires, 1952; Tufts, 1962: 120–121; Vickery, 1978a: 176, 1983: 156).

Cape Sable Island
Occasionally autumn flights of Broad-winged Hawks and Merlins are seen over Cape Sable Island (Christie, 1981a: 7).

Seal Island
Seal Island is an isolated outpost located southwest of Yarmouth County. During autumn various autumn hawk migrations are seen over the island, sometimes containing hundreds of birds, including Turkey Vultures, Sharp-shinned Hawks, Cooper's Hawks, American Kestrels, Merlins, Peregrine Falcons, and occasionally a Gyrfalcon (Burrows, 1981: 25–29; Bagg and Emery, 1966: 14; Finch, 1972: 33; Vickery, 1979a: 155; Christie, 1983b: 16). A few Sharp-shinned Hawks also are reported over the island during spring (Christie, 1981a: 5).

CHAPTER 3

Western Canada Hawk
Migrations

Knowledge of hawk migrations in western Canada—Alberta, British Columbia, Manitoba, the Northwest Territories, Saskatchewan, and the Yukon Territory—is less complete than that for eastern Canada and very much less complete and detailed than for the eastern United States through which many birds of prey reared in Canada migrate twice each year (Godfrey, 1979).

ALBERTA

Although fragmentary hawk migration records for Alberta are published, the only systematic field study of these migrations is that reported by Dekker (1970) from a site west of Cochrane.

Black Diamond

Autumn Sharp-shinned Hawk flights sometimes are reported at Black Diamond in southwestern Alberta (Houston, 1971a: 72).

Calgary

Among the migrant hawks that sometimes appear in autumn in the Calgary area are Swainson's Hawks and an occasional Broad-winged Hawk (Harris, 1981: 195; Serr, 1976a: 88).

Cochrane

A spring and autumn hawk migration study was completed by Dekker (1970) on 83 days during the period 1960–1969 at this ranch site west of Cochrane in the foothill country 30 miles east of the Rocky Mountains.

During spring, the most important migrants are Bald Eagles (50 per day in March) and Golden Eagles (50 per day in March) whereas during autumn the most numerous migrants include Bald Eagles, Sharp-shinned Hawks (5–15 per day), and Golden Eagles. Lesser numbers of other species including Ospreys, Northern Harriers, Cooper's Hawks, Red-tailed Hawks, Rough-legged Hawks, and American Kestrels are also reported.

Dekker (1970: 24) noted a complete absence of the Broad-winged Hawk, Peregrine Falcon, and Gyrfalcon as migrants in the foothills west of Cochrane despite migrations of these species east of Cochrane on the plains.

Edmonton

Spring hawk migrations in the Edmonton area include some Broad-winged Hawks and Golden Eagles—the latter appearing after snow disappears from the ground. There also is a Bald Eagle migration east of Edmonton beginning in late March (Dekker, 1970: 22–24; Houston, 1972b: 775).

During autumn various species migrate through the area including Northern Goshawks, an occasional Broad-winged Hawk, Golden Eagles, and an occasional Gyrfalcon (Dekker, 1970: 22; Gammel and Huenecke, 1956: 33; Houston, 1971a: 72, 1973: 76; Krause, 1961: 52; Lister, 1965: 49).

Rockyford

Up to 50 migrating Swainson's Hawks are reported at Rockyford during autumn (Houston, 1973: 76).

BRITISH COLUMBIA

Detailed hawk migration studies are lacking for British Columbia, but at some locations flights of Turkey Vultures, Bald Eagles, and Sharp-shinned Hawks are reported.

Active Pass

Up to 100 Bald Eagles per day are reported at this site in the Gulf Islands along coastal British Columbia during autumn and early winter (Hancock, 1970; Heintzelman, 1979a: 260).

Cypress Provincial Park

Barry Sauppe conducted a 10 September to 9 October 1979 hawk watch at this park and counted 620 migrating raptors—11.3 hawks per hour. Largest hawk flights occur on north winds. Sharp-shinned Hawks, Red-tailed Hawks, and American Kestrels are seen most frequently (Mattocks and Hunn, 1980: 192).

Fraser Valley

As many as 100 Bald Eagles per day sometimes are seen from October to January in the Fraser Valley northeast of Vancouver (Heintzelman, 1979a: 260).

Pacific Rim National Park

From spring through autumn, more than 170 pairs of nesting Bald Eagles can be seen in this park on the west coast of Vancouver Island. In addition, more northern eagles migrate into the park during August (Beebe, 1974: 42–44; Hancock, 1970; Heintzelman, 1979a: 260).

Pemberton

Moderate numbers of American Kestrels migrate past this site in early autumn (Crowell and Nehls, 1973: 107).

Prince Rupert

As many as 12 Bald Eagles per day migrate past Prince Rupert during early autumn (Crowell and Nehls, 1970: 84).

Victoria

Victoria is located at the southeastern tip of Vancouver Island just north of the Washington border. Both spring and autumn hawk migrations occur here.

The spring hawk flights are not well studied, but Bald Eagles and Sharp-shinned Hawks are reported migrating across the Puget Sound area and occasionally a few other species also appear (Beebe, 1974: 64; Crowell and Nehls, 1971: 787; Flahaut and Shultz, 1955: 351).

In comparison, autumn hawk flights at Victoria are much better documented. Hundreds of Turkey Vultures sometimes are seen migrating southward over Victoria during September and October, and additional flights of Bald Eagles and Golden Eagles, Sharp-shinned Hawks, an occasional Broad-winged Hawk, Swainson's Hawks, and Merlins also appear (Beebe, 1974: 64; Crowell and Nehls, 1977: 213; Guiguet, 1952: 11; Hunn and Mattocks, 1979: 207, 1981: 216, 1982: 209, 1983: 215; Flahaut and Schultz, 1956: 50).

MANITOBA

Knowledge of hawk migrations in Manitoba is fragmentary, but Annual Fall Hawk Watch projects now are established (Holland, 1981: 12–13).

Delta

Spring flights (29 hawks per day) of Red-tailed Hawks and lesser numbers of other buteos and Sharp-shinned Hawks, Cooper's Hawks, Northern Goshawks, and Peregrine Falcons are reported at Delta at the southern end of Lake Manitoba (Hochbaum, 1955: 119, 129, 138, 171; Wedgwood, 1982b: 964). In 1968, Haugh (1972) made a detailed study of these migrations and noted the raptors moving northward along the Red River to the confluence with the Assiniboine River in Winnipeg, then westward along the Assiniboine to the Poplar Point area, then northwestward across the prairie toward Delta. At Delta the hawks tend to move northward along the west side of Lake Manitoba although some go north along the lake's east side.

Haugh (1972: 13) determined that the most abundant spring raptor migrants at Delta are Red-tailed Hawks, Sharp-shinned Hawks, and Northern Harriers.

Red River

Spring migrations of Northern Goshawks and Red-tailed Hawks are reported along the Red River south of Winnipeg (Houston and Shadick, 1974: 815).

Saint Adolphe Bridge

Spring migrations of Red-tailed Hawks, and lesser numbers of Bald Eagles, Northern Harriers, Sharp-shinned Hawks, Cooper's hawks, Northern Goshawks, Rough-legged Hawks, and American Kestrels occur regularly at this site south of Winnipeg. Occasionally Golden Eagles and Merlins also appear (Cleveland, et al., 1980: 16).

Whiteshell Provincial Park

Autumn Bald Eagle migrations occur at this park on the Ontario border in southeastern Manitoba (Serr, 1979: 189).

Winnipeg

Spring Red-tailed Hawk migrations in Winnipeg can contain several hundred birds per day, as many as 50 American Kestrels have been seen, and other species including Broad-winged Hawks also are occasionally reported (Houston, 1971b: 759; Lister, 1964: 461; Nero, 1962b: 424; Renaud, 1973: 786; Serr, 1976b: 856).

Autumm hawk migrations are less well known, but recent hawk counts indicate that Northern Harriers, Sharp-shinned Hawks, Red-tailed Hawks, and American Kestrels appear most frequently (Holland, 1981, 1982; Zoch, 1983). Occasionaly Broad-winged Hawks also appear (Harris, 1983: 192; Holland, 1981, 1982: Zoch, 1983).

NORTHWEST TERRITORIES

Little detailed information is reported on hawk migrations in this region. Renaud (1973: 786) reported a May sighting of a Broad-winged Hawk at Fort Simpson, and Kuyt (1961) reported more than 100 migrating Rough-legged Hawks "in late summer, 1960" at Contwoyto Lake, Mackenzie District, N. W. T.

Kuyt (1967) also reported that a Golden Eagle released in September 1965 near Fort Smith was recovered a month later 1,000 miles away 20 miles east of Roy, Montana. A nestling Peregrine Falcon banded in late July 1965 along the Thelon River, N. W. T., was recovered four months later approximately 9,000 miles south at India Muerta, Chaco Province, Argentina. Enderson (1965) also demonstrated that some Peregrine Falcons from this region migrated to the West Indies, Central America, and South America. Hunt, Rogers, and Slowe (1975) also suggested that Peregrines studied along the Texas coast during autumn originated in northern Canada including this region, and that the Rocky Mountains might serve as a flyway.

SASKATCHEWAN

Both spring and autumn hawk migrations occur yearly in this province. Northern Harriers banded in Saskatchewan, for example, migrate to the Gulf Coast and Mexico although birds from eastern Saskatchewan go to Georgia and Florida (Houston, 1968a). Red-tailed Hawks from this province migrate southward along a narrow corridor through eastern North Dakota, southern Minnesota, Iowa, eastern Kansas, Missouri, eastern Oklahoma, and Arkansas to the Texas Gulf Coast eastward to Louisiana, Mississippi, and Alabama (Houston, 1967: 111). Swainson's Hawks disperse southward to the Dakotas, Montana, Oklahoma, and Texas as well as north-central Argentina (one bird flew more than 6,500 miles from its banding site) and Central and South America (Houston, 1968b, 1974). Migrating Rough-legged Hawks tend to use routes during spring and autumn away from lakes and rivers whereas Bald Eagles prefer aquatic-associated routes (Gerrard and Hatch, 1983: 148).

Color-marked juvenile Bald Eagles in the vicinity of Besnard Lake in north-central Saskatchewan migrated southward during October to the Dakotas, Missouri, and Wyoming. Routes near rivers and lakes are used by 79.3 percent of the eagles whereas 20.7 percent use routes away from those aquatic areas (Gerrard and Hatch, 1983: 148).

Northward spring Bald Eagle migrations occur during March and April with 27.7 percent of the birds following open terrain and 72.3 percent using routes near rivers and lakes (Gerrard and Hatch, 1983: 148). The association with aquatic habitats was due to the need for roost sites near rivers. Those birds moving over open country, however, prefer irregular or hilly terrain because of deflective updrafts and thermals that aid the raptors in flight (Gerrard and Hatch, 1983).

Besnard Lake
As just indicated, autumn Bald Eagle migrations at this lake occur during October whereas the spring migrations occur during March, April, and May (Gerrard, et al., 1974; Gerrard, et al., 1978).

Candle Lake
Some premigratory or migratory Bald Eagles are reported at Candle Lake during September and early October (Serr, 1979: 189).

Indian Head
Spring hawk migrations at this site include Bald Eagles, and flights of as many as 2,542 Red-tailed Hawks during April (Gerrard and Hatch, 1983: 150; Houston, 1972b: 775; Houston and Shadick, 1974: 815; Serr, 1976b: 856, 1978b: 1022).

September migrations of small numbers of Turkey Vultures also are reported (Harris, 1981: 195).

Pathlow

April migrations of large numbers of Red-tailed Hawks, along with Northern Harriers, Rough-legged Hawks, Golden Eagles, and American Kestrels, are reported at this site 180 miles north of Regina (Beveridge, 1954).

Regina

Spring hawk migrations at Regina include Sharp-shinned Hawks, Red-tailed Hawks, and Rough-legged Hawks. A few Broad-winged Hawks also occur (Houston and Shadick, 1974: 815; Lister, 1964: 461; Nero, 1961: 420; Serr, 1977b: 1014; Wedgwood, 1982b: 864).

Autumn hawk migrations, between late August and the end of October, include Turkey Vultures, Cooper's Hawks, Swainson's Hawks, Rough-legged Hawks, Peregrine Falcons, and Prairie Falcons (Harris, 1980: 173; Hatch, 1968: 55; Houston, 1972a: 78; Krause, 1961: 52; Lister, 1965: 50; Nero, 1962a: 48, 1963: 42).

Rouleau

September flights of hundreds of Swainson's Hawks are reported at Rouleau (Serr, 1977a: 192).

Round Lake

September migrations of Turkey Vultures are reported at this site (Harris, 1981: 195; Serr, 1976a: 88).

Saskatoon

Spring hawk migrations at Saskatoon can be impressive. Bald Eagles and Red-tailed Hawks are especially well-represented (Gerrard and Gerrard, 1982; Houston, 1971b: 759, 1972b: 775; Wedgwood, 1982b: 864).

Autumn hawk migrations include Turkey Vultures, Bald Eagles, Swainson's Hawks, Rough-legged Hawks, American Kestrels, and Prairie Falcons (Hatch, 1966: 62–63, 1967: 49; Houston, 1971a: 72, 1973: 76; Nero, 1962a: 48; Serr, 1976a: 88).

Squaw Rapids Dam

Late autumn Bald Eagle migrations are reported at this dam with more than 50 birds counted at times (Houston, 1971a: 72–73; Serr, 1979: 189; Wedgwood, 1982a: 189).

Tisdale

Moderate April flights of Red-tailed Hawks are reported at Tisdale (Houston, 1972b: 775).

Tullis

Hundreds of migrating Swainson's Hawks are reported at Tullis in September (Boon, 1961: 23).

White Bear

September flights of Swainson's Hawks are reported at White Bear (Houston, 1973: 76).

CHAPTER 4

Great Lakes Hawk Migrations

In contrast to many areas of North America, the shorelines of the Great Lakes are major spring and/or autumn hawk migration routes. During spring these flights concentrate at the tips of peninsulas extending northward into the lakes, and/or along the southern shorelines of the lakes. In autumn, however, the largest hawk migrations occur along the northern or western shorelines although some also concentrate at peninsular tips extending southward into the lakes (Haugh, 1972; Heintzelman, 1975, 1979a; Pettingill, 1962).

ONTARIO (CANADA)

Ontario is a major hawk migration region during autumn, especially along the northern shorelines of the Great Lakes which serve as major diversion-lines (leading-lines) for Broad-winged Hawks and some other species which tend to avoid crossing large bodies of water.

Aldershot

Thousands of migrating Broad-winged Hawks have been seen at Aldershot during mid-September (Woodford and Lunn, 1962: 28).

Amherstburg

During September, at the Malden School near Amherstburg, thousands of migrating Broad-winged Hawks were reported by Millie Reynolds (Heintzelman, 1975) and observations there by other observers produced similar results (Bennett, Mitchell, and Gunn, 1958: 27; Goodwin, 1968: 32; Woodford, 1963: 30). A variety of other migrant raptors also pass through the area including Northern Harriers, Sharp-shinned Hawks, Red-tailed Hawks, and American Kestrels but in much smaller numbers than the Broad-wings.

56

Beamer Point Conservation Area

Good spring hawk migrations occur at this site (25.1 hawks per hour), on the Niagara Peninsula at the southwestern end of Lake Ontario a few miles southeast of Hamilton near the town of Grimsby (and frequently referred to as Grimsby in the hawk migration literature). Bird watchers and ornithologists from the Buffalo Ornithological Society, and Canadian associates from Hamilton and Toronto, now conduct careful spring counts at the Beamer Point Conservation Area (Copeland, 1977c; Klabunde, 1976; Klabunde and Burch, 1980). Turkey Vultures, Northern Harriers, Sharp-shinned Hawks, Cooper's Hawks, Red-shouldered Hawks, Broad-winged Hawks, Red-tailed Hawks, and American Kestrels pass the site in largest numbers (Copeland, 1978a; Klabunde, 1979a, 1981a, 1982a, 1983a, 1984a).

Cobourg

Pettingill (1962: 45) reported "heavy" hawk flights during autumn from Cobourg westward.

Great Duck Island

An autumn 1978 hawk count at this location demonstrated that Sharp-shinned Hawks, Broad-winged Hawks, and American Kestrels are the most abundant species, but many other raptors also are seen in smaller numbers (Goodwin, 1979a).

Grimsby

See Beamer Point Conservation Area.

Hamilton

During September there are thousands of migrating Broad-winged Hawks reported in the Hamilton area and a variety of other migrant hawks (386.5 birds per day) also appear in much smaller numbers (Goodwin, 1976a: 60; Woodford, 1963: 30). The most common autumn hawk migrants are Sharp-shinned Hawks, Broad-winged Hawks, Red-tailed Hawks, and American Kestrels (Goodwin, 1977a).

During spring, modest flights of Broad-winged Hawks also are seen at Hamilton, but generally these flights contain less than 1,000 birds (Burton and Woodford, 1960: 384; Gunn, 1958: 350; Woodford, 1962: 406).

Harsen's Island

A flight of 18,000 Broad-winged Hawks was seen on 17 September 1960 at this island near the mouth of the St. Clair River (Woodford and Burton, 1961a: 37).

Hawk Cliff

Hawk Cliff, along the northern shore of Lake Erie near Port Stanley, is one of the best (432.5 hawks per day) autumn hawk concentration areas in the Great Lakes region (Cameron, 1964; Clendinning, 1954; Dales, 1959; Elkins, 1956; Haugh, 1972; W. D. Sutton, 1956). Enormous flights of

Broad-winged Hawks sometimes are seen there during September as, for example, 70,889 Broad-wings on 16 September 1961 (Woodford and Lunn, 1962: 28). The most common species are Turkey Vultures, Northern Harriers, Sharp-shinned Hawks, Broad-winged Hawks, Red-tailed Hawks, and American Kestrels, although smaller numbers of other migrant hawks also are reported (Field and Field, 1979, 1980a, 1980b; Field and Rayner, 1974, 1976, 1977, 1978; Fowler and Fowler, 1981, 1982; Haugh, 1972).

Three separate flight-lines are used at Hawk Cliff. Falcons generally follow the lake shoreline and cliff closely, accipiters occur one-quarter mile inland along the edge of a wooded ravine, and buteos and harriers use varied flight-lines (Haugh, 1972).

The most detailed field studies of autumn hawk migrations at Hawk Cliff are those by Haugh (1972) and the various raptor banders now operating there (Field, 1970, 1971; Field and Field, 1979, 1980a, 1980b; Field and Rayner, 1974, 1976, 1977, 1978; Fowler and Fowler, 1981, 1982). Duncan (1981, 1982b, 1983) provided particularly detailed analyses of Cooper's Hawk, Sharp-shinned Hawk, and Red-tailed Hawk banding data from this site.

Holiday Beach Provincial Park
This park is located about 10 miles southeast of Amherstburg. It is noted for large autumn hawk flights (832.1 hawks per day). The most common species reported are Turkey Vultures, Northern Harriers, Sharp-shinned Hawks, Cooper's Hawks, Red-shouldered Hawks, Broad-winged Hawks, Red-tailed Hawks, and American Kestrels, but various other raptors also pass in much smaller numbers (Goodwin, 1975, 1977a, 1979a, 1980, 1982a; Kleiman, 1981; Weir, 1983a). At times the numbers of Broad-winged Hawks counted is impressive. On 16 September 1970, for example, 34,700 of these hawks were seen (Goodwin, 1971a: 51).

Killarney Provincial Park
Good autumn hawk flights are reported at this remote park along the northern edge of Georgian Bay in Lake Huron, but full details are unavailable (Heintzelman, 1979a: 232).

Kingston
Spring migrations of Broad-winged Hawks and Rough-legged Hawks are reported at Kingston (Goodwin, 1965: 467), and in autumn Rough-legged Hawk flights also are reported (Goodwin, 1968: 33).

Marathon
Marathon is located on the northeastern shore of Lake Superior. A variety of autumn hawk migrations pass this site (1.1 hawks per hour) with Sharp-shinned Hawks, Red-tailed Hawks, Rough-legged Hawks, and American Kestrels being the most commonly observed raptors (Goodwin, 1980, 1981). The site is particularly noted for the large numbers of migrating Rough-legged Hawks reported including 1,000 Rough-legs seen on 18 October 1949 (Korkoerle, 1960a). Minor spring hawk flights also pass the

site (Korkoerle, 1960b; Weir, 1983b: 864).

Meldrum Bay
Goodwin (1982a: 173) stated that "significant" hawk flights sometimes occur near the Mississagi Light on Manitoulin Island in northern Lake Huron.

Pickering
Thousands of migrating Broad-winged Hawks sometimes are reported in September at Pickering (Goodwin, 1969: 44, 1976a: 60, 1982b: 109).

Point Pelee National Park
This splendid park is one of Ontario's most famous birding locations and produces spring and autumn hawk migrations (Gray, 1961; Harrison, 1976: 105–120; Kleiman, 1966; Saunders, 1909; Stirrett, 1960).

Moderate spring hawk flights include Turkey Vultures, a remarkable record of an American Swallow-tailed Kite, several Mississippi Kites, and moderate flights of Red-shouldered Hawks and Broad-winged Hawks (Goodwin, 1970: 596, 1971b: 737, 1978: 998, 1979b: 766; Weir, 1983b: 864).

Autumn hawk flights at Point Pelee include some large flights of Sharp-shinned Hawks and lesser numbers of Turkey Vultures and Red-shouldered Hawks (Goodwin, 1969: 43, 1971a: 51, 1972a: 55; Woodford and Burton, 1961a: 36).

Port Credit
Spectacular autumn hawk flights (416.4 hawks per day) sometimes pass Port Credit southwest of Toronto (Pettingill, 1962: 45). As many as 50,000 Broad-wings have been reported at this site (Woodford, 1963: 30; Woodford and Lunn, 1962: 28).

Port Hope
This is another site where large flights of Broad-winged Hawks sometimes are reported during September (Woodford, 1963: 30; Woodford and Lunn, 1962: 28).

Port Stanley
Generally references to Port Stanley in the ornithological literature refer to Hawk Cliff.

Prince Edward Point National Wildlife Area
This site is located along Lake Ontario about 30 miles southwest of Kingston, at the eastern tip of a peninsula and headland extending into the lake, and serves as a bird-banding station (Copeland, 1977b: 19; Goodwin, 1982b: 157–158; Weir, 1974).

Some good autumn hawk flights (113.8 hawks per day) are reported at this location with Northern Harriers, Sharp-shinned Hawks, Broad-winged Hawks, Red-tailed Hawks, and some American Kestrels the species seen in largest numbers (Goodwin, 1977a).

Spring hawk flights also pass this site but fewer birds (81.7 per day) are reported than during autumn. Northern Harriers, Red-tailed Hawks, and

Rough-legged Hawks are moderately well represented in the spring flights (R. B. Stewart, 1975).

Rondeau Provincial Park

This park is located on the north shore of Lake Erie roughly midway between Hawk Cliff and Point Pelee. Some excellent autumn hawk flights are seen there, especially movements of Turkey Vultures, Sharp-shinned Hawks, and Broad-winged Hawks (Copeland, 1978b: 27–28; Goodwin, 1976a: 60; Heintzelman, 1979a: 233; Kleiman, 1979: 37).

Sarnia

Several thousand migrating Broad-winged Hawks have been reported at Sarnia during spring (Goodwin, 1977b: 994).

Thunder Bay

Autumn migrations of accipiters, buteos, falcons, and other raptors are reported in the Thunder Bay region along the northern side of Lake Superior but details are lacking (Anonymous, 1954).

Toronto

Autumn migrations of Broad-winged Hawks, sometimes numbering thousands of birds, occur in the Toronto area as do lesser numbers of Sharp-shinned Hawks and various other species (Dales, 1959; Goodwin, 1982b: 123).

Tobermory

Excellent spring hawk migrations are reported at Tobermory at the extreme northern tip of the long Bruce Peninsula extending far northward into Lake Huron (Heintzelman, 1979a: 233).

Vineland

Spring hawk migrations at Vineland on the Niagara Peninsula include some large flights of Broad-winged Hawks and smaller numbers of other species including Turkey Vultures, Red-shouldered Hawks, and Red-tailed Hawks (Goodwin, 1972b: 756, 1982b: 79–80).

Vinemount

This is a bird-banding station located about five miles west of the Beamer Point Conservation Area hawk lookout on the Niagara Peninsula (Duncan, 1982a). The species seen there in largest numbers during the spring hawk migration (32.7 hawks per hour) are Turkey Vultures, Sharp-shinned Hawks, Broad-winged Hawks, Red-tailed Hawks, and lesser numbers of various other species (Klabunde, 1979a, 1981a, 1982a, 1983a, 1984a).

ILLINOIS

Autumn hawk migrations in Illinois were studied since late in the last century (Brown, 1966; Cryder, 1928; Ferry, 1896; Ingram, 1966; Wasson and Shawvan, 1953), but David Johnson (1977) arranged the first organized autumn hawk watch in northeastern Illinois at Illinois Beach State Park. Thus began the state's modern era of hawk migration study.

Chicago

Spring and autumn hawk migrations cross Chicago including flights along the lake front (Sanders and Yaskot, 1977).

Spring hawk migrations include moderate numbers of Sharp-shinned Hawks and Broad-winged Hawks (Kleen, 1979b: 776; Mumford, 1959b: 374).

Autumn hawk flights, however, are larger than those in spring and sometimes include flights of Ospreys, Northern Harriers, Sharp-shinned Hawks, Broad-winged Hawks, Peregrine Falcons, and other species (Brown, 1966; Kleen, 1977: 184, 1978a: 212). In 1977, for example, a flight of 47 Peregrine Falcons on 1 October passed along the Chicago lake front at Northwestern University with a seasonal total of 64 Peregrines counted there between 24 September and 15 October (Kleen, 1978a: 212).

Fox River

A September 1977 flight of more than 7,000 Broad-winged Hawks was reported at this location (D. Johnson, 1977: 35).

Illinois Beach State Park

Some autumn hawk flights (8.9 hawks per hour) pass this park along the southwestern shore of Lake Michigan near Waukegan (Heintzelman, 1979a: 165; Sanders and Yaskot, 1977: 6). The species that are seen in largest numbers include Northern Harriers, Sharp-shinned Hawks, Red-tailed Hawks, and American Kestrels (D. Johnson, 1977; Peacock and Myers, 1982b: 30).

Jacksonville

Several moderate to fairly large September flights of Broad-winged Hawks are reported at Jacksonville (Peterjohn, 1982a: 183).

Lake Forest

Autumn hawk migrations at Lake Forest were observed as long ago as 1893 (Ferry, 1896) and various observers in this century also noted hawk flights there. Northern Harriers, Sharp-shinned Hawks, and Broad-winged Hawks seem to pass this site in largest numbers (Ferry, 1896; Nolan, 1953: 19, 1955: 29).

Lawrenceville

A flight of 47 Northern Harriers was seen at Lawrenceville on 12 March 1982 (Peterjohn, 1982b: 858).

Port Byron

Autumn migrations of Red-shouldered Hawks, Broad-winged Hawks, Red-tailed Hawks, and Peregrine Falcons are reported at Port Byron suggesting a flight-line there (Mumford, 1959a: 35).

Quincy

Autumn hawk flights at Quincy include hundreds of Sharp-shinned Hawks, Cooper's Hawks, various buteos, and American Kestrels (Mumford, 1961a: 45).

Thatcher Woods

Thatcher Woods is a part of the Cook County Forest Preserves. Late September migrations of Bald Eagles, Sharp-shinned Hawks, a Red-shouldered Hawk, Broad-winged Hawks, Red-tailed Hawks, several Peregrine Falcons, and other raptors suggest a flight-line there (Wasson and Shawvan, 1953).

Waukegan

Both spring and autumn hawk migrations occur along the lake shore at Waukegan. Sometimes several hundred Broad-winged Hawks are seen during spring (Kleen, 1979b: 776) whereas in autumn considerable numbers of Northern Harriers plus a variety of accipiters, buteos, falcons, and other species are reported (Petersen, 1971: 65; Sanders and Yaskot, 1977: 6).

INDIANA

A broad, general accounting of the migratory status of birds of prey in Indiana was provided by Keller, Keller, and Keller (1979: 107–114). In addition, Hicks (1935) leisurely drove across Indiana in an east-west direction between Fort Wayne, Indiana, and Danville, Illinois, and reported at least one hawk every 15 minutes between dawn and dusk but no raptor concentrations indicating a broad front migration rather than concentrations along diversion-lines or leading-lines. In all, 174 raptors of 11 species were tabulated. Most numerous were Northern Harriers, Cooper's Hawks, Red-tailed Hawks, and American Kestrels.

During the autumns of 1951 and 1952, Cope (1953) directed the Indiana portion of the federal government's international effort to study hawk migrations launched by Chandler S. Robbins. Few observers were available, and few birds seen, in 1951 whereas in 1952 10 persons observed on two September weekends and saw modest numbers of Cooper's Hawks, Broad-winged Hawks, and various other species.

Allen County

September flights of Turkey Vultures are reported in Allen County (Kleen, 1980a: 167).

Brookville

Moderate flights of Broad-winged Hawks are seen in September at Brookville but at most a few hundred birds are counted (Cope, 1953: 15; Nolan, 1956: 28).

Connersville

Moderate October and November flights of Turkey Vultures are reported at Connersville (Kleen, 1979a: 183; Mumford, 1960: 40).

Eel River

Some autumn hawk migrations pass over the vicinity of the Eel River in Wabash County including Turkey Vultures, Red-shouldered Hawks, Broad-winged Hawks, and Red-tailed Hawks (D. B. Snyder, 1961).

Indiana Dunes State Park/National Lakeshore
Good to excellent spring hawk migrations are reported at this state-federal park complex (Heintzelman, 1979a: 166), including some large flights of Red-tailed Hawks (Peterjohn, 1983b: 875).

According to Raymond Grow (letter of 22 January 1979), the spring hawk flights follow a flight-line parallel to the Lake Michigan shoreline in a regular east-to-west direction. Most commonly observed are Red-tailed Hawks, but many Sharp-shinned Hawks also appear. As many as 1,500 hawks were observed during a single day, but more typically 800 to 1,200 hawks are counted during an entire spring migration season (late March to early May). Autumn hawk migrations, however, are rare at this site.

Michigan City
Spring hawk flights at Michigan City sometimes include moderate numbers of Northern Harriers, Sharp-shinned Hawks, and various other species (Noland, 1958b: 357). Few autumn raptor migrants appear, but occasionally some Rough-legged Hawks are noted (Mumford, 1959a: 35).

Miller Beach
On 8 October 1983 a flight of about 21 Peregrine Falcons passed Miller Beach near the extreme southern tip of Lake Michigan. Ten birds moved east early in the morning on north-northeast winds whereas later in the morning 11 of the falcons flew in east and west directions. The flight was ephemeral, restricted to a few hours after a weather front passed, and produced unprecedented numbers of Peregrines for Indiana (Brock, 1984).

Wabash County
Autumn migrations of Turkey Vultures and buteos are seen in Wabash County (Kleen, 1980a: 167; Snyder, 1961).

MICHIGAN

Michigan is an important state for both spring and autumn hawk migrations and a varied assortment of studies and observations are reported (Barrows, 1912; Baumgartner, 1979, 1980, 1981, 1982, 1983; Cook, 1893; Isaacs and Hennigar, 1980; Magee, 1922; Payne, 1983; Pettingill, 1962; Postupalsky, 1976a; Sheldon, 1965; Tyrell, 1934; Wood, 1951).

Beaver Island Group
Both spring and autumn hawk migrations are known to cross these islands, in northeastern Lake Michigan at the northwestern end of the Lower Peninsula (Payne, 1983; Sheldon, 1965: 80). Unfortunately detailed information is not available about these flights.

Berrien County
This county is located in extreme southwestern Michigan bordering the southeastern shore of Lake Michigan. Important spring and autumn hawk

migrations cross the county, probably using the lake shoreline as a diversion-line or leading-line.

In spring, hundreds of migrant Sharp-shinned Hawks and Broad-winged Hawks sometimes are reported as are lesser numbers of many other birds of prey (Bernard, 1968: 529; Soulen, 1970: 606). Some of these migrants may be birds seen in nearby Indiana.

During autumn, thousands of migrating Broad-winged Hawks sometimes are reported in September in this county, and lesser numbers of other raptors including Red-tailed Hawks and Rough-legged Hawks also appear (Maley, 1974a: 57, 1975: 62; Soulen, 1971: 62, 1972: 67).

Brockway Mountain

Important spring hawk migrations (41.9 hawks per hour) occur at this Keweenaw Peninsula site. Sharp-shinned Hawks, Broad-winged Hawks, Red-tailed Hawks, Rough-legged Hawks, and American Kestrels are reported in largest numbers but many other species also are a part of the migrations (Binford, 1965; Isaacs and Hennigar, 1980; Kelley, 1965a, 1965b; Peacock and Myers, 1982a, 1983a; Perkins, 1975; Postupalsky, 1976a; Sheldon, 1965; Weaver, 1977; Wood, 1933).

Delta County

Occasionally September flights of Broad-winged Hawks numbering thousands of birds are reported from this Upper Peninsula county on the northern shoreline of Lake Michigan (Kelley, 1972a: 4, 1975: 26).

Detroit Region

Large September migrations of Broad-winged Hawks sometimes cross the Detroit region—thousands of hawks forming some of the flights (Merriam, 1954a, 1956; Miller, 1952).

Escanaba

During autumn, as many as 9,000 migrating Broad-winged Hawks have been reported at Escanaba, although some flights of these birds contain only several hundred individuals (Green, 1964a: 39; Kelley, 1975: 24, 26).

Isle Royale

Sharp-shinned Hawk migrations are reported at this island in northwestern Lake Superior from mid-August to the third week in September (Wood, 1951).

Binford (1965), Peet (1905), and Wood (1933) speculated that an established autumn flight-line exists between the island and the Keweenaw Peninsula to the south.

Lake Superior South Shore

Postupalsky (1976a) reported a spring Bald Eagle migration, along with some other birds of prey, along Michigan's south shore of Lake Superior in the Porcupine Mountains area.

Luce County

According to Beebe (1933), "immense numbers" of migrating accipiters, buteos, and falcons were seen prior to 1909 in the area near Newberry.

Among the species seen were Cooper's Hawks, Red-shouldered Hawks, Broad-winged Hawks, American Kestrels, and Peregrine Falcons.

Manitou Island
Wood (1933) watched thousands of hawks arriving at this island east of the tip of the Keweenaw Peninsula from late April to mid-June 1931, but Isaacs and Hennigar (1980: 31) did not see any hawks leaving the eastern tip of the peninsula during their spring hawk migration study.

Muskegon
Wood (1951) reported thousands of Sharp-shinned Hawks migrating past Muskegon during September 1929 as noted by Frank Antisdale.

Ottawa County
This county is located along the eastern shoreline of Lake Michigan west of Grand Rapids. An autumn 1973 hawk watch there by James Ponshair produced numbers (93.3 hawks per hour) of Sharp-shinned Hawks, Broad-winged Hawks, and American Kestrels and lesser numbers of various other species (Kelley, 1974: 33).

Pointe Mouille State Game Area
Systematic autumn hawk counts at this site were made in 1982 and demonstrated that excellent hawk flights (178.3 hawks per hour) pass the lookout. The most numerous species reported are Turkey Vultures, Sharp-shinned Hawks, Broad-winged Hawks, Red-tailed Hawks, and American Kestrels, but a variety of other species also are seen in much smaller numbers (Kleiman, 1983: 28–29).

Porcupine Mountains
Bald Eagles, and some other hawks, migrate westward in spring along the Porcupine Mountains along the Lake Superior south shoreline in northern Michigan but full details of these flights are unavailable (Kelley, 1976: 10; Postupalsky, 1976a).

Port Huron
Spring hawk migrations at Port Huron at the southern end of Lake Huron across the international border from Sarnia, Ontario, are substantial. At times up to 5,000 Broad-winged Hawks are observed and small numbers of other raptors also are reported (Bernard, 1966b: 511–512; Green, 1963: 406; Woodford and Burton, 1961b: 407).

Saint Ignace
William N. Grigg's observations of autumn hawk migrations crossing the Straits of Mackinac at St. Ignace demonstrate that dozens of migrant Red-tailed Hawks, and various other species, sometimes occur at this site (Kelley, 1974: 33, 1976: 10).

Saint Joseph
On 30 October 1969, a migration of 150 Rough-legged Hawks passed St. Joseph (Robbins, 1970: 51).

Sault Ste. Marie
Broad-front autumn hawk migrations are reported at Sault Ste. Marie and

many of these birds probably reach sites such as Cedar Grove, Wisconsin, on the western side of Lake Michigan, as they continue their southward migrations (Mueller and Berger, 1967b).

Schoolcraft County

Prior to 1909, Beebe (1933: 118) reported huge autumn hawk flights crossing this county on the Upper Peninsula bordering the north shore of Lake Michigan so that "the earth was appreciably darkened at times as the hordes passed over." Cooper's Hawks, Northern Goshawks, Red-shouldered Hawks, Broad-winged Hawks, and Red-tailed Hawks were identified. Thousands of migrating Broad-winged Hawks also are reported in mid-September near Manistique (Kelley, 1967: 19).

Seney National Wildlife Refuge

Spring Bald Eagle migrations pass this refuge on the Upper Peninsula with peak numbers of birds noted in mid-April (Postupalsky, 1976a: 100).

Straits of Mackinac

Sheldon (1965) demonstrated that considerable numbers of migrant hawks pass across the Straits with Turkey Vultures, Broad-winged Hawks, Red-tailed Hawks, and small numbers of other species being noted. Little information is available on autumn hawk flights at this location, although Kelley (1974: 33, 1976: 10) noted flights of Red-tailed Hawks there.

Whitefish Point

Whitefish Point, at the tip of a broad peninsula that extends into Lake Superior and points toward mainland Ontario about 20 miles away (Wallace, 1960), is one of North America's major spring hawk migration concentration areas (46.9 hawks per hour). As at many other hawk migration sites in the United States, large numbers of migrating hawks were shot at this location earlier in this century (Beebe, 1933: 119; Magee, 1922; Tyrell, 1934).

Systematic observations by the Whitefish Point Bird Observatory, and various observers at the Point prior to the establishment of the Observatory, demonstrate that Northern Harriers, Sharp-shinned Hawks, Broad-winged Hawks, Red-tailed Hawks, Rough-legged Hawks, and American Kestrels are the most numerous species reported, although small numbers of other raptors also are seen (Baumgartner, 1979, 1980, 1981, 1982, 1983; Baumgartner and Kelley, 1981; Bernard, 1966b: 512, 1967: 511, 1968: 529; Eckert, 1980: 779, 1981: 826; Janssen, 1975: 856; Maley, 1972: 764, 1974b: 805; Schneider 1978a, 1979; Soulen, 1970: 606).

Limited autumn hawk migrations also pass Whitefish Point but these flights are meager compared with the spring migrations (Magee, 1922: 258; Wallace, 1960).

A spring Sharp-shinned Hawk banding project also operates at the Point and produced information on Sharp-shin age, eye color, molt, sex, and behavior (Kelley, 1972b: 69).

MINNESOTA

The general status of hawk migrations in Minnesota was documented by Hatch (1892), Roberts (1932), and Green and Janssen (1975). Beer (1966) also studied the state's Merlin records and demonstrated that the spring migration begins in late March, peaks in April, and ends in early May in a scattered or broad-front distribution. During autumn a scattered route also is used from mid-August to late October.

The best known and most completely documented hawk flights in Minnesota, however, are the large autumn migrations that annually pass the Hawk Ridge Nature Reserve in Duluth (Hofslund, 1952, 1954a, 1954b, 1955, 1958, 1962, 1966).

Hawk Ridge Nature Reserve

Interest in hawk flights at Duluth began as a conservation effort designed to stop the shooting of migrating raptors at that site and only later turned to more ornithological aspects of the migrations (Bronoel, 1948, 1954; Lakela, 1946, 1948). Today the Hawk Ridge Nature Reserve conducts the hawk counts at this major autumn concentration site (81.5 hawks per hour). Data from that effort, and earlier efforts, demonstrate that Turkey Vultures, Northern Harriers, Sharp-shinned Hawks, Northern Goshawks, Broad-winged Hawks, Red-tailed Hawks, Rough-legged Hawks, and American Kestrels are the species seen in largest numbers, although other species also are seen in smaller numbers (Green, 1962a, 1962b, 1964b, 1965a, 1965b; Gunderson, 1951; Hofslund, 1952, 1954a, 1954b, 1955, 1958, 1962, 1966; Kohlbry, 1980a; Kuyava, 1958; Peacock and Myers, 1982b; Sundquist, 1973).

A raptor banding station also operates at the Hawk Ridge Nature Reserve in autumn and gathers information on raptor weights, sizes, colors, feeding habits, and engages in some educational programs (Anonymous, 1973; Evans, 1974, 1975, 1976, 1978, 1979, 1980; Olson, 1952; Rosenfield and Evans, 1980).

Homer

During late autumn as many as 100 Bald Eagles are reported migrating southward over bluffs beside the Mississippi River, and small numbers of hawks also use this flight-line (Reese, 1973).

Knife River

An autumn banding station at this site 20 miles northeast of Duluth produces a variety of migrant hawks (11 per hour) with Sharp-shinned Hawks, Northern Goshawks, Broad-winged Hawks, Red-tailed Hawks, and Rough-legged Hawks appearing in largest numbers and small numbers of other species also reported (Peacock and Myers, 1983b: 30–31).

Lake Superior North Shore

Observations of autumn hawk migrations some 60 to 80 miles north of Duluth along the shore of Lake Superior demonstrate that Broad-winged

Hawk flights there are much smaller than those seen at Duluth. Osprey flights, however, are as good as or better than those at Duluth, some fine flights of Bald Eagles and Golden Eagles also are seen, and large numbers of American Kestrels are seen (Green, 1965b).

Minneapolis
During September small migrations of Broad-winged Hawks, generally not more than a few hundred per day, are seen at Minneapolis (Hofslund, 1954b: 97; McIntosh, 1957).

Reno
Small autumn migrations (9.3 hawks per hour) of Bald Eagles, Sharp-shinned Hawks, and Red-tailed Hawks are reported at Reno (Peacock and Myers, 1983b: 30–31).

Route 169
Migrating hawks sometimes are seen over high bluffs adjacent to the Minnesota River and Route 169, southward as far as Nicollet County (Huber, 1964: 23).

Two Harbors
Hofslund (1954b: 96) stated that large autumn hawk migrations sometimes occur at Two Harbors along Lake Superior.

NEW YORK (GREAT LAKES AREA)

New York is a very large state in which hawk migrations occur at numerous locations from the shorelines of Lakes Erie and Ontario, to the state's interior, to the coast of the Atlantic Ocean (Drennan, 1981; Heintzelman, 1975, 1979a; Pettingill, 1977). Hawk migrations passing along the Great Lakes shorelines, and immediately adjacent areas, are discussed in this section.

Allegany Road
This road is located near the southeastern end of Lake Erie a few miles southwest of Buffalo. Spring hawk flights (542.6 hawks per day) are reported at this site with Sharp-shinned Hawks and Broad-winged Hawks the most common species noted but various other species also appearing in small numbers (Anonymous, 1965: 52–54; Drennan, 1981: 466–467; Seeber, 1969: 116–118; Wagner, 1973: 58).

Arkwright Hills
Modest spring hawk flights sometimes are reported from the Arkwright Hills above Lake Erie between Portland and Westfield. Sharp-shinned Hawks and Broad-winged Hawks seem to occur in largest numbers but a few other raptors also may appear at times (Seeber, 1970: 66–67).

Bailey Road
This is an auxiliary lookout for the Braddock Bay Hawk Watch and is located a few miles inland from that major site. It is used when northerly or easterly winds occur, because the flight-line tends to move inland in a wide

or narrow band under those conditions (Moon and Moon, 1980b: 52, 1984b: 53).

Braddock Bay Park
Braddock Bay Park, near Rochester along the south shoreline of Lake Ontario, is one of North America's best spring hawk watching sites (62.7 hawks per hour). Systematic hawk counts made there demonstrate that Turkey Vultures, Northern Harriers, Sharp-shinned Hawks, Red-shouldered Hawks, Broad-winged Hawks, Red-tailed Hawks, Rough-legged Hawks, and American Kestrels are the most numerous species seen, but a variety of other raptors also are counted in smaller numbers (Dodge, 1981; Listman, Wolf, and Bieber, 1949; Moon, 1977a–1983; Moon and Moon, 1979a–1984c; Wolf, 1984). Recently Taylor (1984) established a raptor banding station at this site.

Cassadaga Valley
Spring migrations of Ospreys, Red-shouldered Hawks, and Rough-legged Hawks are reported in this valley near Stockton but these flights seem to be modest in size (Rew, 1971: 42).

Derby Hill
Derby Hill is a major spring hawk watching location near the southeastern end of Lake Ontario. Large numbers of migrant hawks (543.3 per day) pass the site with Turkey Vultures, Ospreys, Northern Harriers, Sharp-shinned Hawks, Cooper's Hawks, Red-shouldered Hawks, Broad-winged Hawks, Red-tailed Hawks, Rough-legged Hawks, and American Kestrels being the most abundant species reported (Anonymous, 1984; Baker, 1983; Haugh, 1972; Muir, 1978, 1979, 1980, 1981, 1982; Peakall, 1962; Smith, 1973; Smith and Muir, 1978).

Dunkirk
Various spring hawk migrations pass Dunkirk, but Sharp-shinned Hawks and Broad-winged Hawks are the species reported in largest numbers— generally a few hundred birds on prime days (Grzybowski, 1966: 63–64).

Four Mile Creek State Park
This park is located along Lake Ontario's shoreline in Niagara County. Some fine spring hawk flights are seen at this site with a good mix of species reported but generally Sharp-shinned Hawks and Broad-winged Hawks the most numerous (Grzybowski, 1966: 63; Seeber, 1969: 116–118, 1970: 65–68).

Frisbee Hill
This is an auxiliary hawk lookout used in association with the Braddock Bay Hawk Watch near Rochester. It is located 1.5 miles southwest of the main lookout in Braddock Bay Park and is used when the wind blows from northerly or easterly directions (Moon and Moon, 1980b: 52).

Hamburg
Fine spring hawk migrations are known to pass the Hamburg area with a good assortment of species represented, Sharp-shinned Hawks and Broad-

winged Hawks being the most numerous birds (Anonymous, 1965: 53–54; Grzybowski, 1966: 63–64; Klabunde, 1984b: 12–13; Ulrich, 1942: 4–5; Woodford, 1962: 406; Woodford and Burton, 1961b: 407).

Hamlin

According to Moon and Moon (1979a: 59), thousands of migrating hawks were seen 20 miles west of Braddock Bay Park at Hamlin on 25 April 1979—presumably part of a large, broad-front hawk migration that day in the region. It was speculated that most of these hawks eventually passed the Braddock Bay Hawk Watch and were counted there.

Hilton-Northwoods School

Located 2.6 miles from the main hawk watching lookout at Braddock Bay Park near Rochester, Hilton-Northwoods School is an adjunct lookout for the Braddock Bay Hawk Watch. Sometimes it is used when northerly or easterly winds force migrating hawks inland away from the Lake Ontario shoreline (Moon and Moon, 1980b: 52).

Irondequoit Bay Area

This site is located in the Rochester area one or two miles south of Lake Ontario and 10 miles east of Braddock Bay Park. Good spring hawk flights are known from the area (Moon and Moon, 1979b: 60).

Lake View

Some interesting spring hawk migrations are reported from Lake View just inland from the south shore of Lake Erie a few miles south of Buffalo. Although data are fragmentary, the most abundant species reported there to date include Northern Harriers, Sharp-shinned Hawks, Red-shouldered Hawks, and American Kestrels (Anonymous, 1965: 52–53).

Lyndonville

An excellent spring flight of hawks was seen on 21 April 1966 at Lyndonville with Sharp-shinned Hawks, Cooper's Hawks, and Broad-winged Hawks the most numerous species but various other raptors also included in the flight (Grzybowski, 1966: 64).

Pecor Road

This road is located in Portland Township, Chautauqua County. Some worthwhile flights of Sharp-shinned Hawks and Broad-winged Hawks, plus a few other species, are reported here in April (Seeber, 1969: 116–118).

Pinehurst

Drennan (1981: 467) stated that this site, the edge of a cliff overlooking Lake Erie near Buffalo, is an outstanding spring hawk watching lookout.

Point Peninsula

This site, almost an island, extends into the northeastern end of Lake Ontario. Some large concentrations of Rough-legged Hawks were reported there in November and December although large rodent populations may have been responsible for the numerous raptors observed (Belknap, 1962: 22).

Portland
Hundreds of migrating Broad-winged Hawks, and small numbers of other raptors, sometimes are reported in spring in the Portland area slightly inland from the south shoreline of Lake Erie and a few miles southwest of Dunkirk (Anonymous, 1965: 54; Woodford and Burton, 1961b: 407).

Shadigee
Modest spring flights of Sharp-shinned Hawks, Broad-winged Hawks, and a variety of other raptors sometimes are reported at Shadigee (Seeber, 1970: 65–68).

Youngstown
During spring limited numbers of migrant hawks are reported at the Creek Road near Youngstown. Sharp-shinned Hawks and Broad-winged Hawks are the species appearing in largest numbers (Anonymous, 1965: 54).

OHIO

The general status of hawk migrations in Ohio was discussed by Trautman and Trautman (1968) and Thomson (1983). More detailed hawk migration observations, however, extend back as far as 1858 (Deane, 1905) and continue to the present (Henderson, 1962, 1963; Klabunde, 1980a: 12, 1984a: 21).

Cedar Point National Wildlife Refuge
This federal refuge east of Toledo along Lake Erie is usually closed to public use. Nevertheless, spring hawk flights pass the site almost daily with Turkey Vultures, Northern Harriers, Sharp-shinned Hawks, Cooper's Hawks, Red-shouldered Hawks, Broad-winged Hawks, Red-tailed Hawks, and American Kestrels seen most frequently, but various other raptors also appearing at times (Thomson, 1983: 10–11).

Cleveland
Spring hawk migrations in Cleveland and its vicinity received as much attention and study as any place in Ohio. There are numerous locations in the area that serve as spring hawk watching lookouts, with varying degrees of success, including the Brecksville Reservation, Cleveland Heights, the Federal Building, Lake Erie Junior Science Center, Lakewood, Portage Escarpment, Rocky River Reservation, the Terminal Tower, and Warren Heights (Henderson, 1962, 1963; Henderson and Margolis, 1959: 15; Hoffman, 1979: 18–19; Newman, 1977; Thomson, 1983). Henderson (1962, 1963) also suggested a possible hawk migration flight-line from west to east across Hilliard Bridge-Perkins Beach-Terminal Tower-Fenn College-Lakeview Cemetery-Sherman Road.

A variety of species appear during these spring hawk migrations, but Turkey Vultures, Sharp-shinned Hawks, and Broad-winged Hawks are most numerous (Henderson and Margolis, 1959: 15; Kleen, 1975b: 860,

1979b: 776, 1980b: 781; Mumford, 1959b: 374, 1961b: 414; Petersen, 1965: 45, 1968b: 532, 1970b: 614; Temple, 1962: 40).

Columbus

Various spring hawk migrations cross the Columbus area with Ospreys, Northern Harriers, Sharp-shinned Hawks, Cooper's Hawks, Red-shouldered Hawks, Broad-winged Hawks, Red-tailed Hawks, and American Kestrels reported most frequently (Mumford, 1959b: 374; Thomson, 1983: 72–75, 97).

Crane Creek State Park

This park is located along the south shoreline of Lake Erie east of Toledo. Some fine spring hawk migrations are reported at this park with Turkey Vultures, Northern Harriers, Sharp-shinned Hawks, Red-shouldered Hawks, and Red-tailed Hawks observed in largest numbers—up to several hundred per day (Kleen, 1980b: 781–782).

Little Miami River Overlook

Spring migrations of Ospreys and Broad-winged Hawks are reported from this site near Cincinnati (Thomson, 1983: 143).

Lorain

Spring migrations of Turkey Vultures, Sharp-shinned Hawks, and Red-tailed Hawks are reported from Lorain (Klabunde, 1984a: 20–21; Kleen, 1979b: 776, 1980b: 781).

Mercer County

This county in western Ohio is the location for the state's oldest record of hawk migrations. On 17 September 1858, G. C. Coney observed a continuous stream of migrating hawks moving from the north-northeast to the south-southwest at Grand Lake (Deane, 1905: 13–14). Presumably the birds were Broad-winged Hawks. The report is based upon a letter sent to John Woodhouse Audubon and Victor Gifford Audubon—sons of John James Audubon.

Ottawa National Wildlife Refuge

Some fine spring hawk migrations (40.9 hawks per hour) pass over this refuge 20 miles east of Toledo along Lake Erie. Species observed in largest numbers tend to be Turkey Vultures, Ospreys, Northern Harriers, Sharp-shinned Hawks, Red-shouldered Hawks, Broad-winged Hawks, Red-tailed Hawks, Rough-legged Hawks, and American Kestrels (Klabunde, 1980a: 12, 17, 1982a: 19; Thomson, 1983: 49).

Painesville

Small spring flights of Broad-winged Hawks and Rough-legged Hawks were reported at Painesville near the south shoreline of Lake Erie (Doolittle, 1919: 568).

Sandusky

This city is midway between Cleveland and Toledo and south of a series of islands that act as stepping stones across the western end of Lake Erie to

Point Pelee, Ontario. During the spring of 1981, a hawk watch in the Sandusky area produced small numbers (3.2 hawks per hour) of migrating Turkey Vultures, Northern Harriers, Red-tailed Hawks, and a few other species, but only 3 Broad-winged Hawks (Klabunde, 1982a: 19).

South Bass Island
South Bass Island is part of a chain of islands extending into Lake Erie north of Port Clinton, Ohio. Some spring hawk migrations are reported over the island—especially Sharp-shinned Hawks, Cooper's Hawks, Red-shouldered Hawks, Broad-winged Hawks, and Red-tailed Hawks (Heintzelman, 1979a: 199; Thomson, 1983: 57–58). During autumn some migrant hawks, especially Rough-legged Hawks, also are reported (Thomson, 1983: 58).

Tiffin
Occasionally spring flights of Turkey Vultures are seen at Tiffin (Petersen, 1968b: 532), and during autumn migrations of Broad-winged Hawks occasionally are reported (Kleen, 1977: 184).

Toledo
Occasionally small flights of Broad-winged Hawks are seen in spring over Toledo (Nolan, 1958b: 357).

Turtle Creek State Park
Edward and Cheryl Pierce report some fine spring hawk migrations at this park along the shoreline of Lake Erie. Red-shouldered Hawks and Red-tailed Hawks are seen in largest numbers (Klabunde, 1979a: 13).

Virginia Kendall Metropolitan Park
High cliffs in this Akron park serve as spring and autumn hawk watching lookouts. Among the species reported are Turkey Vultures, Ospreys, Northern Harriers, Sharp-shinned Hawks, Cooper's Hawks, Red-shouldered Hawks, Broad-winged Hawks, Red-tailed Hawks, American Kestrels, and others (Thomson, 1983: 60–61).

PENNSYLVANIA (GREAT LAKES AREA)

Pennsylvania is a major state in which to study hawk migrations (Heintzelman, 1975, 1979a). This section deals with hawk migrations seen along the south shoreline of Lake Erie and immediately adjacent areas.

Erie County
Spring hawk migrations occur along the south shoreline of Lake Erie (Heintzelman, 1979a: 206–208). The flights in Erie County were known as long ago as 1888 when Samuel E. Bacon reported hundreds of hawks migrating on mild spring days although, by 1903, a decrease in the numbers of hawks observed was suggested (Todd, 1904: 491). Recent studies of spring hawk flights in this county demonstrate that many species (54.7 hawks per hour) sometimes occur, but Turkey Vultures, Northern Har-

riers, Sharp-shinned Hawks, Red-shouldered Hawks, Broad-winged Hawks, Red-tailed Hawks, and American Kestrels appear in largest numbers (Baxter, 1969, 1970a, 1970b, 1971, 1972; Stull and Stull, 1966: 62–64, 1967: 50, 1968).

Lake City

Some good spring hawk migrations are reported at Lake City just south of the Lake Erie shoreline (Heintzelman, 1979a: 206). Field studies conducted there by James and Carolyn Baxter demonstrate that a variety of species of raptors form the flights, but the most numerous species tend to be Turkey Vultures, Ospreys, Northern Harriers, Sharp-shinned Hawks, Broad-winged Hawks, and American Kestrels (Stull and Stull, 1966: 62–64, 1967: 50, 1968: 52–53).

Presque Isle

Todd (1904: 552–556, 1940: 122–167) was perhaps the first to publish some general details on spring and autumn hawk migrations at Presque Isle. The site now is a recognized hawk watching location (Heintzelman, 1979a: 207).

Spring hawk flights there are modest in numbers, but six or more species can be seen at times with Sharp-shinned Hawks and Broad-winged Hawks the most numerous (Burton and Woodford, 1960: 384; Stull and Stull, 1966: 63, 1967: 50, 1968: 52).

Some small hawk migrations also occur at Presque Isle during September and October, but they are meager compared with the northward spring hawk flights (Todd, 1904, 1940).

WISCONSIN

Cory (1909) provided some general comments regarding autumn hawk migrations in Wisconsin, but it was the work of Jung (1935, 1964) and particularly the autumn field studies of Mueller and Berger (1961–1973) and Mueller, Berger, and Allez (1977, 1981) at the Cedar Grove Ornithological Station that fully demonstrated the importance of this state as a center of hawk migration study. In addition, the extraordinary late autumn and spring Bald Eagle (and other hawk) migrations reported at the Eagle Valley Nature Preserve make that site the finest eagle migration observation point in North America (Ingram, Brophy, and Sherman, 1982: 106–112).

Arpin

Modest autumn hawk migrations (3.1 hawks per hour) are reported at Arpin, with Broad-winged Hawks, Red-tailed Hawks, Rough-legged Hawks, and American Kestrels the most plentiful species (Green, 1975: 48–49).

Bark Point

A systematic spring hawk migration study (18.7 hawks per hour) began in 1982 at Bark Point near Herbster, Wisconsin—a site some 45 miles east of

Duluth, Minnesota. A good variety of species was reported, but Bald Eagles, Northern Harriers, Sharp-shinned Hawks, Broad-winged Hawks, Red-tailed Hawks, and Rough-legged Hawks were the most numerous species (Peacock and Myers, 1983a: 18).

Cedar Grove Ornithological Station

This ornithological research station is located near Cedar Grove along the west shoreline of Lake Michigan (Berger, 1954). It is the site of one of the most intensive, long-term hawk migration field studies in North America. Particular emphasis is on banding migrating raptors. Major research papers resulting from the work at this site include studies of weather and autumn hawk migrations (Mueller and Berger, 1961), summer movements of Broad-winged Hawks (Mueller and Berger, 1965), wind drift, leading-lines, and hawk migrations (Mueller and Berger, 1967a), autumn Sharp-shinned Hawk migrations (Mueller and Berger, 1967b; Mueller, Berger, and Allez, 1981), sex ratios and measurements of migrant Northern Goshawks (Mueller and Berger, 1968), periodic Northern Goshawk invasions (Mueller and Berger, 1967c; Mueller, Berger, and Allez, 1977), and daily rhythms of migrant raptors (Mueller and Berger, 1973). A very controversial paper dealing with *Accipiter* identification in North America also was published (Mueller, Berger, and Allez, 1979), but it unfortunately caused great confusion in the hawk watching community because it contained information of little or no value or use to hawk watchers making flight identifications of accipiters.

Generally the most common species observed in autumn at this site are Ospreys, Northern Harriers, Sharp-shinned Hawks, Cooper's Hawks, Broad-winged Hawks, Red-tailed Hawks, American Kestrels, Merlins, and Peregrine Falcons (Mueller and Berger, 1961), whereas during the spring when limited hawk flights occur the most commonly reported species are Northern Harriers, Cooper's Hawks, and American Kestrels (Berger, 1954).

Eagle Valley Nature Preserve

Unequalled in North America as a Bald Eagle migration area in autumn and spring, this preserve near Cassville, Wisconsin, is a high bluff overlooking the Mississippi River (Heintzelman, 1979a: 227–228). As many as 1,500 migrating Bald Eagles have been reported at this site during one week in late December in addition to another 1,455 Bald Eagles counted earlier in the season (Ingram, Brophy, and Sherman, 1982: 106–109; T. N. Ingram, letter of 27 July 1984).

The most numerous species reported in the autumn migrations (13.5 hawks per hour) at this location are Turkey Vultures, Bald Eagles, Sharp-shinned Hawks, Broad-winged Hawks, and Red-tailed Hawks, although smaller numbers of other species also are seen (Brophy, 1982: 1, 3; Ingram, Brophy, and Sherman, 1982: 106–112).

Grant County
Modest autumn hawk flights (9.8 hawks per hour) sometimes are seen in Grant County along the Mississippi River where William A. Smith noted a variety of species. Most numerous are Sharp-shinned Hawks, Broad-winged Hawks, and Red-tailed Hawks, although no species is seen in very large numbers (Green, 1975: 48–50).

Harrington Beach State Park
Good autumn hawk migrations are sometimes seen at this park on the shoreline of Lake Michigan near Port Washington (Heintzelman, 1979a: 228; Pettingill, 1977: 654–655).

Little Suamico Hawk Station
Some large autumn flights of Broad-winged Hawks, at times numbering more than 13,000 birds, occasionally are reported at this site in Brown County (Tessen, 1976: 73, 1981: 185).

Milwaukee
Jung (1935) reported autumn hawk migrations along the bluffs forming the waterfront of Milwaukee—many of the flights seen at considerable heights. September flights of as many as 4,000 Broad-winged Hawks are known (Tessen, 1976: 73).

Moccasin Mike Road
Worthwhile spring hawk migrations (55.3 hawks per hour) are reported from this site along the south shore of Lake Superior two miles east of Superior, Wisconsin. The most numerous species reported included Northern Harriers, Sharp-shinned Hawks, Broad-winged Hawks, Red-tailed Hawks, and Golden Eagles (Schneider, 1978a).

Sauk County
Spring hawk migration studies conducted by Randy Hoffman at Ferry and adjacent bluffs in the southern part of Sauk County produced a variety of migrant hawks (30 birds per hour), of which the most numerous were Turkey Vultures, Ospreys, Sharp-shinned Hawks, Broad-winged Hawks, Red-tailed Hawks, Rough-legged Hawks, and American Kestrels (Hoffman, 1982: 123–124).

Sheboygan County
Jung (1964) stated that this county was on an important autumn hawk migration flyway, and Bernard (1966a: 50) reported more than 2,000 migrant Broad-wings there in late September.

CHAPTER 5

New England Hawk Migrations

Although observations of hawk migrations in New England were published as early as 1756 (Goldman, 1970), it was not until a century or more later that ornithologists studied autumn hawk flights in these states (Allen, 1871; Merriam, 1877; Trowbridge, 1895).

Since then great progress in field studies of New England hawk migrations developed during the first 70 years of this century (Anonymous, 1953; Bagg, 1947, 1949, 1950; Bagg and Eliott, 1937; Clement, 1958; Elkins and Elkins, 1941; Forbush, 1927; Hagar, 1937a, 1937b; Hanley, 1968; May, 1950; Powell, 1957; Schumacher, 1952). More recently still, sustained and organized interest in studying hawk migrations in New England produced large amounts of information and many new ideas as a result of various private persons and the Connecticut Audubon Council, Hawk Migration Association of North America, New England Hawk Watch, and the Vermont Institute of Natural Science (Currie, 1978, 1979; Currie and Cote, 1981–1983; Currie and MacRae, 1979, 1980; Elkins, 1974; Fischer, 1979; Floyd, 1983; Goldman, 1974; Heintzelman, 1975, 1979a; Hopkins and Mersereau, 1971–1976; Norse, 1979a–1983b; Pistorius, 1977, 1978; Roberts, 1977a, 1977b, 1978; Robinson, 1978, 1981; Rowlett, 1977; Trichka, 1981–1984; Will, 1974–1976).

CONNECTICUT

Merriam (1877) and Trowbridge (1895) were the first to document annual autumn hawk migrations in Connecticut. Since 1970, however, increasingly detailed field studies of spring and autumn hawk migrations were completed in this state as documented in this section and elsewhere throughout this book.

77

Ansonia
Mid-September flights of several hundred migrating Broad-winged Hawks are reported at this southern Connecticut location (Hopkins and Mersereau, 1975b).

Bald Peak
Bald Peak is a well known autumn hawk watching site (28.6 hawks per hour) near Salisbury (Heintzelman, 1979a: 159). Species that appear in largest numbers in autumn are Sharp-shinned Hawks, Broad-winged Hawks, and Red-tailed Hawks, although many other species also are reported in limited numbers (Currie, 1978, 1979; Currie and Cote, 1982; Currie and MacRae, 1979).

Barkhamstead
Mid-September migrations of up to several hundred Broad-winged Hawks sometimes appear at this location (Hopkins and Mersereau, 1975b).

Beacon Falls
In mid-September, flights up to several hundred Broad-winged Hawks are sometimes reported at this site northwest of Waterbury (Currie and MacRae, 1979, 1980).

Beech Rock
More than 500 migrating Broad-winged Hawks are reported at this northwestern Connecticut site in mid-September (Currie, 1978).

Bethlehem
In mid-September, up to more than 900 migrating Broad-winged Hawks are reported at Bethlehem in western Connecticut (Currie, 1978; Hopkins and Mersereau, 1975b).

Bluff Head
This rocky bluff in North Guilford is a well known autumn hawk watching lookout (Heintzelman, 1979a: 159–160). In mid-September, as many as 4,761 Broad-winged Hawks are reported although only hundreds of these birds sometimes appear and in spring poor hawk flights are noted (Heintzelman, 1979a: 159–160; Hopkins and Mersereau, 1975b; Proctor, 1978: 63–64).

Booth Hill
At this site near West Hartland, counts of migrating Broad-winged Hawks have numbered over 1,600 birds during mid-September (Currie, 1979; Currie and MacRae, 1980).

Branford
Several hundred migrating Broad-winged Hawks sometimes are seen in autumn at this southern Connecticut site (Currie and MacRae, 1980; Hopkins and Mersereau, 1974b).

Bridgewater
Hundreds of migrating Boad-winged Hawks sometimes are reported in September over this city (Currie, 1978, 1979; Currie and MacRae, 1979, 1980).

Brookfield
From dozens to several hundred Broad-winged Hawks sometimes appear in September at this western Connecticut site (Currie, 1978, 1979; Currie and MacRae, 1979, 1980).

Canaan
Mid-September migrations of more than 1,100 Broad-winged Hawks are occasionally reported at Canaan (Hopkins and Mersereau, 1974b).

Danbury
Autumn hawk migrations at Danbury sometimes produce thousands of migrating Broad-winged Hawks along with dozens of Ospreys, Sharp-shinned Hawks, and other species (Currie and Cote, 1982).

East Hartland
In recent years, some very large mid-September flights of Broad-winged Hawks were reported at this northern Connecticut location, including more than 4,500 Broad-winged Hawks counted in 1981 at Pine Mountain (Currie and Cote, 1981, 1982; Currie and MacRae, 1979, 1980).

Easton
September migrations of Broad-winged Hawks range from dozens to several thousand birds being counted (Currie and MacRae, 1979, 1980).

East Rock Park
This park is located in New Haven (Proctor, 1978: 35–39). During autumn counts of several hundred to several thousand migrating Broad-winged Hawks are reported (Currie, 1978; Currie and MacRae, 1979; Hopkins and Mersereau, 1975b).

Goshen
During September hawk watchers at this location have recorded more than 8,400 migrating Broad-winged Hawks in addition to small numbers of other raptors (Currie, 1978, 1979; Currie and Cote, 1982, 1983; Currie and MacRae, 1979, 1980).

Greenwich Audubon Center
The Audubon Center near Greenwich frequently is used as an autumn hawk watching lookout, often with excellent results (70.3 hawks per hour) (Heintzelman, 1979a: 158–159). Of the various raptors observed migrating past that site the most numerous generally are Ospreys, Sharp-shinned Hawks, Broad-winged Hawks, Red-tailed Hawks, and American Kestrels (Currie and Cote, 1981, 1982, 1983). During some seasons more than 9,500 migrant Broad-winged Hawks are reported (Currie and Cote, 1981).

Guilford
Proctor (1978: 48) reports that fine autumn hawk migrations pass Guilford. Northern Harriers, Sharp-shinned Hawks, American Kestrels, and Peregrine Falcons form many of the flights.

Hammonasset State Park
During September and October large flights of Sharp-shinned Hawks, and fine numbers of American Kestrels, Merlins, and Peregrine Falcons are

reported at Madison along the coast. So, too, are Ospreys, Broad-winged Hawks, and Red-tailed Hawks (Proctor, 1978: 52).

Kent

September migrations of more than 900 Broad-winged Hawks sometimes pass Kent, although generally 100 or less are counted (Hopkins and Mersereau, 1972, 1974b, 1975b).

Larsen Sanctuary

This sanctuary is owned and operated by the Connecticut Audubon Society and is located in Fairfield. Autumn hawk counts made there between 1980–1983 (38 hawks per hour) demonstrate that the most numerous species passing the site are Ospreys, Sharp-shinned Hawks, Broad-winged Hawks, Red-tailed Hawks, and American Kestrels, but smaller numbers of other raptors also are reported (Currie and Cote, 1982; Trichka, 1981, 1982, 1984). Included in the seasonal totals are thousands of migrant Sharp-shinned Hawks and Broad-winged Hawks (Trichka, 1981, 1982, 1984).

Laurel Reservoir

During September counts of migrating Broad-winged Hawks at this site range from hundreds to thousands of birds reported (Currie and MacRae, 1979; Hopkins and Mersereau, 1974b, 1975b). The reservoir is located near Stamford.

Lighthouse Point Park

This New Haven park is Connecticut's finest autumn hawk watching lookout—especially for migrant Sharp-shinned Hawks and American Kestrels as well as lesser numbers of other raptors (Heintzelman, 1979a: 160; Proctor, 1978: 41–44).

Early observations of hawk flights at this site (or its vicinity) extend into the last century when Merriam (1877) and Trowbridge (1895) made observations there. The latter study was especially important.

Autumn hawk counts at the park since 1975 (39.9 hawks per hour) demonstrate that the most abundant species seen there are Ospreys, Northern Harriers, Sharp-shinned Hawks, often Broad-winged Hawks, and American Kestrels, of which thousands of Sharp-shins, Broad-wings, and American Kestrels are noted (Currie, 1979; Currie and Cote, 1982, 1983; Currie and MacRae, 1979, 1980; Mersereau, 1976; Trichka, 1982).

Litchfield

At times during September observers on Chestnut Hill report migrating Broad-winged Hawks crossing the Litchfield area range from hundreds up to occasionally 4,400 (Currie, 1978; Currie and Cote, 1982; Heintzelman, 1979a: 160; Hopkins and Mersereau, 1974b, 1975b).

New Fairfield

During the 1950s and 1960s, Frances J. Gillotti observed a variety of hawks migrating through the area during autumn (474.4 hawks per day) of which thousands of Broad-winged Hawks were noted (Heintzelman, 1975:316).

New Haven Airport
On 29 September 1977, at this airport one mile north of Lighthouse Point Park in New Haven, more than 4,800 migrating Sharp-shinned Hawks and American Kestrels were counted during five hours of observation (Currie, 1978).

Newtown
Hawk watchers stationed on Palestine Hill and/or Whippoorwill Hill sometimes report large September hawk migrations at Newtown in southwestern Connecticut. Counts of Broad-winged Hawks range from hundreds to several thousand birds (Currie, 1978, 1979; Currie and Cote, 1982; Currie and MacRae, 1979, 1980; Hopkins and Mersereau, 1975b).

Old Lyme
Observers stationed at Griswold Point near Old Lyme sometimes report hundreds of migrating Sharp-shinned Hawks and American Kestrels in an autumn season—but far fewer birds (6.3 hawks per hour) than appear at Lighthouse Point (39.9 hawks per hour) farther south along the coast (Currie and Cote, 1982).

Oxford
September migrations of Broad-winged Hawks at Oxford sometimes number hundreds of birds and occasionally reach counts of more than 1,600 hawks (Hopkins and Mersereau, 1975b).

Sharon
There is a record of a flight of 1,300 migrating Broad-winged Hawks at Sharon in mid-September (Hopkins and Mersereau, 1972).

South Windsor
Autumn hawk migrations at South Windsor (19.3 hawks per hour) sometimes produce hundreds of Broad-winged Hawks in September (Currie and Cote, 1982, 1983; Currie and MacRae, 1980).

Stamford Museum
A mid-September 1979 hawk count at the Stamford Museum produced an assortment of more than 5,000 hawks (Currie and MacRae, 1979).

Stratford
In September up to 2,000 migrating Broad-winged Hawks are reported at Stratford, but generally far fewer are seen there (Hopkins and Mersereau, 1974b, 1975b).

Talcott Mountain
Spring hawk migrations at Talcott Mountain west of Hartford include some good counts of Ospreys seen by Rt. Rev. Robert M. Hatch (Heintzelman, 1975: 72–73) and more recently almost 1,000 migrating hawks of various species (Currie, 1979).

September hawk counts there produce only limited numbers of hawks including approximately 100 Broad-winged Hawks (Heintzelman, 1975: 72–73; Hopkins and Mersereau, 1974b).

Torrington
September flights of up to 1,750 migrating hawks, probably mostly Broad-wings, are reported at Torrington (Currie, 1978; Currie and MacRae, 1980).

Washington
As many as 1,505 migrating hawks, presumably mostly Broad-wings, were reported at Washington during four observation days (Currie and Cote, 1982).

Waterbury
More than 3,400 migrating hawks, probably Broad-wings, were reported during two observation days at Waterbury but other observations there produced less than 100 birds (Currie and MacRae, 1979, 1980).

Watertown
The campus of the Taft School in Watertown is an autumn hawk watching site (Heintzelman, 1979a: 161). Thousands of migrating Broad-winged Hawks sometimes are reported there during September, but only very limited numbers of other hawks (90.1 hawks per hour) are known to pass the site (Currie, 1978, 1979; Currie and Cote, 1981, 1982, 1983; Currie and MacRae, 1979, 1980).

West Hartland
Spring hawk flights at West Hartland produce as many as 624 hawks per day in April (Currie and MacRae, 1979), suggesting a spring flight-line there.

Autumn hawk migrations at this site sometimes produce as many as 2,770 hawks (mostly Broad-wings) in a single day, although several hundred birds often are reported (Currie, 1978; Currie and Cote, 1982; Currie and MacRae, 1979, 1980; Nopkins and Mersereau, 1973b, 1974b, 1975b).

Westport
Autumn hawk counts (80.9 hawks per day) made by Mortimer F. Brown at Westport between 1953–1962 produced a varied assortment of hawks with Sharp-shinned Hawks and Broad-winged Hawks the most numerous (Heintzelman, 1975: 73).

Weston
September flights of Broad-winged Hawks have contained up to 1,689 birds per day but at other times only a few hundred or less (Currie, 1979).

West Rock Park
This park near New Haven produces some fine autumn hawk flights (26.5 hawks per hour). The species most numerous are Ospreys, Sharp-shinned Hawks, Broad-winged Hawks, and American Kestrels, but limited numbers of other raptors also are reported (Currie, 1979; Currie and Cote, 1983; Currie and MacRae, 1979, 1980).

Willington
Willington is in northeastern Connecticut where few autumn hawk counts

have been completed. In 1977, however, 49 hours of observation there produced 610 hawks (Currie, 1978).

MAINE

Until about 1970, knowledge of spring and/or autumn hawk migrations in Maine was very fragmentary and isolated (Cruickshank, 1941; Hebard, 1960; Packard, 1962; Palmer, 1949; Powell, 1957). During recent years, however, organized hawk migration studies were established resulting in a good deal more information on these raptor movements in this state.

Beech Mountain
Recently limited studies of autumn hawk migrations (27.2 hawks per hour) were made at Beech Mountain near Southwest Harbor on Mount Desert Island. The species that are most numerous are Ospreys, Northern Harriers, Sharp-shinned Hawks, American Kestrels, and Merlins (Currie and Cote, 1982; Currie and MacRae, 1980; MacRae, 1979a: 10).

Bradbury Mountain State Park
The summit of this park near Freeport is a good autumn hawk watching lookout (Heintzelman, 1979a: 167–168; MacRae, 1979a: 10).

Cape Rosier
As many as 900 migrating Broad-winged Hawks are reported in mid-September at Cape Rosier (Packard, 1962: 163).

Casco Bay Area
During the late 1940s gunners shot migrating hawks during autumn in the Casco Bay area (Appell, 1975: 8–11), but the area now is used as a good hawk migration study site (37.2 hawks per hour) near Harpswell (Heintzelman, 1979a: 168).

Field studies conducted during autumn by George N. Appell during the period 1965–1983 indicate that the most numerous species observed are Ospreys, Northern Harriers, Sharp-shinned Hawks, Broad-winged Hawks, American Kestrels, and Merlins (Appell, unpublished data; Cote, 1982, 1983; Currie and Cote, 1983; MacRae, 1978; Phinney, 1977).

Cutler Area
Pierson and Pierson (1981: 145) state that autumn hawk migrations sometimes are seen along Route 191 between Cutler and West Lubec. Northern Harriers, Sharp-shinned Hawks, Northern Goshawks, Red-tailed Hawks, and Rough-legged Hawks are noted.

East Boothbay
Aaron Bagg and his wife observed more than 100 migrating Broad-winged Hawks in mid-September over East Boothbay—part of a flight of Ospreys, Broad-wings, and American Kestrels he observed that day along the Damariscotta River in Lincoln County (Packard, 1962: 163).

Fletcher Neck
East Point on Fletcher Neck near Biddeford is reported to be an excellent

site to observe autumn hawk flights including Merlins and Peregrine Falcons (Pierson and Pierson, 1981: 34).

Fort Island

A mid-September flight of 125 Broad-winged Hawks and a few Ospreys and American Kestrels was seen at Fort Island in the Damariscotta River by Aaron Bagg and his wife (Packard, 1962: 163).

Freeport

Nearly 100 migrating Broad-winged Hawks were reported in mid-September following a flight-line approaching Mast Landing Sanctuary in Freeport (Packard, 1962: 163).

Hawk Mountain

Despite its unfortunate name, which can cause confusion with the famous Pennsylvania site, some autumn hawk flights (6.2 hawks per hour) are reported at Hawk Mountain, Maine, according to Penny Richards who made autumn field studies there during 1980–1983. Counts of most species were meager, but somewhat larger numbers of Sharp-shinned Hawks, Broad-winged Hawks, Red-tailed Hawks, and American Kestrels were noted (Currie and Cote, 1981, 1982; Currie and MacRae, 1980).

Hermit Island

A late September flight of nearly 400 Sharp-shinned Hawks was reported at Hermit Island near Bath (Finch, 1976: 30).

Kittery Point

A late September flight of more than 400 migrating Sharp-shinned Hawks was reported at Kittery Point in southwestern Maine (Finch, 1976: 31).

Matinicus Rock

Located at the southeastern corner of the Fox Islands in Penobscot Bay, Matinicus Rock occupies a strategic position and attraction for migrating land birds during the autumn migrations. Sprucehead, 18 miles northwest of the island, is the nearest mainland point (Mazzeo, 1955).

When northwest winds prevail, various autumn hawk flights are noted at the Rock where some stop to rest briefly. Sharp-shinned Hawks, American Kestrels, and Merlins, in particular, are sometimes seen there—then generally head toward Monhegan Island 18 miles west of the Rock (Mazzeo, 1955: 351–352).

Monhegan Island

Both spring and autumn hawk migrations are noted at Monhegan Island off the tip of Pemaquid Point in Muscongus Bay.

Among the autumn migrants seen at Monhegan Island (2.6 hawks per hour) are Ospreys, Northern Harriers, Sharp-shinned Hawks, American Kestrels, Merlins, Peregrine Falcons, and a few other species on rare occasions (Cruickshank, 1941: 114; Currie and Cote, 1983; Heintzelman, 1979a: 168; Vickery, 1982: 153, 1983: 156). A Sharp-shinned Hawk and Merlin flight-line seems to exist there.

Some spring hawk migrations also appear at Monhegan Island, especially in the Lobster Cove area (Pierson and Pierson, 1981: 100).

Mount Agamenticus

This site in southern Maine is one of the state's most active autumn hawk migration lookouts. Field studies conducted by Rena Cote during the period 1979–1983 (20.9 hawks per hour) indicate that Ospreys, Sharp-shinned Hawks, Broad-winged Hawks, American Kestrels, and Merlins are the most numerous species observed but various other species also appeared in limited numbers (Cote, unpublished data, 1982; Currie and Cote, 1982, 1983; MacRae, 1981).

Spring hawk migrations at this site are meager. Sharp-shinned Hawks and Broad-winged Hawks are the most numerous species seen (Rena Cote, unpublished data).

Mount Cadillac

Some fine autumn hawk migrations are reported from Mount Cadillac in Acadia National Park (Heintzelman, 1975, 1979a). Sharp-shinned Hawks seem to be the most numerous, but moderate numbers of Ospreys, Northern Harriers, and American Kestrels also are reported, and occasionally other species also appear (Heintzelman, 1975; Howard Drinkwater, personal communication; George H. Kelly, letter of 9 October 1975).

Stockton Springs

A mid-September flight of more than 1,200 migrating Broad-winged Hawks was reported at Stockton Springs near the mouth of the Penobscot River at its entrance into Penobscot Bay (Currie and Cote, 1983).

Swan Island

Powell (1957) reported an early September flight of 76 Broad-winged Hawks crossing the Kennebec River between Swan Island and Dresden Neck.

West Penobscot Bay

Mid-August to early September migrations of Northern Harriers were observed following a flight-line westward across Penobscot Bay, and south toward Little Green Island from Mark Island (Hebard, 1960: 5). There also are mid-August sightings of Golden Eagles in the West Penobscot Bay area (Hebard, 1960: 5).

MASSACHUSETTS

During this century some very distinguished field studies of hawk migrations were conducted in Massachusetts (Bagg, 1947, 1949, 1950; Bagg and Eliot, 1937; Elkins and Elkins, 1941; Fischer, 1979; Forbush, 1927; Gooley, 1978; Hagar, 1937a, 1937b; May, 1950; Robinson, 1981; Schumacher, 1952). As a result of these and other efforts, a variety of hawk migration lookouts now are identified in the state (Heintzelman, 1979a: 172–176; Roberts, 1977b).

Ashburnham
During September and early October, five days of observation at this site produced more than 600 migrating hawks (Currie and Cote, 1982).

Ashby
More than 400 migrating hawks are reported in mid-September at Ashby (Currie and Cote, 1982).

Ashley Falls
Mid-September migrations of up to 1,597 migrating hawks (mostly Broad-wings) are reported at this location (Hopkins and Mersereau, 1974b; 1975b).

Berry Mountain
Berry Mountain is located at Hancock in northwestern Massachusetts. Various migrating hawks are reported there during autumn (16.4 hawks per hour), the most numerous species being Sharp-shinned Hawks, Broad-winged Hawks, and Red-tailed Hawks (Currie and Cote, 1981, 1982, 1983).

Blanford
Occasionally hundreds of migrating Broad-winged Hawks are reported in mid-September at Blanford (Hopkins and Mersereau, 1975b).

Blueberry Hill
This lookout, one mile north of West Granville, produces excellent autumn hawk flights (47.5 hawks per hour) but relatively poor spring flights (Heintzelman, 1979a: 172; Roberts, 1977b: 107, 109). The species reported in largest numbers in autumn are Sharp-shinned Hawks, Broad-winged Hawks, and American Kestrels (Currie, 1979, 1980; Currie and Cote, 1982, 1983).

Bowditch Hills
Moderate autumn hawk flights (22.3 hawks per hour) pass this site with Broad-winged Hawks being the species seen in largest numbers (several hundred) but few individuals of other species (Currie and Cote, 1982).

Crane's Reach
This coastal site at Ipswich is reported to be an autumn migration flight-line for Ospreys, eagles, Northern Harriers, accipiters, and falcons (Roberts, 1977b: 108).

Fisher Hill
Fisher Hill in Westhampton produces some fine autumn hawk migrations (91.6 hawks per hour). Seasonal counts of nearly 4,500 Broad-winged Hawks and lesser numbers of other species are reported occasionally (Bagg, 1949: 135; Currie, 1978, 1979; Currie and Cote, 1981, 1982, 1983; Currie and MacRae, 1979, 1980; Hopkins and Mersereau, 1974b, 1975b). Poor spring hawk migrations are reported at this site (Heintzelman, 1979a: 173).

Fort Devens

Autumn hawk migrations at Fort Devens at Lancaster sometimes contain hundreds of Sharp-shinned Hawks and more than 1,000 Broad-winged Hawks as part of the seasonal totals (Currie and Cote, 1981; Currie and MacRae, 1979).

Framingham

Hundreds of migrating hawks are reported passing Framingham in mid-September (Currie and MacRae, 1979).

Granby

During mid-September more than 700 migrant hawks are seen at Granby (Currie and Cote, 1982).

Greenfield

Flights of more than 1,300 migrating Broad-winged Hawks occasionally are seen at Greenfield (Hopkins and Mersereau, 1974b).

Hatfield

More than 4,400 migrating Broad-winged Hawks, and a few other raptors, were reported at this location during several days of autumn observation (Currie and Cote, 1983).

Lenox Mountain

As many as 900 Broad-winged Hawks are occasionally seen during autumn from Lenox Mountain above the Pleasant Valley Sanctuary near Lenox (May, 1950: 240), but other hawk watching efforts produced only a few dozen birds (Currie and MacRae, 1979).

Littleton

In mid-September more than 1,200 migrating hawks (probably Broad-winged Hawks) are reported at this northeastern Massachusetts site (Currie and MacRae, 1979).

Martha's Vineyard

During autumn limited numbers of migrating Peregrine Falcons pass along the southeastern end of this island in the area between Squibnocket Point to Zacks Cliffs, and some birds also appear over Felix Neck (Heintzelman, 1975, 1979a: 173).

Minnechoag Mountain

In mid-September, more than 500 migrating hawks were reported during several days of observation at this site near Hampden (Currie and Cote, 1981).

Montgomery

More than 1,000 migrating Broad-winged Hawks occasionally are reported in mid-September at Montgomery (Hopkins and Mersereau, 1974b, 1975b).

Mount Everett State Reservation

Excellent autumn hawk migrations (36.2 hawks per hour) are reported at Mount Everett State Reservation near Sheffield (Heintzelman, 1979a: 173).

Some seasonal hawk counts contain more than 2,500 Broad-winged Hawks
and a fine selection of other species in lesser numbers (Currie, 1978, 1979;
Currie and Cote, 1981, 1982, 1983; Hopkins and Mersereau, 1974b,
1975b).

Mount Holyoke

Occasionally in mid-September as many as 475 migrating hawks are reported at Mount Holyoke (Currie and MacRae, 1979), but at other times
few birds are seen (Hopkins and Mersereau, 1975b).

Mount Lincoln

Up to 650 migrating Broad-winged Hawks occasionally are seen in mid-
September at Mount Lincoln in central Massachusetts (Hopkins and Mersereau, 1974b, 1975b).

Mount Toby

Mount Toby, in Sunderland, sometimes produces good autumn hawk
flights containing several hundred birds (Currie and Cote, 1982; Currie
and MacRae, 1979; May, 1950: 240).

Mount Tom State Reservation

Of all of New England's hawk watching lookouts none has a reputation
more celebrated than Mount Tom near Holyoke where autumn hawk
flights (94.2 hawks per hour) are watched from famous Goat Peak (Bagg,
1947, 1949, 1950, 1970; Bates, 1975: 24; Elkins, 1974: 103; Hagar, 1937a,
1937b; Heintzelman, 1975, 1979a: 174; May, 1950; Roberts, 1977b: 107,
109). Some seasonal hawk count totals included more than 11,700 Broad-
winged Hawks plus hundreds of Ospreys, Sharp-shinned Hawks, and
American Kestrels as well as smaller numbers of other raptors (Currie,
1979; Currie and Cote, 1981, 1982, 1983; Currie and MacRae, 1979, 1980;
Fischer, 1979; Heintzelman, 1975: 315; Gagnon, 1974).

Some smaller spring hawk migrations also appear at Mount Tom and
are observed from the Bray Tower or near the old Peregrine cliff two miles
south of the reservation's headquarters (Hagar, 1937a: 6; Elkins, 1974:
103). The most numerous spring migrants are Sharp-shinned Hawks and
Broad-winged Hawks but other species sometimes are seen, too (Fischer,
1979: 132).

Mount Wachusett State Reservation

During the period 1978–1983, field studies at this site (70.1 hawks per
hour) by Paul M. Roberts demonstrate that some very large autumn flights
of Broad-winged Hawks pass Mount Wachusett. In 1983, for example, the
seasonal count for Broad-wings was more than 26,900 birds and hundreds
of Ospreys, Sharp-shinned Hawks, and lesser numbers of other species also
formed part of the migrations (Currie, 1979; Currie and Cote, 1981, 1982,
1983; Currie and MacRae, 1979; Heintzelman, 1979a: 174; Roberts,
1977b: 107, 108, 1981).

Otis
Occasionally hundreds of migrating hawks are reported at Otis in September (Hopkins and Mersereau, 1974b).

Parker River National Wildlife Refuge
During autumn a good flight-line for migrating Ospreys, Northern Harriers, accipiters, eagles, and falcons exists at this refuge along coastal Massachusetts, but spring hawk migrations there are meager (Heintzelman, 1979a: 175; Roberts, 1977b: 108).

Pittsfield
Occasionally during September several hundred migrating hawks are reported at this location—mostly Broad-wings (Currie, 1978, 1979; Currie and MacRae, 1979; Hopkins and Mersereau, 1974b, 1975b).

Plum Island
During spring some migrations of Sharp-shinned Hawks, American Kestrels, and Merlins are reported at Plum Island along the Massachusetts coast (Elkins, 1974: 103; Finch, 1970: 579; Nikula, 1982: 829; Vickery, 1977: 974, 1978b: 978).

During autumn a migration of Ospreys, eagles, Northern Harriers, accipiters, and falcons also is reported at this site (Roberts, 1977b: 108).

Province Lands State Reservation
Fair spring hawk migrations are reported over the outer beach, or the top of High Dune, at this reservation at Cape Cod's outer tip (Heintzelman, 1979a: 175–176; Pease and Goodrich, 1978: 40–47).

Quabbin Reservoir
Good autumn hawk migrations (29.6 hawks per hour) are reported at Quabbin Reservoir in central Massachusetts (Bagg, 1949, 1950; Heintzelman, 1979a: 176; Roberts, 1977b: 107, 109). During some seasons, several thousand migrating Broad-winged Hawks are counted, hundreds of migrant Sharp-shinned Hawks appear, and lesser numbers of various other species also are included in the flights (Currie, 1978; Currie and MacRae, 1979, 1980; Gooley, 1978; Hopkins and Mersereau, 1974b, 1975b).

Round Top Hill
Fair autumn hawk migrations (17.7 hawks per hour) are reported at this site in Athol (Heintzelman, 1979a: 176). Some seasonal hawk counts include hundreds of Sharp-shinned Hawks and Broad-winged Hawks and lesser numbers of other raptors based upon limited coverage (Currie, 1978, 1979; Currie and Cote, 1981, 1982; Currie and MacRae, 1979, 1980; Hopkins and Mersereau, 1974b, 1975b). Poor spring hawk migrations are reported at this location (Heintzelman, 1979a: 176).

Salisbury Beach State Reservation
Roberts (1977b: 108) reported autumn flights of Ospreys, eagles, Northern Harriers, accipiters, and falcons at this coastal Massachusetts site.

Sheffield

Apparently a limited spring hawk migration exists at various Berkshire County locations including Kelsey Road in Sheffield where late March migrations of Turkey Vultures, Ospreys, Northern Harriers, and various other species are reported (Schumacher, 1952: 238–239).

Shelburne Falls

Occasionally mid-September flights of more than 1,100 Broad-winged Hawks are reported at this site (Hopkins and Mersereau, 1974b).

Silver Hill

Occasionally during mid-September up to several hundred migrating Broad-winged Hawks are reported at Silver Hill (Currie and MacRae, 1979; Roberts, 1977b: 108, 110).

Southwick

During recent years some fine autumn hawk migrations (146.6 hawks per hour) are reported at Loomis Street in Southwick. Seasonal counts of more than 17,816 Broad-winged Hawks, hundreds of Sharp-shinned Hawks, and lesser numbers of many other species have been tabulated (Currie, 1979; Currie and Cote, 1981, 1982, 1983).

Springfield

Probably the oldest published report of hawk migrations in Massachusetts is from Springfield where in April 1862 flights of buteos and numerous accipiters were seen. Autumn hawk flights also were noted in the area (Allen, 1871: 173).

Sugarloaf

Occasionally during September more than 100 migrating hawks are observed from Sugarloaf near South Deerfield (Hopkins and Mersereau, 1974b; May, 1950: 240).

Truro

Occasionally during May flights of more than 125 Sharp-shinned Hawks are reported at Truro near Cape Cod's outer end (Nikula, 1983: 846; Vickery, 1977: 974).

Wellesley

During mid-September, flights of several hundred Broad-winged Hawks occasionally are reported at Wellesley (Hopkins and Mersereau, 1975b).

Wellfleet

Forster (1982: 6) reported a late October flight of more than 1,000 Sharp-shinned Hawks in the Marconi Station area at Wellfleet and watched the hawks head north along the adjacent dunes.

West Newberry

In late April small accipiter and falcon migrations occasionally are reported at West Newberry (Elkins, 1974: 103).

Wilbraham

Autumn migrations of more than 2,600 Broad-winged Hawks and lesser numbers of other raptors are reported at Wilbraham (Currie and Cote, 1983).

Williamstown

Elkins (1974: 103) stated that a secondary autumn hawk migration flight-line exists at Williamstown.

NEW HAMPSHIRE

Hawk watching and migration study in New Hampshire is not an activity with a very long history (Anonymous, 1953: 19, 1966; Heintzelman, 1975: 68–69, 314; Hill, 1957; Smart, 1969; Thielen, 1967; Wellman, 1957). During recent years, however, considerably more interest developed in these migrations as a result of a coordinated autumn hawk watch (MacRae, 1979b: 30–31).

Birchwood Ski Area

Autumn hawk flights (24.7 hawks per hour) at this Londonderry site produce seasonal counts of nearly 2,000 Broad-winged Hawks and limited numbers of other raptors such as Ospreys, Sharp-shinned Hawks, American Kestrels, and a few other species (Currie and MacRae, 1980).

Center Ossipee

Mid-September flights of more than 1,000 Broad-winged Hawks are reported at Center Ossipee just south of Lake Ossipee in east-central New Hampshire (Hopkins and Mersereau, 1974b).

Concord

Mid-September flights of up to about 500 Broad-winged Hawks are reported occasionally from Concord (Hopkins and Mersereau, 1974b).

Hancock

During September occasional seasonal counts of more than 2,000 hawks (mostly Broad-wings) are reported at Hancock (Currie and MacRae, 1980).

Kidder Mountain

At least one mid-September flight of more than 3,700 migrating Broad-winged Hawks was reported at Kidder Mountain at the southern end of the Wapack Range (Currie and MacRae, 1980).

Little Round Top

Inspiration Point on Little Round Top, near Bristol, is one of New Hampshire's regularly used autumn hawk watching lookouts (Heintzelman, 1975: 67–68, 1979a: 178–179). The most numerous autumn raptor migrants (53.5 hawks per day) are Broad-winged Hawks, but other species including Ospreys, Sharp-shinned Hawks, and limited numbers of other hawks also form part of the migrations (Currie, 1979; Currie and MacRae, 1980; Hopkins and Mersereau, 1974b, 1975b).

Spring hawk migrations at Little Round Top are relatively poor (41.2 hawks per day) but a few Sharp-shinned Hawks, Broad-winged Hawks, and other species sometimes are noted (Hopkins and Mersereau, 1973a, 1974a; Mersereau, 1977: 4).

Mount Kearsarge

Mount Kearsarge was used as a hawk watching lookout as long ago as 1952 (Anonymous, 1953: 19). More recent autumn hawk counts (18.1 hawks per hour) there produced more than 700 Broad-winged Hawks and a variety of other species during several observation days (Currie, 1979; Currie and Cote, 1982, 1983; MacRae, 1979b: 30).

Mount Uncanoonuc

Good autumn hawk migrations are reported at this site near Goffstown, but poor spring flights are noted (Heintzelman, 1979a: 180). Occasionally mid-September flights of more than 2,500 Broad-winged Hawks are reported (Currie, 1979; Hopkins and Mersereau, 1975b; MacRae, 1979b: 30).

New Ipswich

Wildcat Mountain near New Ipswich is one of New Hampshire's oldest productive autumn hawk watching lookouts (Anonymous, 1953; Heintzelman, 1975: 68–69, 314; Hill, 1957; Wellman, 1957). Cora Wellman's observations there during the autumns of 1952–1960 (99.5 hawks per day) demonstrated that as many as 1,600 Broad-winged Hawks passed the site along with lesser numbers of Ospreys, Sharp-shinned Hawks, American Kestrels, and various other species (Heintzelman, 1975: 314).

Pack Monadnock

Autumn hawk counts at Pack Monadnock produce some good reports (28.2 hawks per hour) of Sharp-shinned Hawks, Broad-winged Hawks, and limited numbers of various other raptors (Currie, 1979; Currie and Cote, 1982, 1983; Currie and MacRae, 1980; Heintzelman, 1979a: 179; Roberts, 1977b).

Peaked Hill

Fair spring hawk migrations (12.2 hawks per hour), but poor autumn flights, are reported at this site near New Hampton (Anonymous, 1966: 98; Heintzelman, 1979a: 179–180; MacRae, 1979c: 8). The most numerous species in the spring migrations are Turkey Vultures, Ospreys, Sharp-shinned Hawks, Broad-winged Hawks, and Red-tailed Hawks (MacRae, 1979c: 8).

Prospect Hill

Mid-September flights of more than 1,400 Broad-winged Hawks occasionally are seen at Prospect Hill (Hopkins and Mersereau, 1975b).

Warner Hill Tower

During September migrations of Broad-winged Hawks numbering as many as 2,400 birds are reported from Warner Hill Tower in East Derry, but other Broad-wing flights contain only a few hundred birds (Currie, 1979; Hopkins and Mersereau, 1975b).

RHODE ISLAND

Interest in hawk migrations in Rhode Island existed back as far as the end of the last century and continued from time to time into this century (Clement, 1958: 118–119; Dunn, 1895: 192; Hanley, 1968). Since the 1970s, however, much more information on hawk migrations in this state was published as part of the overall keen interest in these migrations throughout New England.

Brenton Point

Limited autumn hawk counts (14.6 hawks per hour) at this location produce a variety of migrant hawks, of which Sharp-shinned Hawks are the most numerous but many falcons and other species also appear in limited numbers (Currie and Cote, 1983).

Little Compton

Autumn hawk migrations at Little Compton (39.9 hawks per hour) are dominated by Sharp-shinned Hawks and American Kestrels, but many other species also occur in small numbers (Currie and Cote, 1981).

Napatree Point

The tip of this peninsula is in southwestern Rhode Island and extends into Block Island Sound. Field studies of autumn hawk migrations there (22.8 hawks per hour) indicate that Ospreys, Sharp-shinned Hawks, American Kestrels, and Merlins are the most numerous species observed, but many other species also form part of the flights (Currie and Cote, 1982, 1983; Hopkins and Mersereau, 1974b, 1975b). Clement (1958: 118) stated that autumn hawk flights passed the Point enroute to Fisher's Island in Long Island Sound.

Ninigret Barrier Beach Conservation Area

This barrier beach is the state's longest and is located near Charlestown north of Block Island. Autumn hawk migrations reported there (8.1 hawks per hour) include moderate numbers of Ospreys and Peregrine Falcons (Slack and Slack, 1980: 56–58, 1981: 60–61).

Sakonnet Point

During autumn (27.5 hawks per hour) hundreds of Sharp-shinned Hawks, American Kestrels, and lesser numbers of other hawks sometimes are reported at this spot near Sakonnet (Currie and Cote, 1982, 1983). Clement (1958: 118) stated that Broad-winged Hawks migrate up the estuary of the Sakonnet River during September.

VERMONT

Vermont was the first state in which a hawk flight in North America was documented—a probable flight of some 4,000 Broad-winged Hawks over what is now Brattleboro on 20 or 21 September 1756 (Goldman, 1970). Two centuries passed, however, before more interest developed in the state's hawk migrations (Heintzelman, 1975: 69).

Since 1974 the Vermont Institute of Natural Science coordinates annual spring and autumn hawk watches in the state in the form of short-term hawk migration study projects (Norse, 1979a–1983b; Pistorius, 1977, 1978; Rowlett, 1977; Will, 1974–1976). The autumn migrations are larger and more concentrated than those in spring (Will, 1975), and the latter also tend to fly at higher altitudes than in autumn (Norse, 1979a).

Arrowhead Mountain

Mike Maurer studied spring hawk migrations at this site near Milton in northwestern Vermont but reported relatively meager numbers of birds (6.3 hawks per hour). Sharp-shinned Hawks, Broad-winged Hawks, and Red-tailed Hawks appeared in largest numbers (Norse, 1982a, 1983a).

Bald Mountain

Minor spring hawk migrations pass this site with Sharp-shinned Hawks and Broad-winged Hawks appearing in largest numbers (Norse, 1979a, 1980a, 1983a).

Ball Mountain

Spring hawk migrations at Ball Mountain near South Londonderry produce few birds, but a few dozen Sharp-shinned Hawks and Broad-winged Hawks are reported by Donald Clark (Norse, 1980a, 1981a).

Bluegate

Autumn hawk migrations at Bluegate near Pomfret in central Vermont produce limited numbers of birds (7.8 hawks per hour), Sharp-shinned Hawks, Broad-winged Hawks, and American Kestrels being the most numerous (Norse, 1979b; Pistorius, 1977, 1978; Rowlett, 1977).

Brandon Gap

Some limited (3.4 hawks per hour) autumn hawk migrations are reported at Brandon Gap in west-central Vermont. Broad-winged Hawks and Red-tailed Hawks are the most numerous species observed, but sometimes other species also appeared (Will, 1974, 1975).

Camel's Hump

Camel's Hump is reported as a good autumn hawk watching lookout (Heintzelman, 1979a: 215).

Deer Leap

Autumn hawk migrations passing Deer Leap near Bristol (36.5 hawks per hour) sometimes contain various species, but Northern Harriers, Sharp-shinned Hawks, Broad-winged Hawks, and Red-tailed Hawks appear in largest numbers (Norse, 1981b, 1982b).

Spring hawk flights at this site (21 hawks per hour) produce largest numbers of Turkey Vultures, Ospreys, Sharp-shinned Hawks, Broad-winged Hawks, Red-tailed Hawks, and American Kestrels, although a few other species also appear occasionally (Norse, 1981a, 1982a).

The Dome

Autumn hawk migrations at The Dome near Pownal are modest (8 hawks per hour). Sharp-shinned Hawks and Broad-winged Hawks appear in largest numbers (Will, 1976).

Fuller Mountain

Studies of autumn hawk migrations at Fuller Mountain near Ferrisburg were conducted by John Dye who reported 44 hawks per hour. Broad-winged Hawks and American Kestrels appear in largest numbers (Norse, 1981b).

Poor spring hawk migrations pass the site (10.7 hawks per hour), with Red-tailed Hawks the most numerous (Norse, 1983a).

Gile Mountain

Autumn hawk migrations at Gile Mountain near North Norwich do not produce large seasonal hawk counts (16.3 hawks per hour), but the species reported in largest numbers are Ospreys, Sharp-shinned Hawks, Broad-winged Hawks, Red-tailed Hawks, and sometimes American Kestrels (Norse, 1978, 1979, 1980b, 1981b, 1982b, 1983b; Pistorius, 1978; Will, 1974, 1976).

Spring hawk migrations at Gile Mountain are meager, but Sharp-shinned Hawks, Broad-winged Hawks, and Red-tailed Hawks are most likely to appear (Norse, 1980a, 1981a).

Glebe Mountain

Poor spring hawk migrations pass Glebe Mountain near Londonderry, but the autumn flights (33 hawks per hour) are larger (Heintzelman, 1979a: 215). Species that appear in largest numbers in autumn include Ospreys, Sharp-shinned Hawks, Broad-winged Hawks, Red-tailed Hawks, and American Kestrels (Will, 1974, 1976; Rowlett, 1977).

Grafton

Town Farm near Grafton is one of Vermont's oldest regularly used autumn hawk watching sites. Autumn hawk migrations there are not overly large (14.5 hawks per hour) but the species appearing in largest numbers include Turkey Vultures, Sharp-shinned Hawks, Broad-winged Hawks, Red-tailed Hawks, and American Kestrels (Norse, 1979b, 1980b, 1981b, 1982b, 1983b; Pistorius, 1977, 1978; Rowlett, 1977; Will, 1976).

Hogback Mountain

Autumn hawk flights at Hogback Mountain (13.5 hawks per hour) sometimes produce a few hundred Broad-winged Hawks, a few Sharp-shinned Hawks, and occasionally a few other raptors (Heintzelman, 1975: 69; Norse, 1979b; Pistorius, 1977, 1978; Will, 1974).

Howe Hill

Both spring (7.3 hawks per hour) and autumn (10.2 hawks per hour) hawk migrations are reported at Howe Hill near Pomfret but these flights are relatively meager. During both seasons the species appearing in largest numbers are Sharp-shinned Hawks, Broad-winged Hawks, and American Kestrels, although a few individuals of various other species also pass the site occasionally (Norse, 1979a–1983b).

Interstate 91 Rest Area
Preliminary field checks seem to indicate that some worthwhile spring hawk migrations pass this rest area along Interstate 91 near Exit 6 with several hundred Broad-winged Hawks reported along with a few other species (Norse, 1983a). Autumn hawk migrations at the site are poor (Rowlett, 1977).

Laraway Mountain
During autumn a variety of migrant raptors (17.8 hawks per hour) pass this site near Waterville. Broad-winged Hawks are the most numerous, but small numbers of Ospreys, Sharp-shinned Hawks, and American Kestrels also appear and sometimes a few other species are reported (Norse, 1979b).

Lincoln Hill
Lincoln Hill is located near Hinesburg in northwestern Vermont. Autumn hawk flights there (8.8 hawks per hour) consist mostly of Ospreys, Sharp-shinned Hawks, Broad-winged Hawks, and Red-tailed Hawks, whereas the spring migrations consist mainly of Sharp-shinned Hawks, Broad-winged Hawks, and Red-tailed Hawks (Norse, 1981a–1983b).

Luce's Lookout
Autumn hawk flights at this site (14 hawks per hour) near Lake Lakota contain a variety of species, but Sharp-shinned Hawks and Broad-winged Hawks are the most common (Will, 1976).

Mount Elmore Fire Tower
Mount Elmore is a relatively isolated peak on a 60-mile-long ridge in northern Vermont known as the Worcester Range. Migrations there during autumn (5.9 hawks per hour) contain mostly Ospreys, Sharp-shinned Hawks, Broad-winged Hawks, Red-tailed Hawks, and smaller numbers of various other species (Will, 1976; Rowlett, 1977).

Mount Equinox
Autumn hawk flights (4.8 hawks per hour) at this site near Manchester contain mostly Sharp-shinned Hawks, Broad-winged Hawks, and Red-tailed Hawks, although a few other species sometimes appear (Will, 1974, 1976).

Mount Philo
Spring hawk migrations at Mount Philo near Charlotte produce moderate numbers of birds (9 hawks per hour) with Sharp-shinned Hawks, Broad-winged Hawks, Red-tailed Hawks, and American Kestrels forming the bulk of the flights (Norse, 1980a, 1981a, 1982a, 1983a).

Autumn hawk migrations at this site are smaller (4.7 hawks per hour) with the same species as in spring forming the bulk of the flights (Norse, 1978, 1979b, 1980b, 1981b, 1982b, 1983b).

Owls Head

Norse (1979b, 1980b), Rowlett (1977), and Will (1976) reported that autumn hawk migrations (3.4 hawks per hour) at Owls Head near Marshfield contain mostly Ospreys, Sharp-shinned Hawks, Broad-winged Hawks, and Rough-legged Hawks.

Putney Mountain

Autumn hawk migrations (19.5 hawks per hour) at Putney Mountain near Putney consist mostly of Ospreys, Sharp-shinned Hawks, Broad-winged Hawks, Red-tailed Hawks, and American Kestrels, although sometimes a few other birds of prey appear (Norse, 1977, 1978, 1979b, 1980b, 1981b, 1982b, 1983b; Pistorius, 1977, 1978; Rowlett, 1977; Will, 1974, 1976).

The northward spring hawk flights at this location are very meager but a few Broad-winged Hawks and sometimes other species are reported (Norse, 1983a).

Snake Mountain

Ospreys, Sharp-shinned Hawks, Broad-winged Hawks, Red-tailed Hawks, and American Kestrels are reported in largest numbers during the autumn migrations (10.9 hawks per hour) at Snake Mountain near Addison (Norse, 1979b, 1980b, 1981b, 1982b; Pistorius, 1977, 1978; Rowlett, 1977). Poor spring hawk flights are reported there (Heintzelman, 1979a: 216).

Spencer's Cabin

William J. Norse conducted spring and autumn (4.2 hawks per hour) hawk watches at this site near Winhall but very meager numbers of hawks are seen during both seasons. Most common are Sharp-shinned Hawks, Broad-winged Hawks, and Red-tailed Hawks (Norse, 1979a–1983b; Pistorius, 1977, 1978; Rowlett, 1977).

Sudbury

Autumn hawk migrations at Sudbury (8.4 hawks per hour) contain mostly Turkey Vultures, Ospreys, Sharp-shinned Hawks, Broad-winged Hawks, Red-tailed Hawks, and American Kestrels, but occasionally other species are reported (Will, 1976).

Weathersfield

Skyline Drive in Weathersfield is another of Vermont's regularly used autumn hawk watching sites, although the migrations reported there are relatively meager (5.8 hawks per hour). Seasonal counts indicate that Sharp-shinned Hawks and Broad-winged Hawks are observed in largest numbers but various other species pass in more limited numbers (Norse, 1979b, 1980b, 1981b; Pistorius, 1977, 1978; Rowlett, 1977; Will, 1974).

Westford

The spring migrations of hawks (23 per hour) at Westford, in northwestern Vermont, tend to be somewhat larger than the autumn migrations (9.9

hawks per hour) there. In spring Ospreys, Sharp-shinned Hawks, Broad-winged Hawks, Red-tailed Hawks, and American Kestrels form the bulk of the flights, whereas in autumn Northern Harriers, Sharp-shinned Hawks, Broad-winged Hawks, and American Kestrels are reported in largest numbers (Norse, 1980b, 1981a, 1981b, 1983a; Rowlett, 1977).

White River Junction

The most commonly observed raptors during the spring hawk flights (59 hawks per hour) at this location are Ospreys, Sharp-shinned Hawks, and Broad-winged Hawks, although a few other species appear in very small numbers (Norse, 1981a).

White Rocks

Meager autumn hawk flights (6 hawks per hour) are reported at White Rocks at Wallingford, the most commonly observed species being Ospreys, Sharp-shinned Hawks, Broad-winged Hawks, and Red-tailed Hawks (Norse, 1979b, 1980b, 1981b, 1982b, 1983b; Pistorius, 1977, 1978; Rowlett, 1977).

Meager spring hawk flights (11.5 hawks per hour) also are reported at this site, the most frequently observed species being Sharp-shinned Hawks and Broad-winged Hawks (Norse, 1979a, 1981a, 1982a, 1983a).

C H A P T E R   6

Middle Atlantic Hawk
Migrations

Vast numbers of hawks migrate across the Middle Atlantic states during autumn, often concentrating along long mountain ridges, at the tips of peninsulas, or along barrier beach islands thus providing hawk watchers with some of North America's best autumn hawk watching lookouts along major migration routes and flight-lines. But even in spring, as recent field studies are showing, lesser numbers of hawks migrate across the Middle Atlantic states. Often these spring migrations tend to be dispersed widely, although limited numbers of hawks concentrate at some locations including some of the famous autumn hawk watching lookouts.

DELAWARE

Knowledge of autumn and spring hawk migrations in Delaware remains fragmentary, although Mohr (1969) provided a broad overview of the state's raptor movements and recently autumn field studies at Carpenter Park in northwestern Delaware provided important systematic hawk counts at an excellent location (Dunne, 1982, 1983).

Arundel
A mid-March flight of 60 Red-shouldered Hawks was reported at Arundel by Boyle, Paxton, and Cutler (1982: 834).

Bombay Hook National Wildlife Refuge
Scott and Cutler (1973b: 755) reported an early April flight of 250 American Kestrels at this refuge, suggesting a flight-line there.

Brandywine Creek State Park
A fair autumn hawk migration flight-line exists at this park a few miles

north of Wilmington with hundreds of Broad-winged Hawks sometimes seen at treetop level (Heintzelman, 1979a: 161; Mohr, 1969: 5–6).

A short-term autumn hawk watch (31 hawks per day) was conducted by David Phalen (personal communication) at this site in 1971 and 1972. He reported that Ospreys, Sharp-shinned Hawks, Broad-winged Hawks, and Red-tailed Hawks were seen most commonly, although small numbers of some other species also appeared.

Cape Henlopen
The role of Cape Henlopen in the southward coastal migrations of hawks during autumn was discussed by various observers who discovered that as few as 0.5 percent of the hawk flights reported 13 miles across Delaware Bay at Cape May Point, New Jersey, are seen on the Delaware side at Cape Henlopen (Allen and Peterson, 1936; Dunne, 1980a: 15–16).

During spring, a few Sharp-shinned Hawks and "hundreds" of American Kestrels are reported migrating northward past this site (Boyle, Paxton, and Cutler, 1983: 852; Paxton, Boyle, and Cutler, 1981: 806).

Carpenter Park
James Oliver used this park to study broad-front autumn hawk migrations for the first time in 1981. He discovered that the site near Elkton, Maryland, produces excellent hawk flights (66 hawks per hour), with Turkey Vultures, Ospreys, Northern Harriers, Sharp-shinned Hawks, Broad-winged Hawks, Red-tailed Hawks, and American Kestrels forming the bulk of the flights (Dunne, 1982: 16–17, 1983: 20).

Little Creek
A mid-May flight of 50 Sharp-shinned Hawks was reported at Little Creek (Scott and Cutler, 1975b: 834).

Newark
September flights of as many as 500 Broad-winged Hawks are reported at Newark (Buckley, Paxton, and Cutler, 1978: 185), whereas as many as 75 Broad-winged Hawks are seen there in mid-April (Paxton, Boyle, and Cutler, 1981: 806).

Wilmington
During September and early October, migrating Broad-winged Hawks sometimes pass over Wilmington (Allen and Peterson, 1936: 399; Mohr, 1969: 5; Scott and Cutler, 1963: 19, 1968: 19). A Gyrfalcon even was reported in early November at the city (Scott and Cutler, 1975a: 36).

MARYLAND

During recent decades various important field studies of autumn hawk migrations in Maryland were published, all of which collectively provide an excellent summary of the status of these raptor movements in the state (Hackman, 1954; Hackman and Henny, 1971; Heintzelman, 1975, 1979a:

169–172; Lee, 1977; Lee and Sykes, 1975; Robbins, 1950; Stewart and Robbins, 1958; Titus and Mosher, 1982; Ward and Berry, 1972; Wilds, 1983).

Assateague Island National Seashore

This national seashore is known as an autumn Peregrine Falcon and other falcon migration route. During some years as many as 120 Peregrines are counted there (Berry, 1971; Rice, 1969; Scott and Cutler, 1973a: 37; Ward and Berry, 1972).

Baltimore

As many as 1,000 migrating Broad-winged Hawks are reported in mid-September at Baltimore (Armistead, 1981: 167).

Bay Hundred Peninsula

A fair autumn hawk migration flight-line exists on this narrow peninsula near Fairbank that extends into Chesapeake Bay (Heintzelman, 1979a: 169).

Bellevue

During September, up to 90 Broad-winged Hawks and several Bald Eagles are reported at this site (Scott, 1976: 48).

Chevy Chase

Flights of up to several hundred Broad-winged Hawks sometimes are reported in September at Chevy Chase (Scott and Cutler, 1972a: 43, 1973a: 37, 1974: 34).

Claiborne

As many as 2,000 Broad-winged Hawks are reported migrating over Claiborne in mid-September (Armistead, 1981: 167).

Cove Point

This site is just north of the point where the Patuxent River flows into Chesapeake Bay. On 21 September 1949, George Kelly counted 2,214 migrating Broad-winged Hawks at this location, and noted some of the birds arriving on a broad front moving just south of west from Chesapeake Bay, from Cove Point to Drum Point. Ten days of observation at Cove Point produced a variety of species in addition to Broad-wings, especially modest numbers of Sharp-shinned Hawks (Robbins, 1950).

Dan's Rock

Dan's Rock, at an elevation of 2,895 feet, is on the summit of Dan's Mountain near Midland in western Maryland. Autumn hawk migrations at this site (14.7 hawks per hour) consist mostly of Ospreys, Northern Harriers, Sharp-shinned Hawks, Cooper's Hawks, Red-shouldered Hawks, Broad-winged Hawks, Red-tailed Hawks, and American Kestrels (Finucane, 1976b, 1977b, 1978b, 1979b; Robbins, 1950: 4, 6; Titus and Mosher, 1982).

Spring hawk migrations at Dan's Rock (87.4 hawks per day) consist mostly of Ospreys, Sharp-shinned Hawks, and Broad-winged Hawks (Finucane, 1980a: 12).

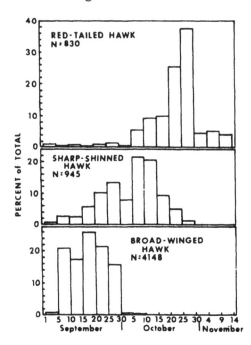

Temporal distribution of three migrant hawk species at Dan's Rock and High Rock, Maryland, during the autumns of 1975–1979. *Reprinted from Titus and Mosher (1982).*

An analysis by Titus and Mosher (1982) of the influence in autumn of seasonality and selected weather variables on Sharp-shinned Hawk, Broad-winged Hawk, and Red-tailed Hawk migrations determined that Sharp-shin flights are associated with following winds and good visibility, Broad-winged Hawk migrations are associated with strong winds, high air temperatures, and good visibility during the afternoon, and Red-tailed Hawk flights are associated with light, opposing, southeasterly winds.

Fort Smallwood

This park is at the mouth of Baltimore's harbor and produces productive spring and autumn hawk migrations (Wilds, 1983: 186).

The autumn flights consist mostly of Ospreys, Northern Harriers, Sharp-shinned Hawks, Red-tailed Hawks, and American Kestrels, although sometimes other species also appear (Wilds, 1983: 186).

Spring hawk migrations at this location may contain as many as 1,000 Turkey Vultures, more than 450 Ospreys, 150 Northern Harriers, over 1,200 Sharp-shinned Hawks, hundreds of Cooper's Hawks, Red-shouldered Hawks, Broad-winged Hawks, and American Kestrels, and dozens of Merlins (Armistead, 1983b: 856).

Foxville Tower
Mid-September hawk migrations at this South Mountain site included more than 300 Broad-winged Hawks plus a few other hawks (Robbins, 1950: 5–6).

High Rock
High Rock is located in western Maryland on Big Savage Mountain west of Dan's Rock. Autumn hawk flights there (25.1 hawks per day) mostly contain Northern Harriers, Sharp-shinned Hawks, Cooper's Hawks, Broad-winged Hawks, Red-tailed Hawks, and American Kestrels (Titus and Mosher, 1982).

Hooper's Island
Hooper's Island is located on the eastern shore of Chesapeake Bay and is located on an autumn hawk migration flight-line on northwest winds. Among the species reported there are Turkey Vultures, Ospreys, Bald Eagles, Northern Harriers, Sharp-shinned Hawks, Cooper's Hawks, Red-shouldered Hawks, Broad-winged Hawks, American Kestrels, Merlins, and Peregrine Falcons (Armistead, 1983a: 165; Scott and Cutler, 1968a: 19; Tyrell, 1935).

Kent Island
This island extends southward into Chesapeake Bay near Stevensville. During autumn various migrating hawks, including hundreds of Broad-winged Hawks, are reported (Heintzelman, 1979a: 170–171; Scott and Cutler, 1966: 24).

Little Cove Point
George Kelly reported a broad-front migration of Broad-winged Hawks, and some other species of raptors, in September over Little Cove Point and adjacent areas near the entrance of the Patuxent River into Chesapeake Bay (Robbins, 1950: 9). See also Cove Point.

Lore's Pond
During September up to 2,100 migrating Broad-winged Hawks are reported at Lore's Pond a short distance north of Solomons in Calvert County (Robbins, 1950: 9–10).

Martin Mountain
Mid-September migrations of several hundred Broad-winged Hawks and smaller numbers of other raptors are reported at this site on the Maryland-Pennsylvania border (Robbins, 1950: 4, 6).

Nicholas Mountain
Robbins (1950: 4, 6) reported that as many as 687 Broad-winged Hawks and a few other hawks were observed at Nicholas Mountain in western Maryland in mid-September.

Rock Run Sanctuary
An early November flight of 59 Red-shouldered Hawks was reported at this site (Scott and Cutler, 1964: 21).

Rothrock Fire Tower
Good autumn hawk migrations (26 hawks per hour) are reported at this fire tower on Backbone Mountain (Heintzelman, 1979a: 171). The species noted in largest numbers include Ospreys, Sharp-shinned Hawks, Broad-winged Hawks, and American Kestrels, but even a few Peregrine Falcons are noted occasionally (Brooks, 1956: 24; Puckette, 1982b: 25, 1983b: 24–25).

Sandy Point State Park
This park near Annapolis produces fair autumn and spring hawk migrations—especially Broad-winged Hawks and Red-tailed Hawks in autumn, and Ospreys, Northern Harriers, Sharp-shinned Hawks, and American Kestrels in spring (Armistead, 1981: 167; Heintzelman, 1979a: 171; Scott, 1976: 48, 1977: 161; Wilds, 1983: 110).

Saverna Park
Late September flights of more than 1,500 Broad-winged Hawks are reported at Saverna Park (Scott, 1978: 190).

Sideling Hill
Several hundred migrating Broad-winged Hawks sometimes are reported in mid-September at Sideling Hill (Robbins, 1950: 6).

Sparks
Autumn hawk migrations at Sparks (30.5 hawks per hour) contain mostly Ospreys, Sharp-shinned Hawks, Broad-winged Hawks, and American Kestrels (Finucane, 1980b, 1981b; Puckette, 1982b, 1983b, 1984b).

Spring hawk migrations at this location are more limited (9.6 hawks per hour) and contain Ospreys, Sharp-shinned Hawks, Broad-winged Hawks, and Red-tailed Hawks in largest numbers (Puckette, 1983a: 15, 1984a: 19).

Tilghman Island
Late September migrations of as many as 160 American Kestrels are reported at Tilghman Island (Scott and Cutler, 1964: 21).

Timonium
Occasionally up to 500 migrating Broad-winged Hawks are reported at Timonium in mid-September (Scott and Cutler, 1970a: 27).

Towson
Low density autumn hawk migrations (32.4 per day) are reported at Towson with Turkey Vultures, Ospreys, Sharp-shinned Hawks, Cooper's Hawks, Red-shouldered Hawks, Broad-winged Hawks, Red-tailed Hawks, and American Kestrels forming the bulk of the flights. This Piedmont site receives a portion of the state's hawk migrations—especially when birds drift from the Appalachian ridges and disperse over a broad front several days after the passage of cold fronts and their associated strong north-westerly winds which tend to produce fine ridge hawk migrations (Lee, 1977; Lee and Sykes, 1975). Mid-September flights of more than 1,400

Broad-winged Hawks occasionally were reported at Towson (Scott, 1977: 161).

Washington Monument State Park

Washington Monument State Park, on South Mountain near Boonsboro, is located on a fair (19.7 hawks per hour) autumn hawk migration flight-line (Heintzelman, 1975, 1979a: 171). The species reported in largest numbers during autumn include Ospreys, Northern Harriers, Sharp-shinned Hawks, Cooper's Hawks, Red-shouldered Hawks, Broad-winged Hawks, Red-tailed Hawks, and American Kestrels, although limited numbers of other species also are reported there (Beaton, 1951; Herbert E. Douglas *in* Heintzelman, 1975: 341; Finucane, 1978b, 1979b, 1980b, 1981b; Puckette, 1982b, 1983b, 1984b).

Spring hawk migrations (16 hawks per hour) at this park consist mostly of Ospreys, Northern Harriers, Sharp-shinned Hawks, Red-shouldered Hawks, Broad-winged Hawks, and Red-tailed Hawks (Puckette, 1984a: 19).

White Marsh

Good autumn hawk migrations (23.3 hawks per hour) are reported at White Marsh northeast of Baltimore. Broad-winged Hawks form the bulk of the flights (Hackman, 1954; Hackman and Henny, 1971).

Wills Mountain Narrows

Occasional mid-September flights of more than 1,200 Broad-winged Hawks are reported above "Lovers Leap" on Wills Mountain (Robbins, 1950: 4).

NEW JERSEY

Hawk migrations in New Jersey, especially during autumn but to a lesser extent during spring, received extensive comment and study making this state one of North America's centers of study of these bird movements (Allen and Peterson, 1936; Clark, 1968–1984; Dunne, 1977–1984; Dunne and LeGrand, 1984; Edwards, 1939; Eynon, 1941; Fables, 1955; Heintzelman, 1972e, 1975; Howland, 1893; Kunkle, 1976; Lang, 1943; Leck, 1975, 1984; LeGrand, 1983a; Marx, 1971; McIntyre, 1982b; Skinner, 1981; Stone, 1909, 1922, 1937; Sutton and Sutton, 1984; Wolfarth, 1952, 1980).

Considerable interest also is developing in educational hawk watching in this state (Dunne, 1984; Heintzelman, 1970e; Keller, 1983a–1984b).

Alpine

Considerable numbers of hawks (59.3 per hour) are reported migrating past this site during autumn with Ospreys, Sharp-shinned Hawks, Broad-winged Hawks, Red-tailed Hawks, and American Kestrels noted in largest numbers (Benz, 1980b; Nichols, 1959: 15).

Avalon

David Ward, Jr., noted an offshore autumn hawk migration at this site with limited numbers (0.17 hawks per hour) of Ospreys, Northern Harriers, American Kestrels, Merlins, and Peregrine Falcons reported (Dunne, 1980: 13).

Bayside

A minor autumn Sharp-shinned Hawk migration flight-line is reported at Bayside at the upper end of Delaware Bay. Hawks cross the bay here from New Jersey to Woodland Beach, Delaware (Kunkle, 1976: 19).

Beach Haven

Some migrating falcons are reported along the outer beaches at Beach Haven in spring and autumn (Edwards, 1939; Potter and Murray, 1953: 9–10).

Bearfort Mountain

Fair autumn (51.4 hawks per day) and spring hawk migrations are reported from the fire tower on Bearfort Mountain in Passaic County (Heintzelman, 1979a: 180–181).

The most numerous species reported during the autumn migrations include Ospreys, Sharp-shinned Hawks, Broad-winged Hawks, Red-tailed Hawks, and American Kestrels (Edwards, 1939; Koebel, 1970). During spring, the bulk of the migrations contain mostly Ospreys, Sharp-shinned Hawks, Red-shouldered Hawks, Broad-winged Hawks, Red-tailed Hawks, and American Kestrels (Marx, 1971).

Boonton

Occasionally during mid-September thousands of migrating Broad-winged Hawks are reported over Boonton (Paxton, Boyle, and Cutler, 1984: 182).

Bowling Green Fire Tower

Spring and autumn hawk flights reported from this fire tower are part of a scattered highlands flight-line (Edwards, 1939; Eynon, 1941). Few birds are observed, but Ospreys, Broad-winged Hawks, and American Kestrels may occur in largest numbers during autumn (Heintzelman, 1975: 102).

Breakneck Mountain

This site in the highlands of northern New Jersey, formerly a hawk shooting area, is another autumn and spring hawk observation post (Edwards, 1939; Eynon, 1941). During the autumn migrations (57.1 hawks per hour), Sharp-shinned Hawks, Broad-winged Hawks, and American Kestrels are reported in largest numbers (Nagy, 1979b). The spring hawk flights, however, consist mostly of Ospreys, Sharp-shinned Hawks, Red-shouldered Hawks, Broad-winged Hawks, Red-tailed Hawks, and American Kestrels (Marx, 1971; Nagy, 1978a).

Brigantine

Edwards (1939) and Eynon (1941) stated that a spring and autumn falcon migration flight-line was located at Brigantine.

Cape May Point

Ornithologists discovered the spectacular autumn hawk migrations at Cape May Point because of the slaughter of thousands of hawks there earlier in this century, and the first efforts at the Point combined conservation activities and hawk count efforts (Allen, 1932, 1933; Allen and Peterson, 1936; Craighead, 1933; Kuerzi, 1937; Rusling, 1935a, 1935b; Saunders, 1931; Stone, 1922, 1937; Tanner, 1936).

After the hawk shooting ended at Cape May Point, occasional hawk migration observations were reported (Brady, 1965; Heintzelman and Mac-Clay, 1974; McLaughlin, 1973). Ernest A. Choate, however, conducted the first systematic autumn hawk counts at the Point in recent years and these now are annual migration study projects (Choate, 1965, 1972; Choate *in* Heintzelman, 1970e; Choate and Tilly, 1973; Clark, 1968–1976; Dunne, 1978b, 1979b, 1980b; Dunne, et al., 1981; Dunne and Clark, 1977; LeGrand, 1984; LeGrand, et al., 1983). These studies (84.7 hawks per hour) demonstrate that Turkey Vultures, Ospreys, Northern Harriers, Sharp-shinned Hawks, Cooper's Hawks, Red-shouldered Hawks, Broad-winged Hawks, Red-tailed Hawks, American Kestrels, Merlins, and Peregrine Falcons form the bulk of the autumn flights with as many as 88,938 birds of prey counted during the autumn of 1981.

Spring hawk migrations at Cape May Point (66.2 hawks per day) only recently received study, but available data indicate that Turkey Vultures, Ospreys, Northern Harriers, Sharp-shinned Hawks, Cooper's Hawks, Broad-winged Hawks, Red-tailed Hawks, American Kestrels, and Merlins form the bulk of the northward flights—although in only a small fraction of the numbers seen during autumn (Sutton, 1980, 1981; Sutton and Sutton, 1984).

Additional aspects of the work with hawk migrations at this site include banding migrants (Clark, 1968–1976), studies of raptor movements in autumn in the Cape May area (Dunne, 1980b; Kerlinger, 1981), variations in individual hawk watchers and the counts they reported (Kochenberger, 1984), several radar studies (Kerlinger, 1983; Kerlinger and Gauthreaux, 1983), and various educational projects (Dunne, 1984; Keller, 1983a–1984b).

Catfish Fire Tower

This fire tower is on a side spur of the Kittatinny Ridge about five miles northeast of Raccoon Ridge. It produces the best autumn hawk flights (213.3 hawks per day) on southerly or easterly winds (Heintzelman, 1972a, 1972b, 1979a: 182).

Seasonal counts of autumn hawk migrants at the Catfish Fire Tower demonstrate that the most commonly observed species are Ospreys, Sharp-shinned Hawks, Broad-winged Hawks, and Red-tailed Hawks (Heintzelman, 1972a: 20).

A comparison of hawk counts from Catfish Fire Tower with hawk counts made on the same days at nearby Raccoon Ridge on the main fold of the Kittatinny Ridge demonstrates that the hawk flights at Catfish vary from about 17.5 to 93.5 percent of the Raccoon Ridge hawk flights (Heintzelman, 1972a, 1972b; Tilly, 1972a, 1972b).

Chimney Rock

Studies of autumn hawk migrations at Chimney Rock (42.3 hawks per hour) indicate that the most commonly seen species are Ospreys, Sharp-shinned Hawks, Broad-winged Hawks, Red-tailed Hawks, and American Kestrels (Benz, 1980b, 1981b, 1982b, 1983b; Edwards, 1939; Nagy, 1979b).

Clifton

Occasionally during September as many as 1,200 migrating Broad-winged Hawks are reported at Clifton (Edwards, 1939).

Culver's Gap

This Sussex County site overlooks the northeastern edge of Culver's Lake. Autumn hawk flights there (65 hawks per day) consist mostly of Ospreys, Northern Harriers, Sharp-shinned Hawks, Broad-winged Hawks, and Red-tailed Hawks (Howard Drinkwater *in* Heintzelman, 1975: 105; William Rusling, unpublished field notes).

Delaware River (Camden to Salem)

The Delaware River is a natural geographic feature forming the border between New Jersey and Pennsylvania. For some hawks it serves as a migration route—at least along some of its length. During a period of ten years (January 1935 to December 1944), surveys of hawks were made during all months between Camden and Salem by Julian K. Potter (1949). The data collected during these surveys exhibit certain peak migration periods for the various raptor species as they migrate along the river route.

Only one Black Vulture was observed, but Turkey Vultures were most common from March through May. Ospreys were most common during April, but few appeared during autumn suggesting the river was not an important autumn migration route—at least at that time. Largest numbers of Bald Eagles appeared in March and again in August, whereas Northern Harriers were seen most commonly during March and September through November.

Too few Sharp-shinned Hawks were seen to determine peak migration periods, but Cooper's Hawks occurred most commonly during March and November. Red-shouldered Hawks peaked in February and October–November, but few Broad-winged Hawks were seen. Migrant Red-tailed Hawks appeared in largest numbers in March and November, whereas Rough-legged Hawks were most common in March and November.

Of the falcons, American Kestrels peaked in numbers in March and October–November, isolated sightings of Merlins occurred in April and October, and Peregrine Falcons were seen in largest numbers in November.

Delaware Water Gap

Autumn hawk flights at Delaware Water Gap, New Jersey (9.3 hawks per hour), where a hawk banding station is located, mostly consist of Ospreys, Northern Harriers, Sharp-shinned Hawks, Cooper's Hawks, Broad-winged Hawks, Red-tailed Hawks, and American Kestrels (Benz, 1981b, 1982b, 1983b)

Egg Island Point

This site extends southward into upper Delaware Bay west of Heislerville. During autumn an American Kestrel migration flight-line exists there, the falcons flying over-the-water on days with northwest winds. The birds presumably come from Cape May Point and its vicinity (Kunkle, 1976: 18–19).

Gandy's Beach

This site is just southwest of Newport. An autumn accipiter and falcon flight-line is reported there, and at times the magnitude of the hawk flights are as large as those reported at Higbee Beach, from whose vicinity (and Cape May Point) these hawks may come (Kunkle, 1976: 18–19).

Greenbrook Sanctuary

This sanctuary near Tenafly is located on an important autumn hawk migration flyway along the Palisades cliffs in northern New Jersey (Serrao, 1976, 1977a, 1977b). The most numerous autumn raptor migrants reported there include Ospreys, Sharp-shinned Hawks, Broad-winged Hawks, Red-tailed Hawks, and American Kestrels (Serrao, 1976, 1977b).

Hardwick

A late April migration of 50 Sharp-shinned Hawks was reported at Hardwick (Paxton, Boyle, and Cutler, 1981: 806).

Higbee Beach Wildlife Management Area

Higbee Beach, along Delaware Bay north of Cape May Point, is an important autumn hawk migration flight-line when northwest winds develop. Under those conditions hawks move up the bay past this site rather than attempting to cross the bay to Delaware from Cape May Point, New Jersey (Heintzelman, 1979a: 182). An autumn 1981 seasonal hawk count at this site produced 120.1 birds per hour of observation with Turkey Vultures, Ospreys, Northern Harriers, Sharp-shinned Hawks, Cooper's Hawks, Red-shouldered Hawks, Broad-winged Hawks, Red-tailed Hawks, American Kestrels, Merlins, and Peregrine Falcons forming the bulk of the flights (McIntyre, 1982a).

High Mountain

Autumn hawk migrations (261.6 hawks per day) at this site northwest of Paterson consist mostly of Ospreys, Sharp-shinned Hawks, Broad-winged Hawks, and American Kestrels (Apps, 1975; Edwards, 1939; Irving H. Black *in* Heintzelman, 1975: 105, 335).

High Point State Park

Fair autumn hawk migrations are reported at this park in extreme north-western New Jersey (Eynon, 1941; Heintzelman, 1979a: 183; Kuser, 1962; Pettingill, 1977: 381).

Island Beach State Park

This park is a barrier beach island along coastal New Jersey and is an autumn migration route (19.9 hawks per day), mostly for Ospreys, Northern Harriers, Sharp-shinned Hawks, American Kestrels, Merlins, and Peregrine Falcons (Dunne, 1978c: 10, 12, 1979b: 17, 20; Heintzelman, 1979a: 183; Scott and Cutler, 1967: 16).

A reverse migration of Sharp-shinned Hawks was noted there on 7 October 1976 when J. C. Miller (1979) and Charles E. Price watched large numbers of these hawks flying north along the west (bay) side of the island early in the morning with some birds headed toward the mainland. Other migrating hawks seen that day headed south in the customary fashion—3 Ospreys, 6 Northern Harriers, 1 American Kestrel, and 2 Merlins. A southwest wind prevailed at 15 miles per hour.

Kearny Marsh

An autumn hawk migration flight-line is reported at this Essex County location for Ospreys, Sharp-shinned Hawks, Merlins, and some Peregrine Falcons (Lawrence and Gross, 1984: 27).

Kittatinny Mountains Raptor Banding Station

This raptor banding station is located on the Kittatinny Ridge in north-western New Jersey some 11 miles south of High Point State Park (Soucy, 1976). The species (13 hawks per hour) forming the bulk of the autumn hawk migrations include Ospreys, Northern Harriers, Sharp-shinned Hawks, Cooper's Hawks, Broad-winged Hawks, Red-tailed Hawks, and American Kestrels (Benz, 1982b, 1983b; Nagy, 1978b; Soucy, 1976) but a Gyrfalcon even appeared there (Soucy, 1983).

Mial Maull Light

During autumn some Northern Harriers sometimes are reported crossing Delaware Bay west of Reed's Beach from New Jersey to Delaware (Kunkle, 1976: 19).

Millbrook Gap

This gap is on the south slope of the Kittatinny Ridge along the Blairstown to Millbrook Road in northwestern New Jersey. Autumn hawk migrations regularly pass the site including some Golden Eagles (Carleton, 1962: 13).

During spring, as many as 1,000 Red-tailed Hawks were observed migrating past the gap in early March and several hundred Turkey Vultures also were reported there (Wolfarth, 1980).

Money Island

On days with northwest winds, heavy autumn migrations of accipiters and falcons sometimes are reported at Money Island west of Newport along upper Delaware Bay (Kunkle, 1976: 18).

Montclair Hawk Lookout Sanctuary

This sanctuary overlooks Montclair in northern New Jersey and is one of the state's best known and most regularly used hawk migration lookout on an excellent (206.1 hawks per day) autumn hawk migration flight-line (Heintzelman, 1979a: 184; Wander and Brady, 1977: 18–19). Prior to its use as a hawk watching lookout, hawk shooting was done there in autumn (Breck, 1960b) but later the formation of the sanctuary put emphasis on hawk watching (Breck, 1960b; Edwards, 1939; Eynon, 1941; Honaman, 1955; Lang, 1943).

Since 1957 annual autumn hawk counts devote particular attention to Broad-winged Hawk migrations, although all species are reported. Although the numbers of birds tend to vary greatly from year to year at this location, the most commonly reported species include Turkey Vultures, Ospreys, Northern Harriers, Sharp-shinned Hawks, Red-shouldered Hawks, Broad-winged Hawks, Red-tailed Hawks, and American Kestrels (Bihun, 1967–1979, 1980b, 1981–1983; Breck, 1962, 1963; Breck and Breck, 1964, 1965, 1966; Redmond and Breck, 1961).

Spring hawk migrations at this site (87.4 hawks per day) were known as long ago as the end of the last century (Howland, 1893), and received some attention earlier in this century (Breck, 1960b; Eynon, 1947), but systematic spring hawk counts were not reported until 1980. These recent field studies demonstrate that Turkey Vultures, Ospreys, Sharp-shinned Hawks, Broad-winged Hawks, Red-tailed Hawks, and American Kestrels are most numerous (Bihun, 1980a, 1984).

Newport

During autumn, or northwest winds, a migration of accipiters and falcons (with some buteos) is reported at Newport near the shore of upper Delaware Bay (Kunkle, 1976: 18).

Plainfield

As many as 750 migrating Broad-winged Hawks were reported in September at Plainfield (Nichols, 1955: 12).

Raccoon Ridge

Raccoon Ridge is New Jersey's most important Kittatinny Ridge hawk migration lookout. Excellent autumn hawk migrations (41.4 hawks per hour) are reported at the site (Edwards, 1939; Heintzelman, 1975: 107–108, 337, 1979a: 184–185; Koebel, 1983). These migrations consist mostly of Ospreys, Northern Harriers, Sharp-shinned Hawks, Red-shouldered Hawks, Broad-winged Hawks, Red-tailed Hawks, and American Kestrels plus small numbers of other species including Bald Eagles and Golden Eagles (Benz, 1980b, 1981b, 1982b, 1983b; Cant, 1946; Heintzelman, 1975: 337; Tilly, 1972a, 1972b, 1972c, 1973, 1979; Wolfarth, 1975).

Spring hawk migrations at Raccoon Ridge contain fewer birds (7.9 per hour) and consist mostly of Ospreys, Northern Harriers, Sharp-shinned Hawks, Red-shouldered Hawks, Broad-winged Hawks, Red-tailed Hawks, and American Kestrels (Dunne, 1977, 1978a).

Rifle Camp Park

Excellent autumn hawk migrations (146.2 hawks per hour) are reported at this park atop the Watchung Mountains near Clifton (Heintzelman, 1979a: 185–186). Species that are observed in largest numbers include Ospreys, Sharp-shinned Hawks, Broad-winged Hawks, and American Kestrels (Nagy, 1977b).

As many as 150 migrating Broad-winged Hawks are reported during mid-April (Paxton, Boyle, and Cutler, 1981: 806).

Sandy Hook

Fair spring hawk migrations (9.7 hawks per hour) are reported at Sandy Hook, a peninsula extending into New York Harbor from the northern end of New Jersey's coast (Clark, 1978a; Heintzelman, 1979a: 186). Northern Harriers, Sharp-shinned Hawks, Cooper's Hawks, Red-shouldered Hawks, American Kestrels, and Merlins are the species seen in largest numbers (Clark, 1978a; Dunne, 1979a; Eynon, 1941: 115; LeGrand, 1983a; McIntrye, 1982b; Nichols, 1953b: 263; Skinner, 1981).

Scott's Mountain Hawk Lookout

Poor spring hawk flights but generally fair to good autumn hawk migrations (82.1 hawks per hour) are reported by Greg Hanisek on Scott's Mountain near Phillipsburg (Heintzelman, 1979a: 186). Despite more than 18,000 Broad-winged Hawks counted at the site on 14 September 1983, generally far fewer numbers of hawks are seen there with the bulk of the flights consisting of Sharp-shinned Hawks, Broad-winged Hawks, and Red-tailed Hawks plus a variety of other species in very limited numbers (Greg Hanisek, field notes).

Skyline Ridge

Autumn hawk migrations (57.6 hawks per hour) at Skyline Ridge near Oakland are excellent (Heintzelman, 1979a: 186–187) and consist mostly of Ospreys, Northern Harriers, Sharp-shinned Hawks, Red-shouldered Hawks, Broad-winged Hawks, Red-tailed Hawks, and American Kestrels plus limited numbers of various other species (Benz, 1980b; Carleton, 1962: 13; Nagy, 1977b, 1978b, 1979b).

Stag Lake

Stag Lake is a former hawk shooting site between Andover and Sparta on the western edge of the Highlands in northern New Jersey (Edwards, 1939; Eynon, 1941). Earlier in this century Justus von Lengerke observed and shot large numbers of migrating hawks there during autumn—especially Sharp-shinned Hawks, Cooper's Hawks, Red-shouldered Hawks, Broad-winged Hawks, and Red-tailed Hawks (Amadon, 1967: 54; Broun, 1949a: 6; Burns, 1911: 234–235; von Lengerke, 1908).

Sunrise Mountain

During autumn, good hawk flights (27.4 hawks per hour) are reported at Sunrise Mountain in Stokes State Forest, Sussex County (Heintzelman, 1979a: 187). The species forming the bulk of these flights are Ospreys, Northern Harriers, Sharp-shinned Hawks, Broad-winged Hawks, Red-tailed Hawks, and American Kestrels, although limited numbers of various other species also are reported (Benz, 1980b, 1981b, 1982b, 1983b, 1984b; Kuser, 1962; Nagy, 1976, 1977b, 1978b, 1979b; Fred Tetlow, field notes).

April hawk migrations at this location sometimes produce more than 1,000 Broad-winged Hawks and smaller numbers of various other birds of prey (Boyajian, 1968b: 509).

Town Bank

As many as 17 migrating Peregrine Falcons were reported during autumn at Town Bank (Scott and Cutler, 1966: 24).

Walnut Valley Bluffs

During early March large flights of migrating Red-tailed Hawks are reported at this site near the south slope of the Kittatinny Ridge near Raccoon Ridge (Dunne, 1977: 25; Wolfarth, 1980: 17).

Watchung Mountains

These mountains are located in northern New Jersey between Paterson and Somerville. Both autumn and spring hawk flights are reported along the mountains (Edwards, 1939; Eynon, 1941; Lang, 1943; Wolfarth, 1952).

The autumn hawk flights, some of which are cross-ridge movements, contain largely Ospreys, Northern Harriers, Sharp-shinned Hawks, Red-shouldered Hawks, Broad-winged Hawks, Red-tailed Hawks, American Kestrels, and Merlins (Eynon, 1941: 114; Lang, 1943: 349; Wolfarth, 1952).

Spring hawk migrations along the Watchungs contain mostly Ospreys, Sharp-shinned Hawks, Red-shouldered Hawks, Broad-winged Hawks, Red-tailed Hawks, and American Kestrels plus small numbers of other species (Wolfarth, 1952).

Wawayanda

Spring hawk flights are reported at Wawayanda in the Central Highlands about 7.5 miles west-northwest of Bearfort Mountain. Ospreys, Sharp-shinned Hawks, Broad-winged Hawks, and Red-tailed Hawks are seen in largest numbers but limited numbers of others raptors also are part of the flights (Marx, 1971).

Windbeam Tower

Some spring hawk migrations are observed from this site in the Central Highlands on a ridge west of the Ramapo Mountains. Ospreys, Sharp-shinned Hawks, Broad-winged Hawks, and Red-tailed Hawks formed the bulk of the flights (Marx, 1971).

NEW YORK

Autumn hawk migrations in New York State received the attention of many observers since the last century (Andrews, 1942; Bailey, 1967, 1969; Bull, 1964, 1974; Darrow, 1963; Eaton, 1910, 1914; Elliott, 1960; Ferguson and Ferguson, 1922; Giraud, 1844; Mills and Mills, 1971; Murphy, 1933; Rogers, 1971; Single, 1980; Snyder, 1945; Thomas, 1971a, 1971b; Townsend, 1892; Ward, 1958, 1960a, 1960b, 1963). As a result of the work of these observers, and others, numerous hawk watching lookouts now are identified in this state (Drennan, 1981: 459–481; Heintzelman, 1979a: 188–195).

Bay Ridge

Bay Ridge, on Long Island, is a western section of Brooklyn adjacent to The Narrows that separate Upper New York Bay from Lower New York Bay. Townsend (1892) reported late September migrations of Sharp-shinned Hawks, Cooper's Hawks, Broad-winged Hawks, and American Kestrels at this site each autumn when northwest winds prevailed.

Bear Mountain State Park

Good autumn hawk migrations are reported at this park adjacent to the Hudson River west of Peekskill (Heintzelman, 1979a: 188–189). Early November flights of at least 150 Red-tailed Hawks are noted in the park (Bull, 1974: 184).

Bedford

Considerable numbers (83.7 hawks per hour) of migrating hawks are reported in autumn at the Butler Sanctuary at Bedford north of Greenwich. Ospreys, Northern Harriers, Sharp-shinned Hawks, Broad-winged Hawks, Red-tailed Hawks, and American Kestrels are observed in largest numbers (Currie, 1982: 15).

Belfrey Mountain Fire Tower

The fire tower atop Belfrey Mountain in northeastern New York is located on an autumn and spring hawk migration flight-line (Drennan, 1981: 469).

Data are too limited to properly evaluate the importance of the site during autumn, but an assortment of accipiters, buteos, falcons, and other species are noted (Nagy, 1979b: 30).

Spring hawk migrations are perhaps larger than those in autumn but nevertheless somewhat modest in magnitude (6.3 hawks per hour). The species counted in largest numbers are Sharp-shinned Hawks, Red-shouldered Hawks, Broad-winged Hawks, and Red-tailed Hawks (Benz, 1981a: 16, 1983c: 14; Nagy, 1977c: 17).

Bonticou Crag

This bald knob is atop the Shawangunk Ridge near Lake Mohonk past which fair (11.7 hawks per hour) spring hawk migrations occur (Heintzelman, 1979a: 189). Observations by Paul Kerlinger and Keith R.

Hanson indicate that Ospreys, Sharp-shinned Hawks, Broad-winged Hawks, Red-tailed Hawks, and American Kestrels formed the bulk of the flights, although a few other species are observed in limited numbers (Nagy, 1978a: 9).

Bronxville

During September more than 1,100 hawks (probably Broad-winged Hawks) plus lesser numbers of Sharp-shinned Hawks, Cooper's Hawks, Red-shouldered Hawks, Broad-winged Hawks, and American Kestrels were reported passing Bronxville north of New York City in Westchester County (Murphy, 1933). The birds followed a course from Long Island Sound's north shore to the Hudson Valley just north of the City line.

Coot Hill

Spring hawk migrations are reported at Coot Hill overlooking Bulwagga Bay and Crown Point on Lake Champlain (Drennan, 1981: 468–469). The flights there are moderate in number (6.2 hawks per hour), with Sharp-shinned Hawks, Broad-winged Hawks, and Red-tailed Hawks appearing in largest numbers (Benz, 1981a, 1982c, 1983c; Nagy, 1977c, 1978a).

Dutchess Hill

James W. Key reported that hundreds of migrating Broad-winged Hawks and lesser numbers of other birds of prey sometimes are observed during autumn at Rose View Farm along Dutchess Hill Road near Hyde Park (Heintzelman, 1975: 74–75; Nagy, 1978b: 24).

Easthampton

Autumn migrations of dozens of Northern Harriers and Merlins are reported at Easthampton at the southeastern end of Long Island (Bull, 1964: 158, 163).

Far Rockaway

Some fine autumn hawk migrations sometimes are reported at Far Rockaway at the eastern end of Jamaica Bay on western Long Island. Sharp-shinned Hawks, American Kestrels, Merlins, and Peregrine Falcons seem to form the bulk of the flights (Bull, 1964: 150, 154, 159, 162–164).

Fire Island

Fire Island is a long, narrow barrier beach island located off the south shore of Long Island in Suffolk County. It is separated from the Oak Beach section of Jones Beach by the Fire Island inlet. Good to excellent autumn migrations of accipiters and falcons pass along Fire Island with several thousand Sharp-shinned Hawks and American Kestrels sometimes reported as well as lesser numbers of Merlins and Peregrine Falcons (Buckley, Paxton, and Cutler, 1978: 184; Darrow, 1963; Heintzelman, 1979a: 191; Kane and Buckley, 1975a: 30).

Field studies by Darrow (1963) indicate that Peregrine Falcons follow a flight-line from Democrat Point on a bearing that never varies more than 232 ± 4 degrees.

Fishers Island

Fishers Island is located at the eastern entrance to Long Island Sound and serves as a connecting link between Rhode Island and Long Island, New York, during autumn for some migrant hawks. Considerable numbers of hawks migrate across the island from east to west during autumn, and earlier in this century thousands of these birds were shot (Ferguson and Ferguson, 1922). Sharp-shinned Hawks and falcons are well represented in the flights, and some Ospreys and Northern Harriers also appear, but buteos are rare (Ferguson and Ferguson, 1922).

Franklin Mountain

Autumn and spring hawk migrations are reported at Franklin Mountain near Oneonta (Drennan, 1981: 470; Heintzelman, 1979a: 191–192). Additional hawk migrations also pass along north-to-south ridges located east of Franklin Mountain, thus suggesting that there are considerable numbers of migrant hawks moving through central New York (Kerlinger and Bennett, 1977).

Autumn hawk migrations are reported at this location at the mean rate of 17.5 hawks per hour with Ospreys, Sharp-shinned Hawks, Broad-winged Hawks, Red-tailed Hawks, and American Kestrels forming the bulk of the flights, although limited numbers of other species also appear (Kerlinger and Bennett, 1977; Nagy, 1978b: 24).

Spring hawk migrations are reported at the mean rate of 115.8 hawks per hour with Turkey Vultures, Sharp-shinned Hawks, Red-shouldered Hawks, Broad-winged Hawks, and Red-tailed Hawks reported in largest numbers (Kerlinger and Bennett, 1977).

Hamershlag Hawkwatch

Drennan (1981: 476–477) stated that autumn migrations of Sharp-shinned Hawks, Broad-winged Hawks, and American Kestrels are reported at the Butler Sanctuary near Mt. Kisco in Westchester County.

Hancock

During spring and autumn dozens of migrating Turkey Vultures are reported at Hancock (Bull, 1974: 164; Nichols, 1958: 337).

Helderberg Mountains

These mountains near Albany serve as a fair (14.8 hawks per hour) autumn hawk migration route. The species reported in largest numbers include Northern Harriers, Red-shouldered Hawks, Broad-winged Hawks, Red-tailed Hawks, and American Kestrels (Kerlinger, 1980; Nagy, 1978b: 24, 1979b: 31; West, 1947: 93).

Hook Mountain

Hook Mountain, a few miles above Nyack, Rockland County, is a short ridge running east to west beside the Hudson River and produces spring and autumn hawk migrations.

Fair spring hawk migrations (12.1 hawks per hour) are reported at this site, the most numerous species being Ospreys, Sharp-shinned Hawks, Broad-winged Hawks, Red-tailed Hawks, and American Kestrels (Benz, 1981a, 1982c, 1983c; Benz and Dereamus, 1984; Nagy, 1977c, 1978a, 1980).

Good autumn hawk migrations (30.8 hawks per hour) are reported at Hook Mountain. The bulk of the flights consist of Ospreys, Northern Harriers, Sharp-shinned Hawks, Red-shouldered Hawks, Broad-winged Hawks, Red-tailed Hawks, and American Kestrels, but smaller numbers of other species also are noted (Benz, 1980b, 1981b, 1982b, 1983b, 1984b; Mills and Mills, 1971; Nagy, 1976, 1977b, 1978b, 1979b; Single, 1974, 1975; Single and Thomas, 1975; Thomas, 1973).

Hudson Falls
As many as 43 Rough-legged Hawks are reported in late November at Hudson Falls (Kane and Buckley, 1975a: 30).

Hudson River (Albany to Kingston)
Almost unique* in the study of hawk migrations was a Hudson River boat trip on 18 September 1945, between Albany and Kingston, that produced several Ospreys, 13 Bald Eagles, 2 Northern Harriers, 2 Sharp-shinned Hawks, 2 Red-shouldered Hawks, and 1 Broad-winged Hawk (Snyder, 1945: 79). Apparently the river is a migration flight-line for some hawks.

Hyde Park
Mid-September flights of more than 250 Broad-winged Hawks are reported at Hyde Park (Mersereau, 1976: 5).

Jamesville
As many as 30 Rough-legged Hawks are reported in mid-October at Jamesville in Onondaga County (Bull, 1974: 189).

Jones Beach State Park
This park, along Long Island's south shore west of Fire Island, produces some fine falcon and other hawk migrations during autumn (Heintzelman, 1979a: 193) when approximately 87 hawks per day are reported with American Kestrels and Merlins appearing in largest numbers (Boyajian, 1971: 33; Buckley, Paxton, and Cutler, 1978: 184; Ward, 1958, 1960a, 1960b, 1963).

Long Island
Autumn hawk migrations crossing Long Island follow a flight-line along the Sound near Orient at the eastern end of the north fork, but the numbers of migrant hawks decreases westward along the shore of Long Island Sound. Important flights also are seen westward along the barrier beaches on the Atlantic Ocean side of the island including those at Fire

---

*In the Great Lakes, Perkins (1964) observed hawk migrations from a ship, offshore hawk migrations were made from ships in the western North Atlantic (Anderson and Powers, 1979; Kerlinger, Cherry, and Powers, 1983) and eastern North Pacific (Henshaw, 1901), and from a small boat (Kunkle, 1976) and ferry (Dunne, 1980b) in Delaware Bay.

Island and Jones Beach, and additional flights are reported at Bronxville on a flight-line from Long Island's north shore to the Hudson Valley north of New York City (Darrow, 1963; Elliott, 1960; Murphy, 1933; Ward, 1958, 1960a, 1960b, 1963).

Millbrook

Autumn hawk flights at Millbrook (167 hawks per hour) include mostly Broad-winged Hawks, a few dozen Sharp-shinned Hawks, and a few other species (Nagy, 1978b: 24).

Mohawk Hills

Ridges facing south, adjacent to the Mohawk River in eastern New York, serve as the Mohawk Hills hawk lookout where Ron Frazier reported 5 hawks per hour during autumn with Northern Harriers, Sharp-shinned Hawks, Broad-winged Hawks, Red-tailed Hawks, and American Kestrels reported in largest numbers (Nagy, 1976: 12–13, 1977b: 13).

Moses Mountain

Moses Mountain is on Staten Island in New York City (Drennan, 1981: 480–481). There are 21.5 hawks per hour reported during autumn at this site with Ospreys, Northern Harriers, Sharp-shinned Hawks, Broad-winged Hawks, and American Kestrels seen in largest numbers (Dunne, 1980a: 12, 14, 1981: 15, 1982: 17).

Mount Aspetong

Mid-September flights of more than 1,100 Broad-winged Hawks are reported at this southeastern New York site (Mersereau, 1976: 5).

Mount Defiance

Spring hawk migrations (10.2 hawks per hour) are reported at Mount Defiance near Ticonderoga. Turkey Vultures and Broad-winged Hawks appear in largest numbers, but a few other species also are seen (Benz, 1981a: 16; Drennan, 1981: 469–470).

Mount Peter

Mount Peter is one of New York's better known hawk migration lookouts located on Bellvale Mountain near Greenwood Lake (Heintzelman, 1979a: 193).

Autumn hawk migrations there (21.9 hawks per hour) consist mostly of Ospreys, Sharp-shinned Hawks, Broad-winged Hawks, Red-tailed Hawks, and American Kestrels, although a variety of other species also are included in the flights in small numbers (Bailey, 1967, 1969; Cinquina, 1975, 1976, 1977; Cinquina and Martin, 1979, 1980, 1981, 1982, 1983, 1984; Martin, 1978; Rasmussen, 1974; Rogers, 1971, 1972).

Spring hawk flights (86.2 hawks per hour) consist mostly of Ospreys, Sharp-shinned Hawks, and Broad-winged Hawks along with a few other species in small numbers (Cinquina, 1984).

Near Trapps

The Near Trapps is the most northern site on the ridge system that includes Bake Oven Knob and Hawk Mountain in Pennsylvania. The Near

Trapps, however, is located on the Shawangunk Mountain range at the Mohonk Trust near New Paltz (Carroll, 1980; Heintzelman, 1979a: 193–194).

Autumn hawk migrations (16.4 hawks per hour) at this site consist mostly of Sharp-shinned Hawks, Broad-winged Hawks, Red-tailed Hawks, and American Kestrels plus small numbers of various other species (Benz, 1981b, 1983b, 1984b; Carroll, 1980; Nagy, 1976, 1977b, 1978b, 1979b).

New York City

During autumn as many as 2,500 migrating Broad-winged Hawks, and occasionally even Rough-legged Hawks and Golden Eagles are reported over New York City (Carleton, 1963: 16, 1965: 19).

Oneida Lookout

The Oneida Lookout near Sherrill produces autumn hawk migrations at the rate of 27.4 hawks per hour. Sharp-shinned Hawks, Broad-winged Hawks, Red-tailed Hawks, and American Kestrels form the bulk of the flights (Benz, 1980b; Crumb and Peebles, 1977; Nagy, 1978b, 1979b).

Orient

Autumn falcon migrations are reported near Orient along the shore of Long Island Sound on the north fork of Long Island's eastern end (Elliott, 1960). During spring, as many as 50 migrant Ospreys also are noted at this location in early April (Bull, 1964: 159).

Ossining

Large numbers of migrating Broad-winged Hawks are reported during September at Ossining in Westchester County (Murphy, 1933: 321).

Owego

Mid-September flights of as many as 2,300 Broad-winged Hawks are reported at Owego in Tioga County (Bull, 1974: 188).

Pleasant Valley

During mid-September, hundreds of migrating Broad-winged Hawks sometimes are reported at Pleasant Valley in Dutchess County (Boyajian, 1968a: 15).

Port Chester

Up to 1,400 migrant Broad-winged Hawks are reported in mid-September at Port Chester (Boyajian, 1968a: 15, 1969: 24).

Port Jervis Hawk Lookout

The hawk lookout near Port Jervis is a roadside pullover beside the westbound lanes of Interstate 84 two miles east of the town. Good autumn hawk flights (27.1 hawks per hour) are noted at this site, especially of Broad-winged Hawks and Red-tailed Hawks (Benz, 1980b, 1981b, 1982b, 1983b; Heintzelman, 1979a: 194; Nagy, 1977b, 1979b).

Poughkeepsie

As many as 500 migrating Broad-winged Hawks are seen in mid-September at Poughkeepsie (Boyajian, 1968a: 15).

Riis Park

Bull (1964: 164) reported as many as 500 American Kestrels migrating past Riis Park in early October. The site is on a barrier beach island south of Jamaica Bay in New York City.

Rockaway Point

Autumn migrations of falcons, and other raptors, are reported at Rockaway Point south of Brooklyn (Elliott, 1960: 157).

Shawangunk Mountains

During September thousands of migrating Broad-winged Hawks are reported along the Shawangunk Mountains near Ellenville (Barbour, 1908).

Shinnecock Bay

Drennan (1981: 477–478) reported fine autumn migrations of Sharp-shinned Hawks, American Kestrels, and Merlins, plus smaller numbers of other species, at Shinnecock Bay toward the southeastern end of Long Island.

Smith Point

Smith Point, on one of Long Island's barrier beaches facing the Atlantic Ocean, produces some fine autumn migrations of Ospreys, American Kestrels, and Merlins (Ward, 1963: 23).

Storm King Mountain

Good autumn hawk flights, but poor spring flights, are reported at Storm King Mountain near Cornwall-on-Hudson (Heintzelman, 1979a: 194–195).

Thornwood

Good autumn hawk flights (24.2 hawks per hour) are noted at Thornwood north of White Plains. Sharp-shinned Hawks, Broad-winged Hawks, and American Kestrels are seen in largest numbers (Nagy, 1977b: 13).

Tuckahoe

Bull (1964: 153) reported as many as 122 migrant Red-shouldered Hawks during late October at Tuckahoe.

Van Cortlandt Park

This New York City park is located on an autumn hawk migration flightline. Bald Eagles, Sharp-shinned Hawks, Broad-winged Hawks (up to 2,700 birds), and American Kestrels are reported in largest numbers (Bull, 1964: 150–164).

Vischer's Ferry

As many as 50 Broad-winged Hawks are observed migrating past Vischer's Ferry along the Mohawk River during late April (Andrews, 1942).

Westhampton

During autumn as many as 600 American Kestrels, and smaller numbers of Merlins, are reported migrating past the Westhampton area on Long Island (Bull, 1974: 200; Elliott, 1960: 157; Nichols, 1956: 10).

Whitehorse Mountain
Whitehorse Mountain is located some 50 miles north of New York City. Excellent autumn hawk flights (163.5 hawks per day) are reported there with Ospreys, Sharp-shinned Hawks, Broad-winged Hawks, Red-tailed Hawks, and American Kestrels forming the bulk of the flights, whereas spring hawk flights at the site are poor (Anonymous, 1977b; Heintzelman, 1979a: 195; Jeheber, 1976; Nagy, 1978b).

PENNSYLVANIA

Pennsylvania is one of the world's leading centers of hawk migration study with a long and very distinguished history of production of major, original work (Benz, 1980a–1984b; Benz and Dereamus, 1984; Bergey, 1975; Broun, 1935–1979b; Fingerhood, 1984a, 1984b; Fingerhood and Lipschutz, 1982; Frey, 1940, 1943; George and Schreffler, 1984; Heintzelman, 1961–1985; Heintzelman and Armentano, 1964; Heintzelman and MacClay, 1971–1979; Heintzelman and Reed, 1982; Horsfall, 1908; Kranick, 1982, 1984; Morton, 1982, 1983; Nagy, 1967–1984; Paff, 1927; Poole, 1934, 1964; Senner, 1984b; Stone, 1887, 1894; Sutton, 1928a–1931; Todd, 1940; Turnbull, 1869; Warren, 18?0; Wood, 1979).

In addition to scientific studies of hawk migrations, major raptor conservation efforts also occurred in Pennsylvania, some of which served as examples for similar national and international programs (Boyer, 1972; Brett, 1984a, 1984b; Broun, 1935, 1936, 1949a, 1956b; Collins, 1933; Graham, 1984; Heintzelman, 1970e: 20, 1979b: 7–8, 119, 1983b, 1984b, 1985; Johnson, 1984; Luttringer, 1938; Nagy, 1984; Pearson and Warren, 1897; Poole, 1934; Pough, 1932, 1936, 1984; Senner, 1984a; Sutton, 1928b, 1929; Tracy, 1943).

Various advances in educational hawk watching also occurred in Pennsylvania (Brett, 1981; Broun, 1949a; Heintzelman, 1980a, 1982a, 1983d, 1984b), and recreational hawk watching also is a major activity in this state (Broun, 1949a; Heintzelman, 1979a, 1979c, 1983b; Peterson, 1985).

Auburn Lookout
This site is on the Appalachian Trail northwest of Hamburg (Wilshusen, 1983). Excellent autumn hawk flights (38.5 hawks per hour) are reported at the lookout with Sharp-shinned Hawks, Broad-winged Hawks, and Red-tailed Hawks the most common of the various species seen there (Benz, 1980b, 1981b; Nagy, 1979b).

Bake Oven Knob
Bake Oven Knob is a superb hawk watching lookout on the summit of the Kittatinny Ridge about 16 miles northeast of Hawk Mountain Sanctuary.

Prior to 1957 it was used as a major hawk shooting site and thousands of migrating hawks were killed there (Brett, 1984b; Broun, 1955; Heintzelman, 1975, 1983d: 135–138, 1985; Nagy, 1984; Senner, 1984a). The Knob now is used for hawk migration research, conservation, education, and recreation.

Excellent autumn hawk migrations (34.9 hawks per hour) are reported at this site with Ospreys, Bald Eagles, Northern Harriers, Sharp-shinned Hawks, Broad-winged Hawks, Red-tailed Hawks, Golden Eagles, and American Kestrels the most numerous of the species observed (Graham, 1972; Heintzelman, 1963a–1968, 1969b, 1970d, 1982c, 1982i, 1983c, 1984e; Heintzelman and Armentano, 1964; Heintzelman and MacClay, 1972–1979; Heintzelman and Reed, 1982).

Spring hawk migrations, however, are much less intensively studied but available data demonstrate that 108.1 hawks per day pass the site, of which Ospreys, Sharp-shinned Hawks, Broad-winged Hawks, and Red-tailed Hawks are the most numerous (D. S. Heintzelman, unpublished data).

In addition to hawk migration field studies, major conservation, education, and recreation hawk watching studies are conducted at this site (Heintzelman, 1970e: 20, 1979a, 1979b, 1979c, 1980a, 1982a, 1983b, 1983d, 1984b, 1985).

Bald Eagle Mountain

This ridge, across a valley from the Allegheny Front, is the most westerly of the ridges of the Appalachians. Autumn hawk migrations (8.5 hawks per hour) along the ridge contain mostly Sharp-shinned Hawks, Broad-winged Hawks, and Red-tailed Hawks, but occasionally other species are observed (Benz, 1983b: 22; Bittner, 1975: 5–6; Charles E. Trost *in* Heintzelman, 1975: 86).

Beam Rocks

Robert Leberman conducted autumn hawk counts at Beam Rocks near Rector. He reported 5.6 hawks per hour migrating past the site, with Sharp-shinned Hawks, Broad-winged Hawks, and Red-tailed Hawks the most common of the species observed (Benz, 1980b: 21, 1981b: 21; Nagy, 1978b: 24, 1979b: 31).

Bear Rocks

The Bear Rocks lookout is on the crest of the Kittatinny Ridge—sometimes incorrectly referred to as "Baer" or "Baer's" Rocks (see Heintzelman, 1984d; Miller, 1941; Wilshusen, 1983)—and is an important hawk migration observation post.

Excellent autumn hawk flights are seen there (35.2 hawks per hour) with Ospreys, Northern Harriers, Sharp-shinned Hawks, Cooper's Hawks, Red-shouldered Hawks, Broad-winged Hawks, Red-tailed Hawks, Golden Eagles, and American Kestrels forming the bulk of the flights, although small numbers of other species also are noted (Benz, 1980b, 1981b, 1982b,

1983b, 1984b; Alan and Paul Grout *in* Heintzelman, 1975: 321; Nagy, 1977b, 1978b, 1979b).

Spring hawk migrations (29.6 hawks per hour) consist mostly of Ospreys, Sharp-shinned Hawks, Broad-winged Hawks, Red-tailed Hawks, and American Kestrels, plus smaller numbers of a variety of other species (Broun, 1979a: 50; Kranick, 1982, 1984).

Broad Mountain

During September flights of almost 1,600 Broad-winged Hawks were reported over Broad Mountain near St. Clair by John Stasho (Maurice Broun, unpublished data).

Chester

As many as 850 Broad-winged Hawks were seen in mid-September over Chester (Paxton, Richards, and Cutler, 1980: 145).

Chickies Rock

This cliff (sometimes called Chicques Rock) overlooks the Susquehanna River near Marietta. Autumn hawk flights there (36.7 hawks per hour) consist mostly of Broad-winged Hawks, although small numbers of various other raptors also are seen (Edge, 1945: 3; Nagy, 1978b: 24). Fair spring hawk flights also are seen at this site (Heintzelman, 1979a: 202).

Compass

Joe Meloney's autumn hawk counts at Compass produce 20.4 hawks per hour with Sharp-shinned Hawks, Broad-winged Hawks, and American Kestrels forming the bulk of the migrations (Benz, 1983b: 22).

Cornwall Fire Tower

Cornwall Fire Tower in Lancaster County produces autumn hawk migrations at the rate of 17.6 hawks per hour with Sharp-shinned Hawks, Broad-winged Hawks, and Red-tailed Hawks the most commonly observed of the various species reported there (George and Schreffler, 1984: 28).

Spring hawk migrations (7.6 hawks per hour) consist mostly of Ospreys, Sharp-shinned Hawks, Broad-winged Hawks, Red-tailed Hawks, and American Kestrels (Nagy, 1977c: 7, 1978a: 9).

Delaware Water Gap

Autumn hawk migrations (274.4 hawks per day) are reported from the west (Pennsylvania) side of the Gap with Broad-winged Hawks forming the bulk of the flights, although a variety of other species also appear in smaller numbers (Heintzelman, 1973: 2; Horsfall, 1908; Nichols, 1960: 18).

Doylestown

Spring hawk migrations containing as many as 240 Broad-winged Hawks sometimes are reported at Doylestown (Scott and Cutler, 1970b: 586).

Drumore

Autumn flights of Black Vultures, Turkey Vultures, and Ospreys are reported at Drumore (Paxton, Buckley, and Cutler, 1976: 42; Paxton, Richards, and Cutler, 1980: 144; Paxton, Smith, and Cutler, 1979: 160).

Spring flights of Sharp-shinned Hawks and Broad-winged Hawks also are seen at Drumore (Scott and Cutler, 1975b: 934).

Easton

Autumn hawk migrations at Easton occasionally contain as many as 10,000 Broad-winged Hawks, although generally far fewer birds are reported (David Dereamus, field observations; Paff, 1927).

Spring hawk migrations (46.7 hawks per day) also are observed at Easton (Boyle, Paxton, and Cutler, 1983: 852).

Elverson

Mid-September migrations of Broad-winged Hawks sometimes number more than 2,700 birds (Scott and Cutler, 1972a: 43, 1974: 34).

During spring, however, less than 90 Broad-winged Hawks were seen in April at Elverson (Scott and Cutler, 1973b: 755).

Exton

During autumn, hawks migrate past Exton at the rate of 13.8 hawks per hour with Turkey Vultures, Sharp-shinned Hawks, and Red-tailed Hawks reported in largest numbers (Nagy, 1978b: 24).

Geneva

As many as 18 Ospreys were seen in early May at Geneva (Hall, 1965: 471).

Governor Dick Fire Tower

Fair autumn hawk migrations, but poor spring flights, are reported at this tower (Heintzelman, 1979a: 203).

Green Lane Reservoir

Mid-September migrations of up to 1,400 Broad-winged Hawks are reported at this southeastern Pennsylvania reservoir (Paxton, Boyle, and Cutler, 1983: 162).

Havertown

Havertown is located northwest of Philadelphia where autumn hawk migrations are reported at the rate of 5.4 hawks per hour. The species observed in largest numbers include Ospreys, Sharp-shinned Hawks, Broad-winged Hawks, Red-tailed Hawks, and American Kestrels (Benz, 1980b: 21, 1981b: 21, 1983b: 22).

Hawk Mountain Sanctuary

Of all the hawk watching locations in North America, perhaps the world, none is more famous or has a richer or more colorful history than Hawk Mountain Sanctuary atop the Kittatinny Ridge in eastern Pennsylvania. It is the birthplace of North American hawk migration conservation, education, research, and recreation (Broun, 1949a; Heintzelman, 1983b; Senner, 1984b). It also is the world's first wildlife sanctuary established to protect birds of prey (Broun, 1936, 1949a; Collins, 1935).

Prior to 1934, when the Sanctuary was established, an annual autumn slaughter of large numbers of migrating hawks occurred at this site—a slaughter that stopped when Maurice and Irma Broun arrived at the site in

September 1934 and transformed the killing ground into the refuge now known to visitors (Brett, 1984a, 1984b; Broun, 1935, 1949a; Collins, 1933; Harwood, 1973; Heintzelman, 1975: 88–90; Poole, 1934; Pough, 1932, 1984; Sutton, 1928). It was then that the first systematic hawk migration studies were completed during autumn—annual autumn hawk counts that continue still resulting in 28.5 hawks per hour reported at the site (based upon an assignment of 720 observation hours per season). The species reported in largest numbers include Ospreys, Northern Harriers, Sharp-shinned Hawks, Red-shouldered Hawks, Broad-winged Hawks, Red-tailed Hawks, and American Kestrels, plus dozens of Golden Eagles and smaller numbers of other species (Anonymous, 1975b, 1976, 1977a, 1978, 1979c, 1981c; Benz, 1980, 1982a, 1983, 1984a; Broun, 1935–1966; Nagy, 1967, 1968; Sharadin, 1972, 1973, 1974; Wetzel, 1969, 1970, 1971).

Spring hawk migrations (4.4 hawks per hour) include Ospreys, Northern Harriers, Sharp-shinned Hawks, Broad-winged Hawks, Red-tailed Hawks, and American Kestrels in largest numbers, plus more limited numbers of various other raptors (Morton, 1982, 1983).

In addition to conservation and research, educational hawk watching also is an important activity at Hawk Mountain (Broun, 1949a; Brett, 1981; Heintzelman, 1983b; Senner, 1984b) as is recreational hawk watching (Broun, 1949a; Heintzelman, 1983b; Senner, 1984b).

Hooversville

Glen and Ruth Sager's autumn hawk counts at Hooversville in the Allegheny Mountains demonstrate a low-density hawk migration there (3.1 hawks per hour) with the bulk of the flights formed by Sharp-shinned Hawks, Broad-winged Hawks, Red-tailed Hawks, and American Kestrels (Benz, 1980b, 1981b, 1982b, 1983b, 1984; Nagy, 1976, 1977b, 1978b, 1979b).

Hopewell Fire Tower

Autumn hawk migrations (39.4 hawks per hour) at this tower in French Creek State Park consist mostly of Sharp-shinned Hawks, Broad-winged Hawks, Red-tailed Hawks, and American Kestrels, although small numbers of various other raptors also are reported (Nagy, 1978b: 24–25).

Huntington Ridge

Gaylon M. Gerrish operated a banding station at this site along the Susquehanna River's east branch during autumn and reported 17.8 hawks per hour passing the station. The species seen in largest numbers were Sharp-shinned Hawks, Cooper's Hawks, Broad-winged Hawks, Red-tailed Hawks, and American Kestrels (Benz, 1980b, 1981b, 1983b; Nagy, 1976, 1977b, 1978b, 1979b).

Jenkintown

Mid-September flights of as many as 2,000 Broad-winged Hawks are reported at Jenkintown (Scott and Cutler, 1963: 19).

Kimberton

As many as 800 migrating Broad-winged Hawks are seen in mid-September at Kimberton (Scott and Cutler, 1963: 19).

Kutztown

Glider pilot Barney Johnston was aloft over Kutztown on 5 October 1969 and observed a kettle of 2,500 and another kettle of 3,000 Broad-winged Hawks only 300 yards apart whereas no Broad-wings were seen that day at Hawk Mountain (Nagy, 1970: 9), suggesting that large numbers of hawks were drifting cross-country rather than following parts of the Kittatinny Ridge as a flight-line.

Lancaster

An autumn hawk watch at Lancaster produced 37.4 hawks per hour of which Sharp-shinned Hawks, Broad-winged Hawks, and American Kestrels appeared in largest numbers (Nagy, 1977b: 13).

During April as many as 85 migrating Broad-winged Hawks also are reported at Lancaster (Paxton, Boyle, and Cutler, 1981: 806).

Lansdale

Thousands of migrating Broad-winged Hawks sometimes are seen in mid-September at Lansdale (Scott and Cutler, 1963: 19).

Larksville Mountain

Autumn hawk migrations at Larksville Mountain near Wilkes-Barre long ago were known to ornithologists mainly because of the slaughter by gunners of hundreds of hawks at that location (Edge, 1945: 3; Broun, 1957: 4; Tracy, 1943).

A limited autumn hawk watch there by Harry Brown, Edwin Johnson, and William Reid produced 57 hawks per day with the largest numbers of birds being Sharp-shinned Hawks, Broad-winged Hawks, and Red-tailed Hawks (Heintzelman, 1975: 92, 326).

Lehigh Furnace Gap

This Kittatinny Ridge hawk watching site, one of the most intensive hawk shooting locations along the ridge prior to 1957, is upridge a short distance from Bake Oven Knob (Broun, 1949b: 7, 1953: 7, 1954: 13; Heintzelman, 1979a: 206).

Autumn hawk migrations (22 hawks per hour) pass Lehigh Furnace Gap with Ospreys, Sharp-shinned Hawks, Broad-winged Hawks, and Red-tailed Hawks appearing in largest numbers based upon only fragmentary counts (D. S. Heintzelman, unpublished data).

Spring hawk migrations (3.2 hawks per hour) also pass the Gap with Ospreys, Sharp-shinned Hawks, Broad-winged Hawks, Red-tailed Hawks, and American Kestrels reported in largest numbers (Benz, 1982c: 17).

Little Gap

Prior to 1957, Little Gap was one of the worst hawk slaughter grounds on the Kittatinny Ridge upridge from Bake Oven Knob and Hawk Mountain

(Broun, 1951a: 7, 1953: 7, 1954: 13). It is now used for autumn hawk watching purposes and excellent (36.5 hawks per hour) flights are reported there. The species observed in largest numbers are Ospreys, Sharp-shinned Hawks, Cooper's Hawks, Broad-winged Hawks, and Red-tailed Hawks (Benz, 1980b, 1981b, 1982b, 1983b, 1984b; Nagy, 1976, 1977b, 1978b, 1979b).

Little Mountain

Frank and Barbara Haas conducted autumn hawk counts at Little Mountain near the Susquehanna River near Sunbury and report 9.7 hawks per hour. The bulk of the migrations are composed of Ospreys, Sharp-shinned Hawks, Broad-winged Hawks, Red-tailed hawks, and American Kestrels (Nagy, 1976: 13, 1977b, 1978b, 1979b).

Longwood Gardens

Jesse Grantham's spring hawk migration observations produced Ospreys, Northern Harriers, Sharp-shinned Hawks, Broad-winged Hawks, and American Kestrels as the main components of the flights (Grantham, 1973: 29).

Militia Hill

John C. Ward III uses this site near Philadelphia as an autumn hawk watching lookout and reports 16 hawks per hour with Ospreys, Sharp-shinned Hawks, Broad-winged Hawks, Red-tailed Hawks, and American Kestrels the most numerous species observed (Benz, 1980b, 1981b, 1982b, 1983b; Nagy, 1978b, 1979b).

Spring hawk migrations (13.3 hawks per hour) consist mostly of Sharp-shinned Hawks, Broad-winged Hawks, and American Kestrels, plus small numbers of various other species (Benz, 1982c: 17).

Morgan's Hill

This site near Easton is used as an autumn hawk watching lookout by David Dereamus who reports 745 hawks per hour of which Ospreys, Sharp-shinned Hawks, Broad-winged Hawks (13,235), and American Kestrels are seen in largest numbers (Benz, 1984b: 24; Heintzelman, 1983c: 3–4).

The spring hawk flights (13 hawks per hour) at Morgan's Hill consist mainly of Ospreys, Sharp-shinned Hawks, Broad-winged Hawks, Red-tailed Hawks, and American Kestrels, plus smaller numbers of other raptors (Benz, 1983c: 14; Benz and Dereamus, 1984: 18).

Muddy Run Pumped Storage Pond

Observations of Ospreys at this lower Susquehanna River location indicate that 24 percent of the birds appear during spring, whereas 66 percent of the birds are seen during autumn (Schutsky, 1982).

Philadelphia

Various fragmentary hawk migrations observations are reported from Philadelphia. Autumn observations include sightings of Sharp-shinned Hawks, Broad-winged Hawks, American Kestrels, and Peregrine Falcons (Gillespie,

1944; Stone, 1887; Todd, 1970), whereas observations during spring include an American Swallow-tailed Kite and a Mississippi Kite (Brendel and Roby, 1984; Ross, 1953).

Pipersville

Ann Webster made autumn hawk counts at Pipersville in central Bucks County near the Delaware River. She reported 108.6 hawks per day of which Ospreys, Sharp-shinned Hawks, Broad-winged Hawks, and American Kestrels appeared in largest numbers (Heintzelman, 1975: 326).

Pocono Mountains

Hawk migrations crossing the Pocono Mountains of northeastern Pennsylvania are documented in broad terms. Northern Goshawks appeared in invasion numbers in 1934–1937, and in 1954, and September flights of as many as 68 Broad-winged Hawks were seen at Pocono Lake (Street, 1954, 1975).

Port Clinton Fire Tower

Earlier in this century this tower near Port Clinton, not far from Hawk Mountain Sanctuary, was used as a hawk shooting site during autumn (Broun, 1949a: 31–32). It also is used occasionally during autumn as a hawk watching lookout.

Spring hawk migrations (13.7 hawks per hour) consist mostly of Ospreys, Sharp-shinned Hawks, Broad-winged Hawks, Red-tailed Hawks, and American Kestrels (Benz and Dereamus, 1984: 18).

Ridley Creek State Park

Some hawk migrations are reported at this park in southeastern Pennsylvania. During autumn, for example, Sharp-shinned Hawks and Broad-winged Hawks appear most commonly, but occasionally other raptors also are noted.

Spring hawk migrations (3.2 hawks per hour) include Sharp-shinned Hawks, Broad-winged Hawks, and American Kestrels in largest numbers (Benz, 1981a; Haas, 1984; Nagy, 1980).

Riegelsville

On 16 September 1948, George Pyle observed 1,500 Broad-winged Hawks migrating over Riegelsville adjacent to the Delaware River. That same day a flight of at least 11,392 hawks (mostly Broad-wings) passed Hawk Mountain Sanctuary on the Kittatinny Ridge to the northwest (Broun, 1949a: 186), suggesting that large numbers of migrating hawks passed across several parts of eastern Pennsylvania on that date.

Rockville

Bergey (1975) observed autumn hawk migrations from a site on the Kittatinny Ridge near Rockville and reported Northern Harriers, Sharp-shinned Hawks, Broad-winged Hawks, and Red-tailed Hawks in largest numbers.

He also noted varying behavior by various hawks when they reached the intersection of the ridge and river cutting through it. Some Northern

Harriers, Sharp-shinned Hawks, Broad-winged Hawks, and Red-tailed Hawks turned and headed downriver, other Turkey Vultures, Broad-winged Hawks, Red-tailed Hawks, and American Kestrels headed upriver toward another ridge, but most hawks crossed the river and continued along the ridge.

Route 183

Autumn hawk migrations were observed at this site on the Kittatinny Ridge for many years by Earl L. Poole, Robert and Anne MacClay, and others. A passage rate of 16.9 hawks per hour is reported for this location with Ospreys, Northern Harriers, Sharp-shinned Hawks, Broad-winged Hawks, Red-tailed Hawks, and fine numbers of Golden Eagles observed (Benz, 1983b, 1984b; Heintzelman, 1975: 327; Robert MacClay, field notes).

Roxborough

Autumn hawk flights at Roxborough (27.3 hawks per hour) include mostly Sharp-shinned Hawks, Broad-winged Hawks, and American Kestrels, plus small numbers of various other birds of prey (Benz, 1982b).

Smith's Gap

This gap is one of several hawk watching locations on the Kittatinny Ridge between the Delaware and Lehigh Rivers. Large numbers of hawks were shot there during autumn prior to 1957 (Edge, 1956: 1; Broun, 1962: 10) indicating that considerable numbers of autumn raptor migrants pass the site.

Spring hawk flights at the Gap include fine numbers of Red-tailed Hawks (Kranick, 1982: 5).

Sterrett's Gap

This gap on the Kittatinny Ridge west of the Susquehanna River produces excellent hawk migrations (37 hawks per hour). The species observed in largest numbers are Sharp-shinned Hawks, Broad-winged Hawks, and Red-tailed Hawks, although a good variety of other birds of prey also are seen in smaller numbers (Benz, 1983b; Frey, 1940, 1943; Nagy, 1978b).

Tott's Gap

Tott's Gap is a small wind gap on the Kittatinny Ridge a few miles southwest of the Delaware River. Fair autumn hawk migrations (65.4 hawks per day) are reported at this site, based upon field studies by Howard Drinkwater and my own work, with Ospreys, Sharp-shinned Hawks, Broad-winged Hawks, Red-tailed Hawks, and American Kestrels the most commonly observed species (Heintzelman, 1973, 1975: 96–97, 1979a: 208–209).

Tri-County Corners

C. J. Robertson operated a banding station at this spot on the Kittatinny Ridge between Bake Oven Knob and Hawk Mountain and reported Sharp-shinned Hawks, Cooper's Hawks, Nothern Goshawks, Broad-winged Hawks, and Red-tailed Hawks the most trapped species (Heintzelman, 1975: 304).

Tuscarora Mountain

The Pulpit, on the summit of Tuscarora Mountain between Chambersburg and McConnellsburg, is used for both autumn and spring hawk watching (Heintzelman, 1979a: 209).

Autumn hawk flights at The Pulpit are fair (12.5 hawks per hour), the bulk of the flights being formed by Ospreys, Sharp-shinned Hawks, Broad-winged Hawks, and Red-tailed Hawks, although a good variety of other birds of prey also are seen in more limited numbers (Benz, 1980b, 1981b, 1982b, 1983b, 1984b; Nagy, 1976, 1977b, 1978b, 1979b).

Spring hawk migrations (8.1 hawks per hour) contain Ospreys, Sharp-shinned Hawks, Broad-winged Hawks, and Red-tailed Hawks in largest numbers, plus smaller numbers of various other raptors (Benz, 1981a, 1982c, 1983c; Benz and Dereamus, 1984; Nagy, 1978a, 1979a, 1980).

Unionville

During September, flights of as many as 4,500 Broad-winged Hawks are reported at Unionville (Buckley, Paxton, and Cutler, 1978: 185).

Waggoner's Gap

Waggoner's Gap (also spelled Wagner's Gap) is the last major hawk lookout on the Kittatinny Ridge before it terminates near Carlisle. D. L. Knohr began systematic use of the site in 1952 and continued making autumn hawk counts there for many years, whereas during recent years various other observers counted from the site. During autumn, hawks pass the site at the rate of 34.5 hawks per hour (Benz, 1980b, 1981b, 1982b, 1983b, 1984b; Heintzelman, 1975: 330–331; Kotz, 1973; Nagy, 1976, 1977b, 1978b). Reported in largest numbers are Sharp-shinned Hawks, Broad-winged Hawks, and Red-tailed Hawks, although considerable numbers of Golden Eagles also are seen along with a variety of other birds of prey.

Spring hawk migrations (7.4 hawks per hour) also are seen at Waggoner's Gap with Ospreys, Sharp-shinned Hawks, Broad-winged Hawks, and Red-tailed Hawks most commonly noted (Benz, 1981a; Benz and Dereamus, 1984).

Wind Gap

Paul Karner operates an observation and raptor banding station on the Kittatinny Ridge near Wind Gap and reports good (22.8 hawks per hour) autumn hawk migrations there. Counted frequently are Ospreys, Sharp-shinned Hawks, Broad-winged Hawks, and Red-tailed Hawks (Benz, 1981b; 1983b).

VIRGINIA

Organized interest in hawk migrations in Virginia began when William J. Rusling made detailed field studies of autumn hawk flights at Cape Charles (Kiptopeke) at the southern end of the Delmarva Peninsula in 1936 (Rusling, 1936; Heintzelman, 1975: 112–113, 249–252, 342). Later field studies at other sites further increased our knowledge of raptor movements in this state—especially at the Mendota Fire Tower in southwestern Virginia near the Tennessee border (Behrend, 1951–1954b; Brooks, 1971; Carpenter, 1949, 1960; Carter, 1978; DeGarmo, 1953; Finucane, 1956–1981b; Fowler, 1982; Turner, 1981). A catalog of some of the state's more active hawk watching lookouts also was published (Heintzelman, 1975, 1979a: 217–223).

Afton Mountain
See Rockfish Gap. This site is designated Afton Mountain in some of the mid-1970's hawk migration literature.

Amherst County
Leonora Wikswo's autumn hawk watch near Turkey Mountain produced good hawk flights (27.7 hawks per hour) with Black Vultures, Turkey Vultures, and Broad-winged Hawks reported in largest numbers (Puckette, 1982b: 25).

Apple Orchard Mountain
Finucane (1980b: 23) reported more than 325 Broad-winged Hawks, plus a few other raptors one autumn day at this site.

Arnold Valley
D. T. Puckette observed more than 1,600 Broad-winged Hawks and a few other raptors during autumn in the Arnold Valley (Finucane, 1980b: 23).

Assateague Island
Considerable autumn Merlin and Peregrine Falcon flights are reported at Assateague Island in both Virginia and Maryland (Berry, 1971; Heintzelman, 1979a: 169; Scott and Cutler, 1970a: 28; Ward and Berry, 1972; Wilds, 1983; 154).

Bald Knob
Autumn hawk migrations (27.3 hawks per hour) are reported at Bald Knob with Sharp-shinned Hawks and Broad-winged Hawks seen in largest numbers (Finucane, 1977b: 17, 1978b: 26).

Bear Church Rock
L. W. Wood observed a few Sharp-shinned Hawks, Red-shouldered Hawks, and Red-tailed Hawks at this site during autumn (Finucane, 1980b:23).

Big Schloss
Good autumn hawk flights are reported at Big Schloss (Heintzelman, 1979a: 217).

Buffalo Ridge
Myriam Moore reported poor autumn hawk migrations (4.8 hawks per hour) at Buffalo Ridge with Sharp-shinned Hawks observed in largest numbers (Finucane, 1976b: 15).

Bull Run
Autumn hawk migrations at this historic Civil War site are meager (3.5

hawks per hour) with an Osprey, Northern Harriers, Sharp-shinned Hawks, Broad-winged Hawks, and an American Kestrel observed (Finucane, 1981b: 23).

Calf Mountain
The autumn hawk flights (23.7 hawks per hour) at Calf Mountain contain Sharp-shinned Hawks and Broad-winged Hawks in largest numbers, plus small numbers of various other birds of prey (Finucane, 1979b: 35, 1980b: 23, 1981b: 23).

Cape Charles
Autumn hawk migrations at Cape Charles and its vicinity (248.1 hawks per day) are excellent with Ospreys, Northern Harriers, Sharp-shinned Hawks, Broad-winged Hawks, Red-tailed Hawks, American Kestrels, Merlins, and Peregrine Falcons forming the bulk of the flights, although a few other species also are reported (Dunne, 1979b: 17, 20; Heintzelman, 1975: 342; Rusling, 1936).

Carter's Mountain
T. F. Wieboldt observed excellent autumn hawk flights (63.1 hawks per hour) at this site with Broad-winged Hawks forming the bulk of the birds, although small numbers of several other species also appeared (Finucane, 1980b: 23).

Chantilly
November flights of up to 6 Rough-legged Hawks are reported at this site in northeastern Virginia (Scott and Cutler, 1961: 21; Scott and Potter, 1960: 21).

Chincoteague National Wildlife Refuge
A variety of hawks migrate past this barrier beach island refuge along coastal Virginia during autumn (8.7 hawks per hour) of which Ospreys, Northern Harriers, Sharp-shinned Hawks, and American Kestrels are reported in largest numbers, but worthwhile numbers of Merlins and Peregrine Falcons also are seen as are a few other species in small numbers (Dunne, 1978c: 10, 12; Heintzelman, 1979a: 217).

Clifton
Clifton is located not far southwest of Washington, D.C. Although very meager numbers of migrant hawks are seen there in autumn, chiefly Sharp-shinned Hawks and Red-shouldered Hawks, occasionally a few hundred Broad-winged Hawks also are reported (Finucane, 1981b: 23; Scott, 1976: 48).

Cove Mountain
Autumn hawk flights at Cove Mountain (6.5 hawks per hour) contain various species, but Sharp-shinned Hawks and Broad-winged Hawks are the most numerous (Finucane, 1981b: 23).

Crawford Mountain
The autumn hawk flights (10.7 hawks per hour) at Crawford Mountain

consist mostly of Broad-winged Hawks, but very small numbers of a few other raptors also are reported (Finucane, 1976b: 15, 1979b: 35).

Dayton
April migrations of Broad-winged Hawks at Dayton in northwestern Virginia may contain as many as 65 birds on some days (Hall, 1963: 402).

Devil's Backbone
This overlook is located near milepost 145 on the Blue Ridge Parkway. Autumn hawk migrations (32.2 hawks per hour) at this site contain a variety of species of which Sharp-shinned Hawks and Broad-winged Hawks are the most numerous (Finucane, 1980b: 23, 1981b: 23).

Spring hawk flights (7.1 hawks per hour) contain Sharp-shinned Hawks and Red-tailed Hawks in largest numbers but a variety of other birds of prey also are reported (Puckette, 1984a: 19).

Dixie Airport Road
As many as 1,000 Broad-winged Hawks were reported by Lawrence Tyrre over Dixie Airport Road eight miles north of Lynchburg (Finucane, 1976b: 14–15).

Dragon's Tooth
Autumn hawk migrations at this site (32.6 hawks per hour) consist mostly of Sharp-shinned Hawks and Broad-winged Hawks, although small numbers of other raptors sometimes are seen (Finucane, 1980b: 23, 1981b: 23).

Dungannon
E. E. Scott reported that as many as 279 migrating Broad-winged Hawks sometimes are seen during autumn at this location (Finucane, 1977b: 17).

Eagle Rock
B. J. Opengari reported that autumn hawk migrations at Eagle Rock (18.2 hawks per hour) consist mostly of Broad-winged Hawks, but small numbers of other hawks also are noted (Finucane, 1980b: 23).

East River Mountain
Sarah Cromer reported a passage rate of 10.5 hawks per hour during autumn at East River Mountain. Broad-winged Hawks form most of the flight, but a few individuals of several other raptors also appear (Finucane, 1976b: 15).

Ekin Park
According to K. L. Kirkpatrick, autumm hawk flights at this site produce 90.8 hawks per hour—mostly Broad-winged Hawks, but with a few other raptors also seen (Finucane, 1980b: 23).

Fisherman's Island National Wildlife Refuge
This refuge, generally closed to public entry, is located just south of the tip of the Cape Charles peninsula. It receives considerable numbers of migrating hawks (56.1 per hour) during autumn and is the focal point for Eastern Shore hawk flights about to cross the mouth of Chesapeake Bay. Of the various raptors observed, Ospreys, Sharp-shinned Hawks, Broad-winged

Hawks, and American Kestrels are reported by Bill Williams in largest numbers (Dunne, 1978c: 10, 12; Heintzelman, 1975: 249–252, 342; Rusling, 1936).

Flat Top

Autumn hawk migrations (4.2 hawks per hour) at this site consist mostly of Sharp-shinned Hawks and Broad-winged Hawks although a few other species also are noted (Finucane, 1977b: 17, 1978b: 26).

Fleming Mountain

W. R. Murphy's autumm hawk watch (34.5 hawks per hour) at Fleming Mountain produced mostly Broad-winged Hawks, but a few other raptors also were observed (Finucane, 1980b: 23).

Fort Lewis Mountain

Woody Middleton conducted an autumn hawk count (7.4 hawks per hour) at this location and reported Broad-winged Hawks in largest numbers but also a variety of other birds of prey in very small numbers (Finucane, 1976b: 15).

Grassy Hill

George W. Stubbs made autumn hawk counts (69.6 hawks per hour) at Grassy Hill and observed Broad-winged Hawks in largest numbers but various other raptors also were seen in very small numbers (Finucane, 1980b: 23, 1981b: 23).

Spring hawk migrations (79.5 hawks per hour) receive very limited coverage but include various species with the bulk of the flights formed by Broad-winged Hawks (Finucane, 1981a: 16).

Great North Mountain

This mountain runs along the Virginia-West Virginia border south of Mayfield. DeGarmo (1953: 41, 52–53) reported autumn hawk flights at the rate of 609.5 hawks per day. Various species were noted, but the bulk of the flights were produced by Broad-winged Hawks which arrived from flight-lines at the southern ends of North and Sleepy Creek Mountains.

Great Valley Overlook

This overlook is on the Blue Ridge Parkway near milepost 100 not far from Roanoke. Autumn hawk migrations (54.1 hawks per hour) contain mostly Sharp-shinned Hawks and Broad-winged Hawks, plus a few other species in small numbers (Finucane, 1978b: 26).

Green Valley

Hundreds of migrating Broad-winged Hawks sometimes are observed during autumn at Green Valley (Finucane, 1978b: 25, 1981b: 23).

Harkening Hill

John Pancake used this Blue Ridge Parkway site as an autumn hawk watching lookout and reported 90.1 hawks per hour—mostly Broad-winged Hawks but also a few other species in very small numbers (Finucane, 1977b: 17).

Harvey's Knob Overlook

Harvey's Knob Overlook is located along the Blue Ridge Parkway at milepost 95.3 near Buchanan. It is one of the area's better hawk watching lookouts (Heintzelman, 1979a: 217–218).

Good autumn hawk migrations (24.9 hawks per hour) are reported at this site. Ospreys, Sharp-shinned Hawks, Broad-winged Hawks, and Red-tailed Hawks form the bulk of the flights, but a variety of other raptors also are seen in smaller numbers (Finucane, 1976b, 1977b, 1978b, 1979b, 1980b, 1981b; Puckette, 1982b, 1983b, 1984b).

Spring hawk migrations (7.7 hawks per hour) contain various species of which Ospreys, Sharp-shinned Hawks, Broad-winged Hawks, and Red-tailed Hawks are the most numerous (Finucane, 1980a, 1981a; Puckette, 1982a, 1983a, 1984a).

Hawksbill Mountain

Wilds (1983: 83) states that the summit of Hawksbill Mountain is the best autumn hawk watching site in Shenandoah National Park.

High Knob

During the period 1949–1959, short-term autumn hawk counts were made from three fire towers—High Knob, Meadow Knob, and Reddish Knob—on Shenandoah Mountain within a seven-mile section of the ridge. The bulk of the migrations consisted mostly of Ospreys, Sharp-shinned Hawks, Cooper's Hawks, Broad-winged Hawks, and Red-tailed Hawks, although some other species also appeared in very small numbers (Carpenter, 1960).

Humpback Rocks

Autumn hawk migrations (272.2 hawks per hour) at Humpback Rocks on the Blue Ridge Parkway near Charlottesville contain large numbers of Broad-winged Hawks and very small numbers of a few other species (Finucane, 1981b: 23).

Irish Creek Overlook

Myriam Moore reports limited autumn hawk migrations (9.7 hawks per hour) at this overlook along the Blue Ridge Parkway near Buena Vista. Broad-winged Hawks form the bulk of the flights, but a few other species also appear in very small numbers (Finucane, 1976b: 15).

Ivy Area

Autumn hawk flights in the Ivy area (78.8 hawks per hour) consist mostly of Broad-winged hawks, plus small numbers of a few other species (Finucane, 1980b: 23, 1981b: 23).

Kennedy Peak

Kennedy Peak, atop Mount Kennedy near Luray, produces excellent autumn hawk migrations (33.5 hawks per hour). Broad-winged Hawks form the bulk of the flights, although a variety of other hawks also are reported in limited numbers (Finucane, 1976b, 1977b, 1979b, 1980b, 1981b; Puckette, 1982b, 1983b).

Spring hawk migrations (31.7 hawks per hour) contain mostly Broad-winged Hawks and a few other raptors in small numbers (Puckette, 1982a).

Kiptopeke

Kiptopeke is located at the southern end of the Cape Charles or Delmarva Peninsula and, like Cape May Point, New Jersey, receives large numbers (84.8 hawks per hour) of migrating birds of prey during autumn as a result of the funneling effect of the peninsula. The most common autumn migrants are Turkey Vultures, Ospreys, Northern Harriers, Sharp-shinned Hawks, American Kestrels, and Merlins, but a variety of other species including Peregrine Falcons are seen in smaller numbers (Armistead, 1981; Dunne, 1978c, 1979b, 1980a, 1981, 1982, 1983; Heintzelman, 1975: 249–252, 342; Potter and Murray, 1955; Rusling, 1936; Scott, 1977, 1978; Scott and Cutler, 1965, 1966, 1967, 1968a, 1969, 1970a, 1972a, 1974, 1975a; Sutton, 1984; Williams, 1983).

Linden Fire Tower

Autumn hawk migrations (60.1 hawks per hour) reported at the Linden Fire Tower northeast of Front Royal by Kerrie Kirkpatrick contain a variety of species of which Sharp-shinned Hawks, Broad-winged Hawks, and Red-tailed Hawks are the most numerous (Finucane, 1980b, 1981b; Puckette, 1982b, 1983b, 1984b).

Loft Mountain

During autumn, the migrations reported at Loft Mountain (69.4 hawks per hour) consist mostly of Broad-winged Hawks, although small numbers of a variety of other species also appear (Finucane, 1976b, 1977b, 1978b, 1981b).

Luray

J. J. Coyle observed autumn hawk migrations (93 hawks per hour) at his home at Luray and reported that Turkey Vultures and Broad-winged Hawks form the bulk of the flights, although a few other species appear in small numbers (Puckette, 1983b, 1984b).

Mary's Rock

This lookout is at Thornton Gap in Shenandoah National Park (Wilds, 1983: 83). Autumn hawk flights there (6.3 hawks per hour) contain Sharp-shinned Hawks, Broad-winged Hawks, and Red-tailed Hawks in largest numbers, but a few other raptors also are noted (Finucane, 1979b, 1980b, 1981b).

Mendota Fire Tower

Mendota Fire Tower on Clinch Mountain near Hansonville in southwestern Virginia near the Tennessee border now represents the site where the state's longest continuous autumn hawk migration data-base was developed. Primary attention is devoted to Broad-winged Hawk migrations which pass the site at the rate of 437.9 hawks per day (Behrend, 1952, 1953, 1954b; Finucane, 1958, 1959, 1960, 1961a, 1961b, 1963, 1964, 1965, 1966, 1967, 1969, 1971,1973, 1974, 1975a, 1976a, 1977a, 1978c, 1980c).

Mills Gap Overlook
This Blue Ridge Parkway overlook produces excellent autumn hawk migrations (67.6 hawks per hour). Sharp-shinned Hawks, Broad-winged Hawks, and Red-tailed Hawks are reported in largest numbers (Finucane, 1977b: 17).

Monterey Mountain
Autumn hawk migrations at Monterey Mountain west of Monterey in Highland County contain large numbers of Broad-winged Hawks but very small numbers of other raptors (Brooks, 1971).

Morris Knob
Ed Kinser and Sarah Cromer conducted autumn hawk counts at Morris Knob and reported 26.1 hawks per hour passing the site. Turkey Vultures, Sharp-shinned Hawks, Broad-winged Hawks, and Red-tailed Hawks appeared in largest numbers (Finucane, 1977b, 1978b).

Nickelsville
Autumn hawk flights at Nickelsville (187.5 hawks per hour) consisted mostly of Broad-winged Hawks, although a few other species also appeared in very small numbers (Finucane, 1976b, 1981b).

Osborn Ridge
Richard Davis made autumn hawk counts at this site and noted 22.3 hawks per hour with Broad-winged Hawks the most common species (Finucane, 1976b).

Spring hawk flights pass this location at the rate of 8.4 hawks per hour. Broad-winged Hawks again are the most commonly observed birds (Finucane, 1981a).

Potts Mountain
The Potts Mountain autumn hawk migrations (20.6 hawks per hour) contain mostly Sharp-shinned Hawks, Broad-winged Hawks, and Red-tailed Hawks, although some other species appear in limited numbers (Finucane, 1978b, 1980b, 1981b; Puckette, 1982b, 1983b, 1984b).

Spring hawk flights (3.2 hawks per hour) contain mostly Broad-winged Hawks, plus a few other species in meager numbers (Finucane, 1981a; Puckette, 1982a).

Purgatory Mountain Overlook
This Blue Ridge Parkway overlook near milepost 92 is one of the area's better known and more regularly used hawk watching sites (Heintzelman, 1979a: 220–221).

During autumn some excellent (33.8 hawks per hour) Broad-winged Hawk flights are reported, smaller numbers of Sharp-shinned Hawks appear, but other species generally are not very numerous (Beck, 1972; Finucane, 1976b, 1977b, 1978b, 1979b, 1980b, 1981b).

Spring hawk migrations (3.2 hawks per hour) contain Sharp-shinned Hawks and Red-tailed Hawks, plus an occasional other raptor (Finucane, 1978a).

Reddish Knob

Reddish Knob, atop Shenandoah Mountain, has been used as an autumn hawk watching lookout since 1949. These autumn flights (7.1 hawks per hour) contain Sharp-shinned Hawks, Cooper's Hawks, Broad-winged Hawks, and Red-tailed Hawks, as the most commonly observed species, but occasionally other raptors appear in limited numbers (Brooks, 1958; Carpenter, 1949, 1960; Finucane, 1976b, 1977b, 1978b, 1979b).

Red Oak Mountain

Red Oak Mountian, near Griffensburg, produces autumn hawk migrations (30.3 hawks per hour) with Sharp-shinned Hawks and Broad-winged Hawks as the most numerous migrants, plus small numbers of other raptors (Finucane, 1980b, 1981b).

Some spring hawk migrations also are reported at this site (Finucane, 1980a).

Rockfish Gap

This site is the parking lot of a motel along the Blue Ridge Parkway at milepost zero. It is referred to as Afton Mountain in some hawk migration literature.

Excellent autumn hawk flights (54.6 hawks per hour) are reported at Rockfish Gap, the most commonly observed species being Turkey Vultures, Sharp-shinned Hawks, Broad-winged Hawks, and Red-tailed Hawks (Finucane, 1976b, 1977b, 1978b, 1979b, 1980b, 1981b; Puckette, 1982b, 1983b, 1984b).

Scott Mountain

Tom Cabe studied autumn hawk migrations (10.2 hawks per hour) at Scott Mountain and reported Sharp-shinned Hawks and Broad-winged Hawks to be most numerous. A few other raptors also were seen in small numbers (Finucane, 1976b).

Seashore State Park

As many as 12 migrating Sharp-shinned Hawks are reported during April at this park on Cape Henry (Lawrence, 1984: 264; Scott and Cutler, 1975b).

Sharp Top

The autumn hawk flights (21.7 hawks per hour) reported at Sharp Top on the Blue Ridge Parkway contain Sharp-shinned Hawks and Broad-winged Hawks in largest numbers (Finucane, 1977b, 1978b).

Short Hill Mountain

This site is a banding and observation post located 25 miles south-southwest of Washington Monument State Park, Maryland. The autumn hawk migrations there (9.6 hawks per hour) include Ospreys, Northern Harriers, Sharp-shinned Hawks, Broad-winged Hawks, and Red-tailed Hawks in largest numbers, although various other raptors also are seen in smaller numbers (Puckette, 1983b, 1984b).

Sinking Creek Mountain

Poor autumn hawk migrations (10.5 hawks per hour) are seen at this location. Broad-winged Hawks form the bulk of the birds seen, but a few other raptors also appear (Finucane, 1978b).

Spring hawk flights (4.2 hawks per hour) consist mostly of Ospreys, Broad-winged Hawks, and Red-tailed Hawks, (Puckette, 1983a).

Stony Man Mountain

This Shenandoah National Park lookout is a hawk watching lookout (Wilds, 1983: 83) where fair autumn hawk migrations (16.5 hawks per hour) are reported. The bulk of the birds seen are Broad-winged Hawks and Red-tailed Hawks, but some other birds of prey appear in smaller numbers (Finucane, 1981b).

Poor spring hawk flights (4.4 hawks per hour) also are reported. Sharp-shinned Hawks and Broad-winged Hawks form most of the flights (Puckette, 1983a).

Strasburg

As many as 265 Broad-winged Hawks are seen in early May at Strasburg in northwestern Virginia (Hall, 1975: 851).

Sunset Fields Overlook

Excellent autumn Broad-winged Hawk flights (60.7 hawks per hour) are reported at this Blue Ridge Parkway overlook along with a few other raptors (Finucane, 1977b, 1978b).

Tanager Knob

In autumn, 26.8 hawks per hour are reported passing Tanager Knob, the bulk of which are Broad-winged Hawks (Finucane, 1980b).

Tazewell Hospital

Autumn hawk flights (214.2 per hour), mostly Broad-wings, are seen at this site (Finucane, 1976b)

Thunder Ridge Overlook

Excellent autumn Broad-winged Hawk migrations (129.2 per hour), plus small numbers of other raptors, are seen at this Blue Ridge Parkway overlook near Big Island (Finucane, 1976b, 1977b, 1978b, 1981b).

Troutville

Barry Kinzie studied autumn hawk migrations (60.1 hawks per hour) at Troutville and reported that Sharp-shinned Hawks and Broad-winged Hawks appear in largest numbers, although a variety of other raptors also are seen in limited numbers (Finucane, 1976b, 1977b, 1978b, 1979b 1980b, 1981b).

Spring hawk migrations contain mostly Broad-winged Hawks but a few other species also are seen resulting in 38 hawks per hour reported (Puckette, 1982a).

Turkey Mountain

Excellent autumn hawk migrations (39.8 hawks per hour) are reported at

Turkey Mountain near Amherst by Ray and/or Sandra Chandler. Broad-winged Hawks form most of the flights, although small numbers of other raptors are seen occasionally (Finucane, 1976b, 1977b, 1978b, 1979b, 1980b, 1981b).

Virginia Coast Reserve

This reserve, owned by the Nature Conservancy, is a chain of barrier beach islands extending along the Delmarva Peninsula from the mouth of Chesapeake Bay northward to the Maryland-Virginia border. Good autumn hawk flights are reported along some of the islands (Heintzelman, 1979a: 222; Wilds, 1983: 164).

Wallops Island

As many as 463 migrating Sharp-shinned Hawks are seen in early October at Wallops Island (Armistead, 1981: 167).

Waynesboro

Flights of as many as 1,500 Broad-winged Hawks are reported in September at Waynesboro (Hall, 1972: 64).

Wind Rock

Wind Rock is near the War Spur Trail in Giles County and is sometimes used as a hawk watching lookout (Lawrence and Gross, 1984: 312). Autumn hawk migrations there (14.9 hawks per hour) contain mostly Broad-winged Hawks but smaller numbers of Sharp-shinned Hawks and other raptors also are reported (Finucane, 1980b, 1981b; Puckette, 1984b).

Woodstock Tower

Lawrence and Gross (1984: 293) stated that this tower on the summit of Massanutten Mountain near Edinburg is used to observe autumn hawk migrations.

WEST VIRGINIA

There is a long and distinguished history of hawk migration field studies in West Virginia extending back into the late 1940s. As a result of this collective work much now is known about autumn hawk flights in the state (Brooks, 1944; DeGarmo, 1953; Finucane, 1976b, 1978b, 1979b, 1980b, 1981b; Hall, 1964a, 1983a; Heimerdinger, 1974; Heintzelman, 1975, 1979a; Hurley, 1970, 1975; Johnston, 1923; Puckette, 1982b, 1983b, 1984b; Radis, 1976; Shreve, 1970; Tinsley, 1971).

Allegheny Front

The Allegheny Front in West Virginia is a rocky escarpment forming the boundary between the ridge and valley province and the more westerly Allegheny plateau. Since the late 1940s, ornithologists were aware of autumn hawk and passerine migrations along the ridge—especially at Bear Rocks and Pinnacle Rock (DeGarmo, 1953: 41, 42, 46; Hall, 1964a: 30).

Autumn hawk migrations on the Allegheny Front (334.3 hawks per day) contain large numbers of Broad-winged Hawks but smaller numbers of other species such as Sharp-shinned Hawks, Red-tailed Hawks, and other species (DeGarmo, 1953: 47–54).

Backbone Mountain

This mountain forms the western edge of the Alleghenies in Tucker County (DeGarmo, 1953: 41). Autumn hawk flights there (370.1 hawks per day) contain large numbers of Broad-winged Hawks but only a few other raptors (DeGarmo, 1953: 49–54).

Bear Den Knob Fire Tower

This tower is on Canaan Mountain in the center of Tucker County (DeGarmo, 1953: 42, 44; Hall, 1983a: 167). Some excellent autumn hawk migrations (53 hawks per hour) are reported at the site with Broad-winged Hawks forming the bulk of the flights but various other birds of prey also included in small numbers (DeGarmo, 1953: 52; Finucane, 1978b, 1979b, 1980b).

Bear Rocks

Bear Rocks, on the Allegheny Front at the Dolly Sods Scenic Area in Monongahela National Forest, is one of West Virginia's most famous hawk watching lookouts (DeGarmo, 1953; Hall, 1964a; Heimerdinger, 1974; Heintzelman, 1975, 1979a). Autumn hawk migrations there (50.9 hawks per hour) consist mostly of Broad-winged Hawks, although a variety of other raptors also are reported in small numbers (Brooks, 1959; DeGarmo, 1953; Finucane, 1979b, 1980b, 1981b; Hall, 1961, 1964b; Heimerdinger, 1974; Puckette, 1982b, 1983b, 1984b).

Belle

Occasionally during September as many as 1,200 migrating Broad-winged Hawks are reported near Belle (Hurley, 1975: 115).

Blue Ridge

The Blue Ridge is the most easterly of West Virginia's mountains and is an important autumn hawk migration flight-line. The birds seem to arrive from Maryland's Catoctin and South Mountain ridges and perhaps from as far north as Sterrett's Gap on the Kittatinny Ridge in Pennsylvania (DeGarmo, 1953: 40).

Cabin Mountain

This ridge in southeastern Tucker County is a secondary autumn hawk migration flight-line (51 hawks per day). Broad-winged Hawks are reported in largest numbers, but a variety of other raptors also appear (DeGarmo, 1953: 42, 46, 52).

Cacapon Mountain

This northwestern Morgan County ridge is another secondary autumn hawk migration flight-line (12 hawks per day). Broad-winged Hawks are reported in largest numbers (DeGarmo, 1953: 42, 46, 49).

Canaan Mountain

Autumn hawk migrations (190 hawks per day) are reported along this eastern Tucker County ridge. Most of the birds are Broad-winged Hawks, but a variety of other raptors also are seen in very limited numbers (De-Garmo, 1953: 42, 46, 52).

Centennial Park

Excellent autumn hawk flights (250 hawks per day) are seen from a road-side lookout and park near Thomas on Backbone Mountain by Bill Wylie and students from West Virginia University (Wylie, 1976: 116).

Charleston

During September, as many as 3,300 migrating hawks (probably mostly Broad-wings) are reported at Charleston (Hall, 1984: 201; Hurley, 1975: 115).

In April, as many as 150 Broad-winged Hawks also are seen at this city (Hall, 1975: 851).

Charles Town

September flights of as many as 4,000 Broad-winged Hawks are reported at Charles Town (Hall, 1972: 64).

Cheat Mountain

This roadside lookout near Mace is an autumn hawk migration route (96.8 hawks per day) in the Central Alleghenies. Broad-winged Hawks are reported in largest numbers, but various other raptors also are seen in very small numbers (DeGarmo, 1953; J. L. Smith, 1982).

East River Mountain

This ridge is a continuation of Peters Mountain south of Peterstown, West Virginia, or Narrows, Virginia (Hurley, 1970: 82). Excellent autumn hawk flights (41.6 hawks per hour) are reported at the site with Broad-winged Hawks forming the bulk of the migrations but various other raptors also included in small numbers (DeGarmo, 1953; Finucane, 1981b; Puckette, 1982b, 1983b, 1984b).

Gap Mills

DeGarmo (1953) reported as many as 394 migrating Broad-winged Hawks during September at this Peters Mountain location as well as a few other hawks.

Gaudineer Knob

DeGarmo (1953: 42) stated that Maurice Brooks observed autumn hawk flights (89 hawks per day) at Gaudineer Knob on Shaver's Mountain with Broad-winged Hawks forming most of the flight.

Hanging Rocks Fire Tower

This fire tower on Peters Mountain near Waiteville is one of the state's best autumn hawk watching sites and produces excellent autumn hawk migrations (67.9 hawks per hour). Most of the birds reported are Broad-winged Hawks, but a variety of other raptors also are seen in small numbers

(DeGarmo, 1953; Finucane, 1978b; Hurley, 1970, 1975; Puckette, 1982b, 1983b, 1984b; Tinsley, 1971).

Harpers Ferry
Brooke Meanley and Anna Gilkeson reported fair autumn hawk migrations (20.2 hawks per hour) at Harpers Ferry. Broad-winged Hawks formed the bulk of the birds observed, but very limited numbers of various other raptors also were seen (Robbins, 1950: 5–6).

Knobley Mountain
Knobley Mountain in Mineral County sometimes produces important autumn hawk flights (213.3 hawks per day), the bulk of which are formed by Broad-winged Hawks. An assortment of other raptors also are included in some of the migrations in small numbers (DeGarmo, 1953).

Laurel Ridge
Laurel Ridge is located in northwestern Randolph County and produces fair autumn hawk migrations (60.4 hawks per day). Most of the birds observed are Broad-winged Hawks, but some other raptors also are reported in meager numbers (DeGarmo, 1953).

Middle Ridge
During September some very large Broad-winged Hawk flights (up to 3,300 birds) are reported at this site near Charleston, and various other birds of prey also appear in limited numbers (DeGarmo, 1953: 42; Shreve, 1970).

Moncove Lake
More than 130 Broad-winged Hawks are reported during September at Moncove Lake in Monroe County west of Peters Mountain (Hurley, 1975: 114).

North Fork Mountain
North Fork Mountain is located in Pendleton County. Autumn hawk migrations there (88 hawks per day) contain mostly Broad-winged Hawks, but as many as 5 Bald Eagles also are seen (DeGarmo, 1953).

North Mountain
This ridge extends only through the length of Berkeley County and serves as an autumn hawk migration flight-line (235.5 hawks per day) only for a limited distance (DeGarmo, 1953: 40). The raptors reported in largest numbers are Ospreys, Sharp-shinned Hawks, Cooper's Hawks, and Broad-winged Hawks, but various other hawks also are seen in small numbers (DeGarmo, 1953).

North River Mountain
DeGarmo (1953) reported excellent autumn hawk flights (509 hawks per day) at this site in Hampshire County. Although a variety of hawks were seen, most of the migrations consist of Broad-winged Hawks.

Paddy Knob
Autumn hawk migrations (87.5 hawks per day) at Paddy Knob on the

Allegheny Front are part of the flight-line in that area (Hurley, 1970).

Peters Mountain

Peters Mountain early was recognized as an important autumn hawk migration flight-line (DeGarmo, 1953: 42). Years later, Hurley (1970, 1975) made more detailed field studies of autumn hawk flights along the ridge—especially at the Hanging Rocks Fire Tower and further confirmed the importance of the ridge.

Hurley (1970: 84–85) suggested that some hawks migrating along the Allegheny Front and Warm Springs Mountain in Virginia eventually reach Peters Mountain and use it as a flight-line, whereas other raptors use ridges to the east of Peters Mountain or continue farther west of the ridge. He concluded that none of the ridges in the Peters Mountain area is particularly favored, but that hawks tend to follow whichever mountain they find themselves near at any given time. Nevertheless, higher hawk counts reported from Peters Mountain tend to indicate that it receives more hawks than reach some of the adjacent ridges.

Peterstown

Occasionally autumn hawk flights (711 hawks per day) contained large numbers of Broad-winged Hawks and very small numbers of a variety of other raptors (DeGarmo, 1953).

Saint Albans

Occasionally during September flights of more than 3,300 migrating Broad-winged Hawks are reported at St. Albans (Hall, 1970: 48).

Shaver's Mountain

Few autumn hawk counts are reported from Shaver's Mountain in Pocahontas County with the exception of Gaudineer Knob (DeGarmo, 1953). H. Uhlig, however, reported 115 hawks per day at a site at Route 250 on this ridge. Most of the birds were Broad-winged Hawks.

Sleepy Creek Mountain

Excellent autumn hawk flights (200 hawks per day) are seen at this mountain in eastern Morgan County, but some of the flights were cross-ridge movements—apparently from nearby Third Hill Mountain. Broad-winged Hawks appear in largest numbers, but various other raptors also appear (DeGarmo, 1953).

Thorny Mountain

Autumn hawk counts (9.5 hawks per day) reported at Thorny Mountain by R. H. Holderby contained a few Sharp-shinned Hawks, Cooper's Hawks, Broad-winged Hawks, and Red-tailed Hawks (DeGarmo, 1953).

CHAPTER 7

Southern Appalachian Hawk Migrations

The seasonal (especially autumn) migrations of hawks in the South Appalachian states—Georgia, Kentucky, North Carolina, South Carolina, and Tennessee—form a continuation of hawk flights observed farther northeast in North America. Migration routes are used along inland mountains, the coastal plain, and the coastline and barrier beach islands. The most detailed hawk migration studies in this part of North America are reported from Tennessee.

GEORGIA

Knowledge of hawk migrations in Georgia is relatively limited and fragmentary, but some information is available in general terms (Burleigh, 1958; Stoddard, 1978). More detailed reports concerned records of nomadic or seasonal migrations of kites (Chamberlain, 1960b; Fitch, 1965; Henderson, 1941; Hopkins, 1966; Johnston, 1957; Knighton, 1970; Swiderski, 1975; Thomas, 1943; Tomkins, 1962; Wharton, 1941; Williams, 1962). The status of Golden Eagles in the state also was studied (Erwin, 1974; Griffin, 1941; Hopkins, 1968; Kale and Almand, 1963), and Mellinger (1973) described a small autumn hawk flight in Rabun County.

Altamaha and Ocmulgee Rivers

During early June several American Swallow-tailed Kites and 22 Mississippi Kites were observed along the Ocmulgee and Altamaha rivers between Red Bluff and the Altamaha River in Glynn County (Hopkins, 1966).

Augusta Area

During May an occasional American Swallow-tailed Kite and up to 30

145

Mississippi Kites are seen in the Augusta area (Knighton, 1970; Thomas, 1943).

Chattahoochee River National Recreation Area
During late October more than 90 migrating Red-tailed Hawks sometimes are reported at the Chattahoochee River National Recreation Area (LeGrand, 1983b: 168).

Chickamauga Battlefield
As many as 85 migrating Broad-winged Hawks sometimes are seen in autumn at this northwestern Georgia site (Finucane, 1978b: 25).

Cumberland Island National Seashore
Although poor autumn hawk flights generally are reported at this site near St. Marys, as many as 125 Peregrine Falcons sometimes are counted there during autumn (Heintzelman, 1979a: 165; LeGrand, 1980a: 151; Teulings, 1974a, 1977).

Eastman Mountain
During September as many as 63 migrating Broad-winged Hawks sometimes are seen over Eastman Mountain in Rabun County (Mellinger, 1973).

Jekyll Island
As many as 150 Sharp-shinned Hawks were observed moving *north* in early November at Jekyll Island along coastal Georgia, and a few other raptors also were seen (Dunne, 1979b; LeGrand, 1979).

Lookout Mountain
This site in northwestern Georgia is just south of Chattanooga, Tennessee. Autumn hawk flights (136.1 hawks per hour) at Lookout Mountain contain mostly Broad-winged Hawks, but a few other hawks also are seen occasionally (Finucane, 1975b, 1977b).

KENTUCKY

The status of hawk migrations in Kentucky is known very incompletely, although some fragmentary details are published (Brown, 1978; Carpenter, 1934; Mayfield, 1980; Mengel, 1965; Schneider, 1950; Shadowen, 1968; Stamm, 1957, 1961, 1965, 1969, 1972).

Bernheim Forest
April hawk migrations (15 hawks per day) at Bernheim Forest in Bullitt County contain mostly Broad-winged Hawks, but a few other raptors also are seen (Able, 1965).

Bowling Green Area
Autumn migrations of as many as 150 Turkey Vultures are seen near Bowling Green (Wilson, 1925), and in mid-April as many as 45 Broad-winged Hawks and a few other hawks also are noted (Shadowen, 1968).

Bullitt County
During September as many as 50 migrating Broad-winged Hawks some-

times are seen in northwestern Bullitt County (Carpenter, 1934).

Cadiz
Stokes (1980) reported 117 hawks per day in mid-September near Cadiz in western Kentucky. Most of the birds were Broad-winged Hawks, but some Red-tails also were seen.

Cumberland Gap National Historical Park
Autumn hawk migrations (444.3 hawks per day) are seen at this site. Most of the birds are Broad-winged Hawks, but various other raptors also are noted in meager numbers (Hall, 1977, 1979; Stamm, 1972).

Curlew
Martin (1959) reported 157 migrating hawks (probably Broad-wings) in September at Curlew.

Fulton County
As many as 12 Mississippi Kites are seen in late August in Fulton County (Peterjohn, 1982a: 183).

Harlan County
Occasionally more than 225 Broad-winged Hawks are reported in mid-September in Harlan County (Nolan, 1955: 29).

Henderson
Autumn hawk migrations (13.4 hawks per day) appear at Henderson with Red-tailed Hawks and American Kestrels seen in largest numbers (Finucane, 1977b).

As many as 40 Northern Harriers are reported in early April at this location (Rhoads, 1958).

Jefferson County Forest
Anne L. Stamm conducted an autumn hawk watch at the Jefferson County Forest south of Louisville and reported 21.4 hawks per hour passing the site. Cooper's Hawks, Broad-winged Hawks, and Red-tailed Hawks were seen in largest numbers but other hawks also were noted (Stamm, 1957).

Lexington
Occasionally up to 50 migrating Broad-winged Hawks are seen over Lexington during September (Brown, 1978).

Louisville
Autumn hawk flights passing Louisville (53.1 hawks per day) contain mostly Broad-winged Hawks, but occasionally a few other species are seen (Kleen, 1974, 1979a; Peterjohn, 1982a; Petersen, 1965; Schneider, 1950; Stamm, 1961, 1965, 1969).

Mississippi River
Mengel (1965) states that considerable Red-tailed Hawk flights follow the Mississippi Valley southward during autumn.

Morehead
Michael W. Rice's autumn hawk counts at Morehead (2.5 hawks per hour)

contained more Red-tailed Hawks and American Kestrels than other species (Finucane, 1975b).

Pikeville

On 24 September 1979, Mayfield (1980) was in his airplane at 4,500 feet over a spot six miles east of Pikeville when he was surrounded by 3,000 Broad-winged Hawks, some of which rose and disappeared into clouds at 6,000 feet. He later observed scattered kettles of hawks along the entire route to Pikeville. Apparently, large numbers of migrating Broad-winged Hawks can cross parts of Kentucky without being observed by people on the ground.

Reelfoot National Wildlife Refuge

As many as 30 migrating Mississippi Kites are seen in mid-June at a section of this refuge just north of the Tennessee border (Bierly, 1973).

NORTH CAROLINA

Pearson, Brimley, and Brimley (1942) presented a broad picture of hawk migrations in North Carolina, but it was not until 1952 that the first organized effort was made to study hawk flights in this state (Simpson, 1952a, 1952b, 1954). Since then more interest developed in North Carolina hawk migrations as documented in the following site reports.

Bear Wallow Gap

In September at least 210 Broad-winged Hawks are observed at this site on the Blue Ridge Parkway between Little Switzerland and Mount Mitchell (Johnson, 1950: 72–73).

Beaufort

A mid-October flight of Red-tailed Hawks, Merlins, and Peregrine Falcons is reported at Beaufort on the coast (Chamberlain, 1956: 16).

Beech Mountain

Fred H. Behrend reported a flight of 225 Broad-winged Hawks in September at Beech Mountain in Avery County (Chamberlain, 1958: 20).

Blowing Rock

A late August flight of 35 Merlins, and small numbers of Broad-winged Hawks in September, were reported at Blowing Rock on the Blue Ridge Parkway north of Lenoir (Murray, 1946; Simpson, 1952b).

Blue Ridge Parkway

During autumn, migrations of Broad-winged Hawks, and various other raptors including Ospreys, Bald Eagles, Northern Harriers, Sharp-shinned Hawks, Cooper's Hawks, Red-tailed Hawks, American Kestrels, Merlins, and Peregrine Falcons are reported along this parkway (Dapper, 1953; Simpson, 1954).

Bodie Island Lighthouse

Autumn Sharp-shinned Hawk flights (17 birds per day) are reported at the Bodie Island lighthouse (Lee and Lee, 1978; Parnell, 1965).

Buxton Woods
Fair autumn migrations of accipiters are reported near the Buxton Woods area near Cape Hatteras lighthouse. The site also is a migration staging area (Heintzelman, 1979a: 196; Lee and Lee, 1978).

Craggy Gardens
Good autumn hawk flights are reported at Craggy Gardens (Heintzelman, 1979a; Pettingill, 1977).

Doughton Park
September migrations of Broad-winged Hawks (80 birds per day) are reported at Doughton Park on the Blue Ridge Parkway (Simpson, 1952b, 1954; Parnell, 1967).

East Fork Mountain
Mike Tove's autumn hawk watch at East Fork Mountain (16.2 hawks per hour) produced mostly Sharp-shinned Hawks and Broad-winged Hawks, plus small numbers of other raptors (Finucane, 1980b, 1981b).

Fort Fisher
Early October migrations of Sharp-shinned Hawks (124 birds per day) are reported at Fort Fisher (LeGrand, 1981a).

Fort Macon State Park
Autumn hawk flights (122 hawks per day) are reported at this barrier beach island park the bulk of which consist of Sharp-shinned Hawks, although other raptors also are seen in very small numbers (Lee and Lee, 1978; Teulings, 1976a).

Hatteras Island
Excellent autumn hawk migrations (54.2 hawks per hour) are reported at Hatteras Island. Sharp-shinned Hawks and American Kestrels are reported in largest numbers, but various other birds of prey also are noted in small numbers (Teulings, 1976a).

Hensley Ridge
Occasionally as many as 250 Broad-winged Hawks are seen at Hensley Ridge during autumn (Finucane, 1975b).

Highlands
During September as many as 29 Broad-winged Hawks are reported near Highlands (Simpson, 1952b).

Hump Mountain
William Cable reported as many as 600 Broad-winged Hawks in September at Hump Mountain in Avery County (Chamberlain, 1958).

Little Switzerland Area
W. M. Johnson reported more than 2,300 migrating Broad-winged Hawks in September over the Blue Ridge Parkway between Little Switzerland and Mount Mitchell (Simpson, 1954).

Long Beach
There are 23.2 hawks per hour reported during autumn at Long Beach.

The bulk of the birds are Sharp-shinned Hawks and American Kestrels, but several other raptors also are noted (Chamberlain, 1962; Dunne, 1983).

Mount Mitchell

Large September migrations of Broad-winged Hawks (1,351.6 hawks per day) sometimes are seen at this Blue Ridge Parkway location (Johnson, 1950; Simpson, 1954).

Peregrine Falcons also are reported in May and July carrying food, thus suggesting breeding there (Simpson, 1969).

Nags Head

Autumn Sharp-shinned Hawk and Cooper's Hawk migrations are reported near Nags Head. The hawks head toward Oregon Inlet and continue beyond it (Crossan and Stevenson, 1956).

Ocracoke Island

October hawk migrations (81 hawks per day) reported at Ocracoke Island include Ospreys, Cooper's Hawks, and occasionally a few other raptors (Teulings, 1972, 1976a).

Oregon Inlet

Crossan and Stevenson (1956) report that migrating Sharp-shinned Hawks and Cooper's Hawks cross Oregon Inlet during a September flight after arriving from the vicinity of Nags Head.

Outer Banks

The Outer Banks are well known as an autumn hawk migration flight-line (57.5 hawks per day) for Ospreys, accipiters, American Kestrels, and some other birds of prey (Crossan and Stevenson, 1956; Lee and Lee, 1978; Teulings, 1972, 1976a).

Pea Island National Wildlife Refuge

Autumn hawk migrations (6.8 hawks per hour) passing this refuge on the Outer Banks consist of Northern Harriers, Sharp-shinned Hawks, American Kestrels, Merlins, Peregrine Falcons, and occasionally a few other raptors (Dunne, 1978c; Parnell, 1968; Potter and Sykes, 1980; Sutton, 1984; Teulings, 1977).

In late May there were 13 Broad-winged Hawks reported at the refuge (Davis, 1976).

Pilot Mountain

Good autumn hawk migrations (24.6 hawks per hour) are reported at Pilot Mountain by Ramona R. Snavely. Black Vultures, Sharp-shinned Hawks, Broad-winged Hawks, and Red-tailed Hawks are seen in largest numbers but various other birds of prey also are seen in small numbers (Finucane, 1979b; LeGrand, 1982, 1983b; Puckette, 1983b).

Salter Path

Occasionally in mid-October as many as 46 migrating Sharp-shinned Hawks are reported at Salter Path (LeGrand, 1979).

Scotland Neck

In late May, 23 Mississippi Kites are seen at Scotland Neck (Teulings, 1976b).

Stone Mountain State Park

Occasionally during September as many as 446 Broad-winged Hawks are reported at this park (LeGrand, 1981a).

Sullivans Island

Sidney Gauthreaux reported autumn hawk flights (45.5 birds per hour) at Sullivans Island with Ospreys and American Kestrels forming the bulk of the flights, although a few other raptors also appeared (Teulings, 1976a).

Table Rock

Sometimes during September flights of Broad-winged Hawks (189.5 birds per day) are reported at Table Rock in northwestern Burke County on the eastern side of the Blue Ridge (Pratt, 1967; Simpson, 1954).

Thunder Hill

Broad-winged Hawk migrations (341 birds per day) sometimes are reported in September at Thunder Hill on the Blue Ridge Parkway (Parks, 1957; Simpson, 1954).

Topsail Island

October hawk migrations (16 hawks per hour) at Topsail Island on the coast contain mostly Ospreys, but a few other hawks also are noted (Grant, 1967).

Winston-Salem

September hawk migrations at Winston-Salem may include up to 274 Broad-winged Hawks (Chamberlain, 1958).

Wrightsville Beach

Early October Osprey migrations (45 birds per hour) are reported at Wrightsville Beach (Carter, 1974).

SOUTH CAROLINA

Information on hawk migrations in South Carolina is limited (Sprunt and Chamberlain, 1970). The first organized study of autumn hawk flights in the state began in 1951 at the request of Chandler S. Robbins and with the aid of William P. Van Eseltine (Van Eseltine, 1952). The most detailed, informative, and important study of autumn hawk migrations in the state, however, is the work done at Fort Johnson at Charleston Harbor (Laurie, McCord, and Jenkins, 1981).

Camden Area

Concentrations of as many as 40 Mississippi Kites are seen in late May near the Wateree River 10 miles below Camden (Chamberlain, 1961).

Edisto Beach State Park

As many as 21 Merlins are observed in late September at Edisto Beach State Park (LeGrand, 1982).

Folly Beach
October hawk migrations (116 hawks per day) are reported at Folly Beach near Charleston. The species reported in largest numbers are Sharp-shinned Hawks, American Kestrels, and Merlins, but a few other hawks also are noted (Teulings, 1971).

Fort Johnson
Major field studies of autumn hawk migrations (31.9 hawks per hour) are reported at Fort Johnson overlooking the Charleston Harbor. Turkey Vultures, Ospreys, Sharp-shinned Hawks, and American Kestrels are seen in largest numbers, but a variety of other birds of prey also appear in more limited numbers (Dunne, 1979b, 1980a, 1982, 1983; Laurie, McCord, and Jenkins, 1981; Sutton, 1984).

Georgetown
In mid-April as many as 12 American Swallow-tailed Kites are seen at Georgetown (Teulings, 1976b).

Santee River Delta
During May as many as 8 American Swallow-tailed Kites and an occasional Mississippi Kite are reported in the Santee River delta, and as many as 100 American Swallow-tailed Kites are seen there in mid-July (Chamberlain, 1953; LeGrand, 1981b, 1983b).

Sumter County
In mid-August as many as 31 Mississippi Kites are seen soaring above Interstate 95 at the Lynches River in Sumter County (Teulings, 1975).

TENNESSEE

Organized studies of autumn hawk migrations began in 1951 in Tennessee under the leadership of Fred W. Behrend and were coordinated by the Tennessee Ornithological Society (Behrend, 1951, 1952, 1953, 1954a, 1954b). Since then annual autumn hawk counts were made at many sites in the state (Finucane, 1956–1981b; Fowler, 1982; Odom, 1966; Turner, 1981), and details of recoveries in Tennessee (or elsewhere) of several banded Red-tailed Hawks and American Kestrels also were reported (Williams, 1980). As a result of this collective effort, more is known about hawk migrations in this state than elsewhere in the Southern Appalachian states.

Ashland City
Occasionally during October as many as 130 migrating Turkey Vultures are observed near Ashland City 20 miles west of Nashville (Ganier, 1951).

Bald Mountains Tower
This site is located in Unicoi County in Cherokee National Forest on the Tennessee-North Carolina border. Autumn hawk migrations there (8.5 hawks per hour) contain mostly Cooper's Hawks and Broad-winged Hawks (Finucane, 1968; Fowler, 1982).

Bays Mountain

Bays Mountain is located in the Upper East Tennessee Valley near Kingsport. Since autumn hawk counts began there in 1950, some large Broadwinged Hawk flights (311.1 birds per day) are reported as are occasional sightings of other hawks (Behrend, 1950b, 1951, 1952, 1953; Finucane, 1958, 1960, 1964, 1969, 1971).

Birchwood

Occasionally during September flights of Broad-winged Hawks (184.3 hawks per hour) are reported passing down Walden's Ridge from the town of Birchwood in Hamilton County (Fowler, 1982).

Black Oak Ridge

Autumn hawk migrations (1.7 hawks per hour) reported at Black Oak Ridge near Fountain City contain mostly Broad-winged Hawks, but occasionally another raptor is observed (Finucane, 1957, 1958, 1959, 1960).

Brushy Mountain

Turner (1981) reported that autumn hawk flights (25.6 hawks per hour) observed at Brushy Mountain on the eastern edge of the Cumberland Mountains in Campbell County contain Broad-winged Hawks, plus a few other raptors in small numbers.

Camp Creek Bald Fire Tower

This fire tower is located southeast of Greenville. Autumn flights of Broadwinged Hawks (77.1 birds per day) form the bulk of the migrations, although other raptors also are reported occasionally in small numbers (Behrend, 1950b, 1951, 1952, 1953, 1954b; Finucane, 1959, 1961, 1964, 1966, 1967).

Chattanooga Area

Observations of migrating hawks in the Chattanooga area (198.7 birds per day) generally are of Broad-winged Hawks, but as many as 150 Turkey Vultures also are reported (Finucane, 1959, 1961, 1963, 1971, 1973, 1976a; Hall, 1981).

Chickamauga Dam

This site is located near Chattanooga. Autumn hawk flights there (437.3 hawks per hour) consist mostly of Broad-winged Hawks, plus small numbers of other raptors (Finucane, 1969, 1978c, 1980b).

Chilhowee Mountain

This mountain in Blount and Sevier counties is a long ridge bordering the northern side of Great Smoky Mountains National Park. The autumn hawk migrations reported there (91.4 hawks per hour) consist of large numbers of Broad-winged Hawks, lesser numbers of Black Vultures and Turkey Vultures, and small numbers of various other raptors (Finucane, 1982b; Puckette, 1984b; Stedman and Stedman, 1981).

Spring hawk flights at the site (10.5 hawks per hour) mostly contain

Turkey Vultures, Broad-winged Hawks, and Red-tailed Hawks, plus a few other species in minute numbers (Finucane, 1981a; Puckette, 1983a).

Chimney Top

The autumn hawk migrations (13.5 hawks per hour) reported at Chimney Top in Greene County are composed mostly of Broad-winged Hawks, plus minor numbers of other birds of prey (Finucane, 1958, 1961, 1964, 1965, 1966, 1968).

Clinch Mountain Fire Tower

Clinch Mountain extends into Virginia, whereas the fire tower in Tennessee is located north of Rogersville. As many as 225 migrating Broad-winged Hawks sometimes are reported from the site during September (Behrend, 1951; Finucane, 1954b).

Collegedale

Occasionally during late September as many as 2,000 migrating Broad-winged Hawks are reported at Collegedale east of Chattanooga (Finucane, 1961, 1976a).

Copper Ridge

At times during September as many as 150 migrating Broad-winged Hawks are reported at Copper Ridge north of Knoxville (Behrend, 1953; Finucane, 1976a).

Crossville Area

Autumn hawk counts made at a site seven miles north of Crossville contain a raptor passage rate of 33.2 hawks per hour. Most of the birds are Broad-winged Hawks, but a few other hawks also are seen (Finucane, 1973, 1974, 1975a, 1976a, 1977a).

As many as 175 Broad-winged Hawks also are seen there in mid-April (West, 1974).

Cumberland Mountain

In September of 1953, many hawks (probably Broad-wings) were seen near Sparta near the Cumberland Plateau—the first evidence of autumn hawk flights in the area (Behrend, 1953).

Daisy

More than 450 migrating Broad-winged Hawks sometimes are seen at Daisy northeast of Chattanooga during late September (Finucane, 1961).

Dry Gap

A late September flight of 639 Broad-winged Hawks and a few other raptors is reported at Dry Gap near Knoxville (Finucane, 1961).

Dunlap Fire Tower

Autumn hawk migrations at this site near Chattanooga are excellent (36 hawks per hour). The bulk of the flights consist of Broad-winged Hawks, but small numbers of other raptors also are seen (Finucane, 1969, 1971, 1973, 1974, 1975a, 1976a, 1977a, 1978c).

Elder Mountain Fire Tower

Elder Mountain is located northwest of Chattanooga and is divided from Signal Mountain by the Tennessee River. The autumn hawk migrations seen there (176 hawks per day) consist of Broad-winged Hawks and occasionally a few other birds of prey (Behrend, 1954b; Finucane, 1957, 1958, 1959, 1960, 1961a, 1961b, 1963, 1964, 1965, 1966, 1967, 1968, 1975).

Fairfield Glade

Fairfield Glade, in Cumberland County on the Cumberland Plateau some 14 miles northeast of Crossville, is located on an excellent (73.2 hawks per hour) autumn hawk flight-line. Most of the birds seen are Broad-winged Hawks, but sometimes another raptor also appears (Finucane, 1978c, 1980c; Turner, 1981).

Fall Creek Falls State Park

This park is located near Spencer and was used for many years as a hawk watching lookout. The excellent autumn hawk migrations (37.7 hawks per hour) reported at this site consist of large numbers of Broad-winged Hawks and an occasional raptor of another species (Behrend, 1954b; Finucane, 1971, 1973, 1974, 1975a, 1977a, 1978c; Spofford, 1949).

Foothills Parkway

The autumn hawk migrations (17.5 hawks per hour) that pass the Foothills Parkway, along the northern side of the Great Smoky Mountains National Park, contain mostly Black Vultures, Turkey Vultures, Broad-winged Hawks, and small numbers of a variety of other birds of prey (Finucane, 1968; Puckette, 1983b).

Fordtown

Occasionally as many as 200 migrating Broad-winged Hawks are reported at Fordtown south of Kingsport (Behrend, 1950b).

Holston Mountain Radar Station

The autumn hawk migrations (15.6 hawks per hour) reported at the Holston Mountain Radar Station on Holston Mountain in Carter County are composed mostly of Broad-winged Hawks, but small numbers of various other hawks also are observed (Behrend, 1951, 1954b; Finucane, 1956, 1957, 1960, 1961a, 1963, 1964, 1969, 1973, 1974, 1975a, 1976a, 1977a, 1978c).

House Mountain

House Mountain is located 15 miles northeast of Knoxville and is the southwestern end of Clinch Mountain. The autumn hawk migrations seen there (15.4 hawks per day) contain various species of raptors including Sharp-shinned Hawks, Broad-winged Hawks, Red-tailed Hawks, and a few other hawks (Behrend, 1953, 1954b; Finucane, 1956, 1957, 1958, 1960).

Hump Mountain

Hump Mountain, near Shell Creek 18 miles southeast of Elizabethton, was used as a hawk watching lookout for the first time in 1944 (Behrend, 1950a). The autumn hawk flights there (37 hawks per day) consist of considerable numbers of Broad-winged Hawks and a few other species in small numbers (Behrend, 1950a, 1950b, 1953; Finucane, 1960, 1963, 1964, 1971; Herndon, 1949).

Jane Bald

Autumn hawk flights (42.8 hawks per hour) reported at Jane Bald contain Broad-winged Hawks, plus small numbers of various other species of raptors (Finucane, 1957, 1961b, 1963, 1964, 1974).

Joppa

Joppa, on the eastern side of Clinch Mountain in the Richland Valley, produces autumn hawk flights at the rate of 299 hawks per day with Broad-winged Hawks forming the bulk of the migrants (Fowler, 1982).

Kingsport

Sometimes autumn migrations of Broad-winged Hawks and a few other species are reported in the Kingsport area at the passage rate of 56.8 hawks per day (Finucane, 1958, 1961a, 1976a, 1977a, 1979b).

Kingston

Occasionally autumn Broad-winged Hawk flights of as many as 472 birds are reported at Kingston approximately 40 miles west-southwest of Knoxville (Behrend, 1952).

Knoxville Area

Many autumn hawk flights (122.3 hawks per day) are reported in the Knoxville area with Broad-winged Hawks forming the bulk of the migrations. Nevertheless, a variety of other raptors also are seen in small numbers (Behrend, 1953; Finucane, 1957, 1958, 1961b, 1963, 1964, 1965, 1966, 1967; Fowler, 1982; Johnson, 1950).

Laurel Grove Fire Tower

Hawk watchers using the Laurel Grove Fire Tower north of Knoxville report autumn hawk flights (68 hawks per hour) consisting mostly of Black Vultures, Turkey Vultures, and Broad-winged Hawks, although other raptors also are seen in limited numbers (Puckette, 1983b, 1984b).

Mason

Behrend (1951) reported occasional September Broad-winged Hawk flights of as many as 183 birds near Mason northeast of Memphis.

Meadow Creek Fire Tower

The autumn hawk migrations (62.5 hawks per day) reported at this site in Cocke County contain Broad-winged Hawks and a few other hawks in meager numbers (Finucane, 1968, 1976a).

Memphis Area

In the Memphis area, autumn hawk migrations (61.5 hawks per day) contain Broad-winged Hawks and small counts of other raptors including 2

Prairie Falcons (Behrend, 1951; Coffey, 1981; Finucane, 1961b, 1963; Irwin, 1962).

Millington

As many as 850 migrating Broad-winged Hawks were seen in September at the Naval Air Station at Millington north of Memphis (Behrend, 1951).

Missionary Ridge

Missionary Ridge, east of Chattanooga, sometimes produces autumn hawk flights of as many as 357 Broad-winged Hawks and a few other raptors (Finucane, 1973, 1976a).

Mountain Creek

Autumn hawk migrations at Mountain Creek (101.3 hawks per day) contain Broad-winged Hawks plus a few other raptors. An alternative name for this site is Dowler Heights (Finucane, 1977a; Turner, 1981).

Mount Roosevelt Fire Tower

This fire tower (sometimes called Rockwood) is located on Walden's Ridge near Rockwood. The autumn hawk migrations (34.8 hawks per hour) contain Broad-winged Hawks in largest numbers, plus other hawks in limited numbers (Behrend, 1951, 1954b; Finucane, 1963, 1964, 1971, 1973, 1974, 1976a, 1977a, 1980c; Fowler, 1982).

Nashville Area

Occasionally during autumn nearly 100 migrating Broad-winged Hawks and a few other raptors are reported in the Nashville area (Finucane, 1958, 1969).

Oak Ridge

The autumn hawk flights (35.8 hawks per day) in the Oak Ridge area contain Broad-winged Hawks, plus occasional sightings of other raptors in limited numbers (Finucane, 1957, 1959, 1960, 1965).

Powell Mountain

During autumn flights of as many as 150 migrating Broad-winged Hawks sometimes are reported at Powell Mountain north of Knoxville (Finucane, 1958, 1964).

Roan Mountain

Hawk watchers on Roan Mountain report autumn hawk migrations (20.8 hawks per day) consisting mostly of Broad-winged Hawks, although small numbers of various other raptors also are noted (Behrend, 1950b; Finucane, 1956, 1957, 1958, 1963, 1964, 1965, 1976).

Rogersville-Kyles Ford Fire Tower

This fire tower on Clinch Mountain near Edison is one of the state's better hawk watching locations (Heintzelman, 1979a: 211). Excellent autumn hawk migrations (44 hawks per hour) are reported there with the bulk of the birds observed there being Broad-winged Hawks (Finucane, 1959, 1960, 1961a, 1961b, 1963, 1964, 1965, 1966, 1967, 1968, 1969, 1971, 1973, 1974, 1975a, 1976a, 1977a, 1978c; Fowler, 1982; Turner, 1981).

Short Mountain Fire Tower
September hawk flights at this site sometimes contain as many as 228 hawks per day most of which are Broad-winged Hawks, although a variety of other hawks also are noted in small numbers (Finucane, 1973).

Signal Point (in Chickamauga and Chattanooga National Park)
Signal Point, near the town of Lookout Mountain, is an interpretative center on Signal Mountain in Chickamauga and Chattanooga National Park. The autumn hawk migrations there (9.5 hawks per hour) consist of Broad-winged Hawks and occasional views of other raptors in small numbers (Finucane, 1969, 1976a, 1977a, 1978c, 1980c; Fowler, 1982; Puckette, 1983b, 1984b; Turner, 1981).

Spring hawk migrations (2.5 hawks per hour) at this site contain Ospreys, Broad-winged Hawks, and several other raptors in limited numbers (Finucane, 1978a, 1981a).

Sparta
Observations of autumn hawk migrations at Sparta (73.4 hawks per hour) produce mostly Broad-winged Hawks and a few other hawks in small numbers (Finucane, 1981b).

Stepp Gap
September hawk flights at this Pickett County location contain 161.6 hawks per hour—mostly Broad-winged Hawks (Finucane, 1971).

Sunset Rock
Hawk watchers at Sunset Rock west of Bon Air on the Cumberland Plateau in White County report 206.8 hawks per day during autumn with Broad-winged Hawks forming the bulk of the birds observed (Behrend, 1954b; Finucane, 1956, 1958, 1959, 1975a, 1980c; Turner, 1981).

Townsend
Townsend is a town in Tuckaleechee Cove in Blount County in the foothills of Great Smoky Mountain National Park. Occasional September hawk flights include as many as 832 Broad-winged Hawks and a few other raptors including several Golden Eagles (Fowler, 1982).

White Oak Mountain Fire Tower
Observers on this lookout (sometimes referred to as the Ooltewah Lookout) on the borders of Bradley and Hamilton Counties, report autumn hawk flights (43.1 hawks per hour) containing Broad-winged Hawks and a few other raptors in limited numbers (Finucane, 1973, 1975a, 1976a, 1977a, 1978c, 1980c; Puckette, 1982b; Turner, 1981).

CHAPTER 8

Gulf Coast Hawk Migrations

The Gulf Coast states—Alabama, Florida, Louisiana, Mississippi, and Texas—form an important part of the autumn hawk migration routes used by some species including Ospreys, Sharp-shinned Hawks, Broad-winged Hawks, and Peregrine Falcons while also serving as the winter range for some birds of prey. In spring the northward hawk flights also appear at some of the same autumn locations but the current picture of spring hawk migrations tends to suggest a wide dispersal for many species.

ALABAMA

The general status of hawk migrations in several sections of Alabama—the Tennessee Valley, Mountain Region, Piedmont, Upper Coastal Plain, Lower Coastal Plain, and Gulf Coast—was summarized in Imhof (1962, 1976). Various other fragmentary hawk migration observations also are reported from the state.

Birmingham
During September as many as 80 migrating Broad-winged Hawks have been seen at Birmingham (Lowery and Newman, 1953), and a February record of a Prairie Falcon also was reported by Einspahr and Meehan (1975).

Brownsboro
Autumn hawk migrations (52.6 hawks per day) are reported at Brownsboro near Huntsville. Broad-winged Hawks form the bulk of the flights which follow a northeast-to-southwest flight-line along a valley between two mountains (Robinson, 1960).

Castleberry
Occasionally during mid-September as many as 400 Broad-winged Hawks are reported migrating across the Castleberry area (Swindell, 1959).

Dauphin Island
Fair autumn hawk migrations are reported at Dauphin Island near Mobile (Heintzelman, 1979a: 157). Broad-winged Hawks and American Kestrels

159

are reported in largest numbers but various other raptors also are seen occasionally (Duncan and Duncan, 1979b: Newman, 1960).

Fairfield
During mid-September more than 400 migrating Broad-winged Hawks sometimes are reported at Fairfield (Newman, 1960).

Fairhope
As many as 18 Mississippi Kites have been seen in May at Fairhope (Imhof, 1972: 770).

Fort Morgan State Park
The autumn migrations of hawks at this park on the eastern side of Mobile Bay produce fair flights (63.9 hawks per day) with the bulk of the migrations formed by Northern Harriers, Sharp-shinned Hawks, Broad-winged Hawks, and American Kestrels, plus various other species in limited numbers (Duncan, 1980b, 1981; Duncan and Duncan, 1978b, 1979b; Heintzelman, 1979a: 157; Kennedy, 1976b, 1977b; Newman, 1958a; Purrington, 1975; Rogers, 1982, 1983b, 1984).

Spring hawk migrations at this site (13 hawks per day) are meager and consist mostly of some Broad-winged Hawks, plus a few other raptors (Duncan, 1980a).

Marshall County
September flights of as many as 600 Broad-winged Hawks are reported in Marshall County (Purrington, 1975).

Mobile Bay Area
Hawk migrations in the Mobile Bay area during autumn (7.8 hawks per hour) consist of Northern Harriers, Broad-winged Hawks, and American Kestrels, plus limited numbers of various other raptors (Heintzelman, 1975: 128; Newman, 1958a; Parrish and Wischusen, 1980).

As many as 8 American Swallow-tailed Kites also are reported in late March in the area (Imhof, 1967).

Pinnacle State Park
M. D. Floyd observed 17 hawks per day at this park during late September. Ospreys are the most numerous raptors noted (Donohue, 1981a).

Russel Cave National Monument
James Peavy reported fair autumn hawk migrations (14 hawks per hour) at this site. Broad-winged Hawks appear in largest numbers, but a few other raptors also are observed (Finucane, 1975b).

Talladega Mountains
The Talladega Mountains, at the end of the Blue Ridge, produced a spectacular sight of 5,000 roosting Broad-winged Hawks, on 24 September 1957 (Newman, 1958a: 38).

Tensaw River
As many as 88 Mississippi Kites are seen in late April along the Tensaw River (Imhof, 1983: 880; Newman, 1958b).

FLORIDA

Hawk migrations in Florida were studied by various observers who some-times found these raptor movements very complex, important, but often confusing (Darrow, 1983; Duncan and Duncan, 1975; Enderson, 1965; Gillespie, 1960; Henny and Van Velzen, 1972; Howell, 1932; Kennedy, 1973; Pough, 1939; Simons, 1977; Sprunt, 1954).

Broley (1947), for example, demonstrated that Bald Eagles from Florida nest sites engaged in a northward postbreeding dispersal then return again to the state in autumn to resume breeding activities, whereas a few north-ern Bald Eagles also migrate into Florida and adjacent states during winter (Postupalsky, 1976b).

A small immature Broad-winged Hawk population also migrates into Florida during autumn and winters between Miami and Key West (Tabb, 1973; Darrow, 1983), as do a few Swainson's Hawks (Heintzelman, 1979a). Adult Broad-winged Hawks, however, winter in Central and South Amer-ica—including some birds that were banded as immatures in winter in southern Florida (Tabb, 1979).

There also is a remarkable intrastate migration of several hundred Short-tailed Hawks. These birds are distributed in central and southern Florida (rare records in the northern part of the state) from late February through early October. From mid-October through early February the hawks migrate southward to their wintering grounds in the extreme south-ern part of the mainland (especially in or adjacent to Everglades National Park) and occasionally farther into the keys (Ogden, 1974).

A variety of raptors including some Northern Harriers, Sharp-shinned Hawks, Red-shouldered Hawks, Red-tailed Hawks, and American Kestrels also migrate into Florida and winter there (Bohall and Callopy, 1984; Layne, 1982).

Alligator Point
Mid-October migrations of up to 16 Merlins and 20 Peregrine Falcons are reported at Alligator Point on the western side of Apalachee Bay in north-western Florida (Stevenson, 1955).

Amelia Island
Robertson (1972) reported 25 Sharp-shinned Hawks going *north* over Ame-lia Island in early October.

Boca Grande Pass
Edscorn (1977) reported 42 Sharp-shinned Hawks flying *north* at Boca Grande Pass in late October.

Brevard County
Cruickshank and Cruickshank (1980) document some conspicuous autumn migrations of Turkey Vultures, Ospreys, Bald Eagles, Sharp-shinned Hawks, Red-shouldered Hawks, Broad-winged Hawks, Red-tailed Hawks,

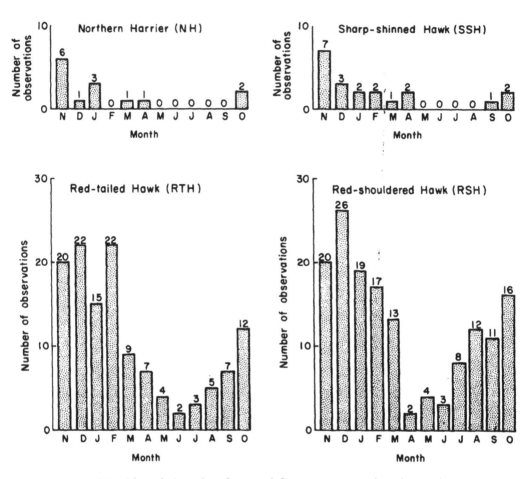

Monthly relative abundance of four raptor species observed along 24 16-kilometer transects in north-central Florida from 2 November 1981 to 31 October 1982. *Reprinted from Bohall and Collopy (1984).*

American Kestrels, Merlins, and Peregrine Falcons in Brevard County on the Atlantic coast.

Somewhat less conspicuous spring flights of Ospreys, Sharp-shinned Hawks, Red-tailed Hawks, American Kestrels, Merlins, and Peregrine Falcons also are noted.

Canaveral National Seashore

Fair to excellent October migrations of Sharp-shinned Hawks and Peregrine Falcons sometimes are reported at this Atlantic coast location (Heintzelman, 1979a: 161–162).

Cape Florida

October hawk migrations (346 hawks per day) reported at Cape Florida south of Miami are composed of Sharp-shinned Hawks and occasionally a few other species in small numbers (Atherton and Atherton, 1981, 1982, 1983). These Atlantic coastal hawk flights can be considered as part of the raptor migrations moving down the keys toward Key West.

Cedar Keys

Byrum W. Cooper reported early October hawk flights (213 hawks per day) at Cedar Keys on the Gulf coast between St. Petersburg and Tallahassee. Sharp-shinned Hawks form the bulk of the migrations, but a few other species also are seen in small numbers (Duncan, 1981).

Dade County

The autumn hawk migrations in Dade County (8.7 hawks per hour) are formed by Ospreys, Sharp-shinned Hawks, Broad-winged Hawks, Merlins, Peregrine Falcons, and several other raptors in limited numbers (Rogers, 1983b, 1984).

Destin

Destin is located in Okaloosa County along the Gulf of Mexico (Heintzelman, 1979a: 162). Eric Lefstad's autumn hawk watch there produced Broad-winged Hawks and a few other raptors generally moving *west* or *northwest*. Most of the Broad-wings were immatures (Duncan and Duncan, 1978b).

Dry Tortugas

The status of hawk migrations in the Dry Tortugas remains unclear and is based entirely upon very fragmentary observations. Nevertheless, most raptor sightings are made in spring rather than in autumn.

During spring, Ospreys, Northern Harriers, Sharp-shinned Hawks, Cooper's Hawks, Red-shouldered Hawks, Broad-winged Hawks, Red-tailed Hawks, American Kestrels, Merlins, and Peregrine Falcons are seen occasionally in the Dry Tortugas but rarely number more than a few individuals (Cunningham, 1966; Kale, 1970, 1973, 1976, 1979, 1980; Paulson and Stevenson, 1962; Stevenson, 1956b, 1960b, 1963). On 1 May 1983, however, 59 Sharp-shinned Hawks were seen on Loggerhead Key in the Dry Tortugas (Kale, 1983: 861).

During autumn, several Broad-winged Hawks also are observed in the Dry Tortugas (Robertson, 1970, 1972).

Everglades National Park

During autumn observers in this park note small numbers of American Swallow-tailed Kites, and as many as 60 Broad-winged Hawks also are seen (Cunningham, 1964; Robertson, 1970). Ogden (1974) also reported small numbers of Short-tailed Hawks migrating into the park during autumn.

Florida Keys

There are various published reports of raptors migrating across the Florida

Keys during both autumn and spring. Vultures, accipiters, buteos, and falcons seem to be particularly well represented in many of these flights.

The autumn migrations (167.3 hawks per day) consist of Turkey Vultures, Ospreys, Northern Harriers, Sharp-shinned Hawks, Cooper's Hawks, Broad-winged hawks, Swainson's Hawks, American Kestrels, Merlins, and Peregrine Falcons, some of which also are occasionally noted in reverse migrations (Atherton and Atherton, 1980, 1983; Cunningham, 1964; Darrow, 1983; Edscorn, 1974, 1976, 1977; Robertson, 1970; Robertson and Paulson, 1961; Stevenson, 1958, 1962, 1973).

Spring hawk migrations (22.7 hawks per day) in the keys contain Turkey Vultures, Northern Harriers, Sharp-shinned Hawks, Red-shouldered Hawks, Broad-winged Hawks, Swainson's Hawks, and Red-tailed Hawks (Kale, 1979, 1980, 1981).

Fort DeSoto State Park
Larry and Jerry Hopkins and others observed autumn hawk flights at this park along Tampa Bay and reported 98 raptors. Sharp-shinned Hawks and Broad-winged Hawks are seen in largest numbers (Duncan, 1980b).

Fort Lauderdale
George and Ruth Breck made autumn hawk counts along the Intracoastal Waterway at Fort Lauderdale and noted a variety of raptors of which Turkey Vultures and Broad-winged Hawks were seen in largest numbers (Duncan, 1980b, 1981).

Fort Pickens
October hawk migrations (87 hawks per day) at Fort Pickens, a northwestern Florida barrier island between the Gulf of Mexico and Pensacola Bay, include Sharp-shinned Hawks as the most numerous species seen (Duncan, 1981; Duncan and Duncan, 1978b).

Fort Pierce
Occasionally in early November as many as 126 Turkey Vultures are reported migrating past Fort Pierce (Stevenson, 1967).

Gasparilla Island
This island is located at Charlotte Harbor. Malcolm M. Simons, Jr., reported Sharp-shinned Hawks (20 birds per hour) heading northward along the island's interior into a brisk northwest wind. The hawks approached across Boca Grande Pass and the direction of Cayo Costa Island, suggesting a reverse migration (Simons, 1977).

Guano Lake
Steven L. Sutton's October hawk watch at Guano Lake, in extreme northeastern Florida on the Atlantic coast near the Georgia border, produced a variety of migrating hawks (90.8 hawks per day) the bulk of which were Black Vultures, Turkey Vultures, Ospreys, Northern Harriers, Sharp-shinned Hawks, and American Kestrels (Duncan and Duncan, 1978b).

Spring hawk migrations (1,210 hawks per day) contain Ospreys, Northern Harriers, American Kestrels (2,000 on one day), and Merlins (Kale, 1972).

Gulf Breeze

Francis M. Weston made the first autumn hawk counts in 1943 at Gulf Breeze, the end of a 20-mile-long peninsula with Pensacola Bay on the north and west and Santa Rosa Sound on the south, and noted that Red-shouldered Hawks were the most abundant migrant hawks there prior to 1965 (Weston, 1965), whereas Broad-winged Hawks were the most common hawks seen there in 1972–1974 (Duncan and Duncan, 1975). Annual autumm hawk counts made at this site between 1972–1982 indicate that the bulk of the migrations consist of Sharp-shinned Hawks and Broad-winged Hawks, plus small numbers of other species (Duncan, 1980b, 1981; Duncan and Duncan, 1975, 1978b, 1979b; Kennedy, 1976b, 1977b; Rogers, 1982, 1983b; Kennedy, 1975).

Hypoluxo Island

Observations of autumn hawk flights at this island near Lantana along the Atlantic coast south of West Palm Beach produce various species of which Sharp-shinned Hawks and American Kestrels are the most numerous (Atherton and Atherton, 1982; Duncan and Duncan, 1979b).

Indian River County Barrier Islands

As many as 40 Peregrine Falcons are reported in autumn along the Indian River County barrier islands (Robertson and Ogden, 1968).

Indiatlantic Beach

A reverse migration of 400 Sharp-shinned Hawks was seen in early November as the birds headed *northeast* along Indiatlantic Beach (Edscorn, 1975: 45).

Jacksonville

March migrations of Turkey Vultures (185.5 birds per day) are reported by Fred Wetzel and others at Jacksonville (Kale, 1972, 1977, 1983).

Key West Area

During autumn some large concentrations of migrating hawks are reported over Key West and its vicinity. It is not clear, however, what happens to many of the hawks seen at this location. Darrow (1982) suggested that some hawks cross the Straits of Florida and head for Cuba or other islands in the West Indies whereas other hawks doubtless reverse direction and return north to rejoin other hawks migrating around the Gulf of Mexico (Heintzelman, 1975: 293–299). In 1977, for example, Sharp-shinned Hawks were seen "everywhere" on 14 October, and these hawks were observed "falling dead out of trees" on 29 October, apparently due to starvation (Edscorn, 1978: 194).

Autumn hawk migrations (109.9 hawks per hour) at Key West include Turkey Vultures, American Swallow-tailed Kites, Sharp-shinned Hawks,

Cooper's Hawks, Broad-winged Hawks, Short-tailed Hawks, and Swainson's
Hawks, plus various other species in limited numbers (Atherton and Atherton, 1980, 1981, 1983; Cunningham, 1965; Edscorn, 1974, 1976, 1977,
1978; Robertson and Ogden, 1968, 1969; Stevenson, 1955, 1956a).

Occasionally during spring a few Broad-winged Hawks and Swainson's
Hawks are reported at Key West or its vicinity (Kale, 1978, 1979).

LaBelle

In mid-August as many as 18 American Swallow-tailed Kites are reported at
LaBelle (Atherton and Atherton, 1982).

Lakeland

In early October, 27 hawks per day have been seen at Lakeland, mostly
Sharp-shinned Hawks (Edscorn, 1976).

Lake Worth

Late October hawk migrations (169 hawks per day) are reported at Lake
Worth. Ospreys, Northern Harriers, Sharp-shinned Hawks, and American
Kestrels are observed (Stevenson, 1960a).

Loxahatchee National Wildlife Refuge

As many as 20 American Swallow-tailed Kites are reported in August at this
refuge (Atherton and Atherton, 1982).

Marquesas Keys

A Broad-winged Hawk was reported in mid-October at the Marquesas Keys
in the Gulf of Mexico between Key West and the Dry Tortugas (Robertson,
1970).

Merritt Island

March raptor migrations (69 birds per day) are reported at Merritt Island.
American Kestrels are observed (Duncan and Duncan, 1978a).

Miami

The autumn hawk migrations in the Miami area (51.5 hawks per day)
include Turkey Vultures, Ospreys, Sharp-shinned Hawks, Broad-winged
Hawks, Merlins, and Peregrine Falcons (Brookfield, 1953; Cunningham,
1965; Rogers, 1984; Stevenson, 1967, 1973).

Ochopee

In mid-May more than 150 American Swallow-tailed Kites are reported at
Ochopee (Kale, 1979).

Pembroke Pines

Autumn hawk migrations reported at Pembroke Pines, in southwestern
Broward County, include Turkey Vultures, Sharp-shinned Hawks, American Kestrels, and small numbers of other species (Duncan, 1981).

Pensacola

Among the autumn raptor migrants reported at Pensacola (57.7 hawks per
day) are Mississippi Kites, Northern Harriers, Sharp-shinned Hawks, Red-
shouldered Hawks, and Broad-winged Hawks. On 15 October 1961, a
flight of Red-shouldered Hawks headed *west* at the same time that a flight of

Northern Harriers and Sharp-shinned Hawks headed *east* (Imhof, 1962; Herman, 1960).

Port Canaveral

Duncan and Duncan (1978b) state that the autumn hawk migrations observed at Port Canaveral near Cocoa (21.5 hawks per day) include Northern Harriers, Sharp-shinned Hawks, American Kestrels, plus some other birds of prey in small numbers.

Port Charlotte Harbor

The autumn hawk flights seen on the west coast at Port Charlotte Harbor include Turkey Vultures, Ospreys, Sharp-shinned Hawks, American Kestrels, and a few other species in small numbers (Duncan and Duncan, 1979b).

Port Saint Lucie

Mid-October migrations of Sharp-shinned Hawks (357 hawks per day) are reported at this Atlantic coast location (Atherton and Atherton, 1982).

Rockledge

Robert Barber's autumn hawk counts along the Intracoastal Waterway at Rockledge contained various species including Turkey Vultures and Sharp-shinned Hawks in largest numbers (Duncan, 1980b).

Saddle Creek Park

The autumn hawk flights (20.6 hawks per day) reported at this park at Lakeland consist of Sharp-shinned Hawks, plus several other species in small numbers (Duncan and Duncan, 1978b).

Saint Augustine

Mid-October migrations of Sharp-shinned Hawks and Cooper's Hawks (296 hawks per day) are reported moving *north* at this location, and migrating Ospreys also are reported there (Edscorn, 1977, 1978).

Saint George Island

The autumn hawk flights reported at Saint George Island (8.1 hawks per day) near Apalachicola contain Northern Harriers, Sharp-shinned Hawks, American Kestrels, plus a variety of other raptors in small numbers (Duncan and Duncan, 1978b; Kennedy, 1977b).

Saint Joseph Peninsula

Excellent autumn hawk migrations (140.7 hawks per day) are reported at this 18-mile-long peninsula along the Gulf of Mexico. The most numerous species are Sharp-shinned Hawks, Broad-winged Hawks, and American Kestrels (Duncan, 1980b, 1981; Duncan and Duncan, 1978b, 1979b; Kennedy, 1976b, 1977b; Rogers, 1982, 1983b).

Saint Marks Light

A mid-September flight of 53 Turkey Vultures and a few other hawks has been reported moving *east* past this location (Atherton and Atherton, 1980).

Saint Petersburg
Early October hawk flights (356.5 hawks per day) are reported at St. Petersburg. Sharp-shinned Hawks and Red-tailed Hawks form the migrations (Edscorn, 1978, 1979).

Santa Rosa Peninsula
Mid-October migrations (63 hawks per day) of Broad-winged Hawks and several other species are reported at this site (Duncan and Duncan, 1978b).

South Allapattah Gardens
Swainson's Hawk migrations (176.6 birds per day) are reported in November and December at this southern Florida location (Brookfield, 1953).

South Ponte Vedra Beach
Poor autumn hawk migrations (25.2 hawks per day) are reported at this Atlantic coast site. Turkey Vultures, Ospreys, Northern Harriers, Sharp-shinned Hawks, and American Kestrels form the bulk of the flights (Duncan, 1980b, 1981; Duncan and Duncan, 1978b, 1979b; Kennedy, 1977b).

Spring hawk migrations (17.6 hawks per day) also are seen at South Ponte Vedra Beach. Black Vultures, Ospreys, and American Kestrels appear in largest numbers (Duncan and Duncan, 1979a; Kennedy, 1976a; Rogers, 1983a).

Straits of Florida
In late November an immature Northern Harrier was reported heading south across the Straits of Florida halfway between Cuba and Cape Sable, Florida (Pough, 1939).

Tavernier
Autumn migrations of Ospreys, Sharp-shinned Hawks, and Broad-winged Hawks (692 hawks per day) are reported at Tavernier (Edscorn, 1976, 1978; Stevenson, 1956).

Tierra Verda Golf Course
An autumn hawk watch at this location in the Tampa Bay area produced Sharp-shinned Hawks in largest numbers but some other hawks in small numbers (Duncan and Duncan, 1979b).

Upper Plantation Key
Fair autumn hawk flights sometimes are reported at Upper Plantation Key (Heintzelman, 1979a).

Wakulla County
As many as 12 American Swallow-tailed Kites have been seen in early August in Wakulla County (Edscorn, 1979).

LOUISIANA

General outlines of hawk migrations in Louisiana were provided by Oberholser (1938) and Lowery (1974), but much remains to be learned about raptor movements in the state. Recent interest in hawk migrations in Louisiana should provide a more complete picture in the future.

Alexandria

As many as 700 migrating Broad-winged Hawks were seen in mid-September over Alexandria (Newman, 1956a).

Avery Island

Avery Island is an uplifted area surrounded by tidal marshes in Iberia Parish five miles southwest of New Iberia. Of 22,600 Black Vultures banded there between 1934 and 1946, only 840 were recovered—740 in Louisiana within a 100-mile radius of Avery Island, and 65 in seven nearby states. One bird reached South Carolina and another reached northeastern Florida. Thus Black Vultures from Avery Island have no true migration, but some wander over a considerable area (Parmalee and Parmalee, 1967). In late September nearly 500 Broad-winged Hawks also were reported over Avery Island (McIlhenny, 1939).

Baton Rouge Area

The autumn hawk flights (93 hawks per hour) seen in the Baton Rouge area contain Broad-winged Hawks and small numbers of various other birds of prey (Rogers, 1983b, 1984).

Cameron

During autumn as many as 500 migrating Broad-winged Hawks are observed at Cameron, and a few other birds of prey also are noted (Newman, 1960; Purrington, 1971; Rogers, 1983b).

Grand Isle

As many as 100 Broad-winged Hawks, and sometimes a few other hawks, are reported in autumn at Grand Isle (Newman, 1960; Rogers, 1983b).

Lafayette

Occasionally in late September more than 325 Broad-winged Hawks have been observed at Lafayette (James, 1966).

Livingston

Rogers (1983b) reported as many as 915 Broad-winged Hawks and several other raptors at Livingston during autumn.

New Iberia

Occasionally during September as many as 500 migrating Broad-winged Hawks are seen at New Iberia (Purrington, 1974, 1976).

New Orleans

Occasionally during October small numbers of migrating Sharp-shinned Hawks and Merlins are reported at New Orleans (Purrington, 1978, 1980).

Spring raptor migrations sometimes include as many as 7 American Swallow-tailed Kites and 150 Mississippi Kites (Imhof, 1967, 1969).

Sabine National Wildlife Refuge

Autumn flights of up to 1,000 Broad-winged Hawks are seen at Sabine National Wildlife Refuge (Lowery and Newman, 1953).

Saint Francisville Area
An autumn hawk watch in the Tunica Hills near St. Francisville produced a passage rate of 946 hawks per hour. Most of the birds observed are Broad-winged Hawks, but a few other species also appear (Purrington, 1982).
Shreveport
Occasional autumn hawk migrations at Shreveport (109 hawks per day) are composed of Broad-winged Hawks (Newman, 1959, 1960; Rogers, 1982).

Mid-March flights of Red-tailed Hawks (13 birds per day) may contain individuals of the *kriderii* race (Newman and Warter, 1959).
Vidalia
As many as 3,000 Broad-winged Hawks are reported in early October at Vidalia (Newman, 1961).
Wakefield
Autumn hawk migrations reported at Wakefield contain Broad-winged Hawks and various other birds of prey in very small numbers (Rogers, 1982, 1983b).

MISSISSIPPI

Very limited information is available regarding hawk migrations in Mississippi, but a few fragmentary reports are discussed here.
Attala County
Occasionally as many as 250 Broad-winged Hawks are reported in late September in this county (Purrington, 1979).
Louisville
In late September more than 1,400 Broad-winged Hawks were seen at Louisville (Purrington, 1983).
Rosedale
Spring hawk migrations at Rosedale (35.3 hawks per day) sometimes contain as many as 80 Mississippi Kites and a few Broad-winged Hawks (Imhof, 1966, 1981; Newman 1956b; Newman and Warter, 1959).
Warren County
In late September as many as 85 migrating Broad-winged Hawks are reported in this county (Purrington, 1979).

TEXAS

Hawk migrations in Texas, especially of Broad-winged Hawks and Swainson's Hawks, are the most spectacular in North America in terms of numbers of birds seen. As many as 500,000 Broad-winged Hawks, for example, are observed in a single day in Nueces County (Fox, 1956; Oberholser, 1974; Webster, 1978a). During recent years greatly increased interest in studying hawk flights developed in this state resulting in much new information on flight-lines and numbers of birds (Donohue, 1978a–1982; Hunt, Rogers, and Slowe, 1975; Kennedy, 1975–1977b; Rowlett, 1978; Sexton, 1982–1984b).

Alice Area

Occasionally during September flights of as many as 1,000 Mississippi Kites are reported in the Alice area west of Corpus Christi (Lasley, 1984; Webster, 1969a, 1976a).

During early April as many as 75,000 migrating Broad-winged Hawks and 500 Swainson's Hawks also are reported in this area (Webster, 1961b, 1974b).

Austin Area

Autumn hawk migrations in the Austin area sometimes include as many as 220 Mississippi Kites, 2,500 Broad-winged Hawks, and 3,000 Swainson's Hawks, plus small numbers of various other hawks (Emanuel, 1982; Webster, 1964a, 1969a, 1970a, 1972a, 1975a, 1980a, 1983a).

The spring hawk flights in the Austin area occasionally include as many as 250 Mississippi Kites, 75,000 Broad-winged Hawks, and 5,000 Swainson's Hawks (Goldman and Watson, 1953b; Webster, 1958b, 1960b, 1961b, 1962b, 1964b, 1965b, 1966b, 1969b, 1970b, 1973b, 1979b).

Bartlett

October migrations of American Kestrels (69.5 falcons per day) sometimes are reported at Bartlett north of Austin (Webster, 1972a).

Baytown

September hawk flights at Baytown at the northern end of Galveston Bay sometimes contain as many as 10,000 Broad-winged Hawks and 400 Swainson's Hawks (Webster, 1963, 1976a, 1980a).

Beeville

The autumn hawk migrations reported at Beeville and its vicinity (1,581.4 hawks per day) include both Broad-winged Hawks and Swainson's Hawks, although the former tend to appear in the largest numbers (Donohue, 1981b; Webster, 1956a, 1960a, 1962a, 1968a, 1969a, 1972a, 1974a).

During the spring hawk migrations in the Beeville area, Mississippi Kites, Broad-winged Hawks, and Swainson's Hawks also are well represented (2,300 hawks per day). As many as 500 Mississippi Kites, 10,000 Broad-winged Hawks, and 1,000 Swainson's Hawks form some of the flights (Arvin, 1982; Webster, 1956b, 1959b, 1960b, 1964b, 1965b, 1967b, 1970b, 1976b, 1977b, 1980b).

Bentsen-Rio Grande Valley State Park

During September, flights of as many as 18,000 Broad-winged Hawks are reported at this park, and occasionally rare species including Hook-billed Kites and a Roadside Hawk appear there (Webster, 1968a, 1973a, 1981, 1983a).

May migrations of more than 300 Mississippi Kites also are reported at the park (Arvin, 1982; Webster, 1961b, 1972b).

Blessings
As many as 500 Mississippi Kites are reported in late April at Blessings (Donohue, 1978a).

Brazos County
In early September as many as 200 Mississippi Kites are seen in this county (Williams, 1971), whereas late April flights contain up to 173 of these kites (Webster, 1960b).

Bryan
Late September migrations of Broad-winged Hawks at Bryan contain as many as 10,000 birds (Donohue, 1981b).

Chambers County
Migrations of as many as 1,000 Mississippi Kites are reported in late August in Chambers County (Emanuel, 1982).

Coastal Bend Area
September and early October Broad-winged Hawk flights in the Coastal Bend area sometimes contain as many as 13,300 birds, and some hawk watchers speculate that more than 136,000 Broad-wings may actually cross the area undetected (Donohue, 1979b; Rowlett, 1978).

College Station
Broad-winged Hawk flights in autumn (4,937.5 hawks per day) in the College Station area occasionally contain as many as 16,250 birds (Donohue, 1981b; Webster, 1965a, 1966a).

Spring hawk migrations (1,600 hawks per day) also contain Broad-winged Hawks (Webster, 1964b, 1965b).

Columbus
Occasionally as many as 3,000 Broad-winged Hawks are seen in early October at Columbus in Colorado County (Webster, 1971).

Copano Bay Bridge Area
Late April hawk migrations (2,615.5 hawks per day) in the Copano Bay Bridge Area contain Mississippi Kites, Sharp-shinned Hawks, and Broad-winged Hawks, plus a few other species in small numbers. The Sharp-shins cross the narrowest section of the bay, at that point about two miles wide (Donohue, 1978a).

Corpus Christi Area
Major autumn hawk migrations sometimes are reported crossing the Corpus Christi Area (11,781.8 hawks per day). As many as 407 Mississippi Kites and 62,000 Broad-winged Hawks are reported in some of the flights (Emanuel, 1982; Webster, 1968a, 1973a, 1975a, 1981, 1983a).

The spring hawk migrations at this location (14,300 hawks per day) sometimes contain as many as 500 Mississippi Kites and 65,000 Broad-winged Hawks (Webster, 1966b, 1973b, 1976b, 1977b).

Cove

The autumn hawk flights reported at Cove (680 hawks per day) consist of Broad-winged Hawks (Webster, 1958a, 1961a, 1967a, 1973a).

April hawk migrations at Cove (130.5 hawks per day) consist of Broad-winged Hawks and occasionally a few other raptors (Goldman and Watson, 1953b; Watson, 1955b).

Cuero

Occasionally in late March as many as 2,000 Broad-winged Hawks are reported at Cuero (Webster, 1965b).

Dallas

Early October migrations of as many as 2,500 Swainson's Hawks are reported at Dallas (Williams, 1968).

Denton \

Late August flights of as many as 800 Mississippi Kites are seen at Denton north of Dallas (Williams, 1982).

Dickinson

A mid-April flight of 15,000 Broad-winged Hawks was reported at Dickinson in Galveston County (Webster, 1969b).

Edinburg

Occasionally in late September as many as 1,000 Broad-winged Hawks are reported at Edinburg in the lower Rio Grande valley (Webster, 1972a), whereas as many as 300 migrating Swainson's Hawks appear in late March (Webster, 1978b).

Eldorado

Early October migrations of up to 400 Swainson's Hawks occasionally are reported at Eldorado (Williams, 1979).

Falcon Dam

Occasional autumn hawk flights at the Falcon Dam include 100 Mississippi Kites and 630 Swainson's Hawks (Webster, 1972a, 1976a). Flights of 228 Mississippi Kites also are reported in early May (Arvin, 1982).

Falfurrias

The autumn hawk migrations reported at Falfurrias (5,621.2 hawks per day) contain various species of which flights of as many as 500 Mississippi Kites, 25,000 Broad-winged Hawks, and 3,500 Swainson's Hawks are reported (Webster, 1963, 1967a, 1969a, 1970a, 1971, 1972a, 1974a, 1975a, 1978a, 1983a).

In spring, hawk migrations at this location (1,200 hawks per day) include flights of up to 150 Mississippi Kites, 1,500 Broad-winged Hawks, and 3,000 Swainson's Hawks (Webster, 1968b, 1973b, 1975b, 1976b, 1983b).

Garner State Park

Occasionally early October migrations of as many as 1,000 Swainson's Hawks are reported at this park south of Leakey (Webster, 1973a).

Goliad

Some autumn Broad-winged Hawk migrations (1,900 hawks per day) are reported at Goliad, and as many as 3,500 birds were noted in one flight (Webster, 1961a, 1973a).

Harris County

August migrations of Mississippi Kites (41.5 per day) in Harris County sometimes contain as many as 65 birds (Webster, 1959a, 1976a).

April raptor migrations (150 birds per day) include up to 200 Mississippi Kites and 100 Broad-winged Hawks (Webster, 1959b, 1983b).

Hays County

October Swainson's Hawk migrations (383.5 birds per day) include as many as 717 birds (Watson, 1955a).

Heard

As many as 1,000 Swainson's Hawks are seen in early October at Heard (Williams, 1984).

Hidalgo County

Broad-winged Hawk migrations (5,866.6 birds per day) reported in autumn in Hidalgo County contain up to 9,200 birds (Webster, 1977a).

Houston Area

Autumn hawk migrations (3,262 hawks per day) reported in the Houston area contain as many as 68 Mississippi Kites and 20,000 Broad-winged Hawks in some flights (Donohue, 1981b; Goldman and Watson, 1953a; Webster, 1959a, 1960a, 1964a, 1965a, 1966a, 1967a, 1968a, 1970a, 1973a, 1978a, 1983a).

In spring the hawk flights in this area (2,216.6 hawks per day) contain up to 10,000 Broad-winged Hawks and 300 Swainson's Hawks on some days (Webster, 1956b, 1958b, 1960b, 1962b, 1974b).

Ingram

As many as 300 migrating Swainson's Hawks are reported in early October at Ingram (Williams, 1981).

Jacinto City

A flight of 4,900 migrating Broad-winged Hawks was reported in late September at this city in Harris County (Kennedy, 1977b).

Karnes City

April raptor migrations at Karnes City occasionally contain as many as 100 Mississippi Kites and 4,000 Broad-winged Hawks (Webster, 1968b, 1975b).

Katy

As many as 5,000 migrating Broad-winged Hawks are reported in September at Katy (Webster, 1981).

Kingsville

Spring raptor migrations at Kingsville (13,398.6 hawks per day) include flights of up to 196 Mississippi Kites and 30,000 Broad-winged Hawks (Arvin, 1982; Donohue, 1981a, 1982a).

Kountze Area

The autumn hawk migrations in the Kountze area (3,192.5 hawks per day) include as many as 10,000 Broad-winged Hawks in some flights (Goldman and Watson, 1953a; Webster, 1970a, 1971, 1972a, 1976a).

Laguna Atascosa National Wildlife Refuge

The spring hawk migrations at this refuge (1,794.5 hawks per day) include up to 8 Mississippi Kites, 10,000 Broad-winged Hawks, and 300 Swainson's Hawks in some of the flights (Goldman and Watson, 1953b; Webster, 1956b, Webster, 1958b, 1960b, 1973b).

La Porte

September flights of up to 4,350 Broad-winged Hawks are reported at La Porte (Webster, 1961a).

Spring hawk migrations (285 birds per day) include up to 700 Broad-winged Hawks in some flights (Webster, 1959b, 1960b, 1961b).

Little Thicket Nature Sanctuary

In mid-September 150 migrating Broad-winged Hawks were seen at this site (Webster, 1956a).

Live Oak County

April migrations of up to 100 Swainson's Hawks are seen in this county (Watson, 1955b).

Luling

As many as 25 Mississippi Kites are reported in early September at Luling in Caldwell County (Webster, 1970a).

Madero

Flights of as many as 2,700 Broad-winged Hawks are reported in late September at Madero (Webster, 1974a).

Matagorda County

As many as 11,000 migrating Broad-winged Hawks are reported in late September in this county (Webster, 1969a).

Mathis

In September as many as 2,000 migrating Broad-winged Hawks are reported in Mathis (Webster, 1972a).

McAllen

Early October Broad-winged Hawk flights of up to 2,000 birds are seen at McAllen (Webster, 1981), and late March flights of up to 25,000 Broad-wings also appear there (Webster, 1976b).

Midland

Williams (1974) reported 140 migrating Swainson's Hawks in late September at Midland.

Mission

The autumn hawk migrations reported at Mission exhibit a passage rate of 1,520.4 hawks per day. Some of these flights include as many as 1,200 Turkey Vultures, 11,600 Broad-winged Hawks, 1,000 Swainson's Hawks, and a variety of other species in small numbers (Kennedy, 1977b; Webster, 1959a, 1972a, 1973a, 1978a).

Spring hawk migrations at Mission are composed of Black Vultures, Turkey Vultures, Broad-winged Hawks, Swainson's Hawks, and various other species in limited numbers (Kennedy, 1977a).

Mitchell County

In early October up to 500 migrating Swainson's Hawks are seen in Mitchell County (Williams, 1974).

Nacogdoches

The autumn hawk migrations reported at Nacogdoches (434.1 hawks per day) include as many as 50 Mississippi Kites and 1,086 Broad-winged Hawks in some flights (Williams, 1974, 1980, 1981, 1984).

Norias

In late March a flight of 50,000 Broad-winged Hawks was reported at Norias (Webster, 1976b).

Nueces County

Enormous autumn Broad-winged Hawk flights are reported in Nueces County—often very high, very narrow, dense, and fast-moving (Donohue, 1979b; Sexton, 1984b). Published records indicate a passage rate of 89,585.7 hawks per day with as many as 500,000 hawks reported in one early October flight (Lasley, 1984; Webster, 1964a, 1966a, 1969a, 1977a, 1978a, 1979a, 1981, 1983a).

During spring as many as 50,000 migrating Broad-winged Hawks also are reported in some flights (Webster, 1964b, 1965b).

Oilrig Platform

A unique autumn hawk watch on an oilrig platform in the Gulf of Mexico 75 miles south and 25 miles east of Galveston produced few birds—1 Sharp-shinned Hawk, 5 American Kestrels, 1 Merlin, and 1 Peregrine Falcon but no buteos (Kennedy, 1977b). The effort demonstrated that few migrating hawks cross that section of the Gulf of Mexico.

Padre Island

October field studies at Padre Island on the Texas coast demonstrated that Peregrine Falcons pass the site in considerable numbers (4.7 falcons per day). Age classes contained 20.3 percent adults and 79.7 percent immatures indicating a differential migration. The birds may have migrated from Alaska and northwestern Canada (Hunt, Rogers, and Slowe, 1975).

Panther Run

Shirley Fellers reported more than 5,600 migrating Broad-winged Hawks between early March and mid-April at Panther Run west of Fort Davis (Donohue, 1981a).

Port Aransas

In early October flights of 50 Ospreys and at least 50 American Kestrels are seen at Port Aransas (Webster, 1981), whereas 5,000 migrating Broad-winged Hawks are reported there in April (Webster, 1983b).

Rancho Chiquito
Autumn hawk migrations (407.4 hawks per hour) reported at Rancho Chiquito two miles west of Mission by Gladys Schumacher Donohue consist of Broad-winged Hawks and very limited numbers of a variety of other species (Donohue, 1979b).

Rio Hondo
Webster (1983b) reported a flight of 10,000 Broad-winged Hawks in early April at Rio Hondo.

Rockport Area
Spring hawk flights in the Rockport area occasionally contain as many as 50 Mississippi Kites, 218 Sharp-shinned Hawks, and 15,000 Broad-winged Hawks (Webster, 1973b, 1977b, 1983b).

Saint Paul
An early October flight of 10,000 Broad-winged Hawks was reported at St. Paul in San Patricio County (Webster, 1972a).

San Angelo
Williams (1981) reported 1,000 migrating Swainson's Hawks in early October at San Angelo.

San Antonio
An early October flight of 25,000 to 100,000 Swainson's Hawks, plus a few other hawks, was reported at Route 90 some 10 miles west of San Antonio (Fox, 1956), and as many as 50 Mississippi Kites also were reported in late August at San Antonio (Webster, 1969a).

San Benito
Occasionally in early October as many as 3,000 Broad-winged Hawks are seen at San Benito (Webster, 1959a), whereas mid-April migrations of up to 4,000 Broad-winged Hawks and 1,000 Swainson's Hawks also are reported (Webster, 1961b).

San Jacinto Battlegrounds
Webster (1958a) reported a late September flight of at least 10,000 Broad-winged Hawks at this location.

San Marcos Area
Occasionally as many as 800 Swainson's Hawks are reported in early October in the San Marcos area (Webster, 1966a).

Santa Ana National Wildlife Refuge
This superb wildlife refuge, one of the jewels in the federal system, is located along the Rio Grande River. Autumn and spring hawk migrations are seen there.

The autumn hawk migrations (1,777.2 hawks per day) sometimes contain flights of as many as 5,000 Broad-winged Hawks and 500 Swainson's Hawks (Webster, 1959a, 1965a, 1966a, 1983a).

Many spring hawk migrations at this refuge are impressive (4,326.1 hawks per day). Some of the flights contain up to 1,500 Mississippi Kites, 40,000 Broad-winged Hawks, and 3,000 Swainson's Hawks (Arvin, 1982;

Goldman and Watson, 1953b; Watson, 1955b; Webster, 1956b, 1958b, 1959b, 1960b, 1961b, 1962b, 1964b, 1966b, 1967b, 1968b, 1969b, 1970b, 1972b, 1973b, 1974b, 1975b, 1977b, 1979b, 1980b, 1983b).

San Ygnacio

Occasionally in mid-March as many as 750 migrating Swainson's Hawks are reported at this Zapata County location (Arvin, 1982).

Sarita

As many as 1,000 Broad-winged Hawks are seen in early April at Sarita (Webster, 1961b).

Tarrant County

Williams (1976) reported 80 migrating Mississippi Kites in late August in Tarrant County.

Travis County

An October flight of 1,200 Swainson's Hawks was seen in Travis County (Webster, 1966a).

Welder Wildlife Refuge

This important wildlife research station is located eight miles north of Sinton. Included in autumn hawk migrations occasionally reported there are September flights of 10,000 Broad-winged Hawks and 1,000 Swainson's Hawks (Webster, 1956a, 1958a, 1967a, 1978a).

Some of the spring hawk flights at the refuge include as many as 5,500 Broad-winged Hawks (Webster, 1958b, 1960b, 1983b).

C H A P T E R 9

Central States Hawk Migrations

The large section of the United States covered by the central states—Arkansas, Iowa, Kansas, Missouri, Nebraska, North Dakota, Oklahoma, and South Dakota—received very uneven field study coverage of annual hawk migrations with few systematic hawk watch efforts organized to date. The most detailed field study efforts thus far are in Iowa, Nebraska, and Oklahoma.

ARKANSAS

Limited organized study of hawk migrations exists in Arkansas, but a few of the larger hawk flights seen in the state are summarized here.

Calion Area
Occasionally during September as many as 1,000 migrating Broad-winged Hawks are reported in the Calion area (James, 1965).

Chicot
Some September flights of Broad-winged Hawks in the Chicot area contain as many as 258 birds (Purrington, 1983), and in mid-April up to 50 Mississippi Kites are reported there (Imhof, 1981).

Hempstead County
Occasionally as many as 2,000 migrating Broad-winged Hawks are reported in September in Hempstead County (Purrington, 1975).

Little Rock
September and early October migrations of Broad-winged Hawks at Little Rock contain as many as 4,300 birds (Purrington, 1973, 1975, 1976).

Petit Jean State Park
Pettingill (1981) stated that Sharp-shinned Hawk and Broad-winged Hawk migrations in autumn are reported at this park near Morrilton.

Pinnacle Mountain State Park
September hawk migrations (2.6 hawks per hour) contain Ospreys and a few other hawks, whereas April hawk flights (1.6 hawks per hour) are composed of Turkey Vultures and an occasional Northern Harrier, Sharp-shinned Hawk, or Red-tailed Hawk (Marty Floyd, letter of 10 May 1980).

Saint Francis Forest
As many as 25 Mississippi Kites are reported at this site in mid-May (Imhof, 1980).

Scott County
Occasionally as many as 1,200 Broad-winged Hawks are seen in Scott County in late September (Purrington, 1980).

Union County
Occasional September flights of up to 1,061 Broad-winged Hawks are reported in Union County (Purrington, 1973, 1974).

IOWA

Hawk migrations in Iowa, well summarized by Dinsmore, et al. (1984), were documented as early as 1875 at Marshalltown (Anonymous, 1875) and generally received more study and documentation than in many other central states (Bailey, 1912; Black, 1969a, 1969b; Dumont, 1934a, 1934b; La Mar, 1937; Proescholdt, 1961; Spiker, 1933; Varland, 1982; Williams, 1941; Youngworth, 1935).

Backbone State Park
Occasionally as many as 140 migrating Broad-winged Hawks are observed at Backbone State Park in Delaware County (Spiker, 1933).

Booneville
At times mid-November hawk flights at Booneville in Dallas County contain as many as 15 Rough-legged Hawks and a few other raptors (La Mar, 1937).

Cedar Rapids
Bailey (1912) reported a September hawk flight at Cedar Rapids containing 3,000 Northern Harriers, Cooper's Hawks, Broad-winged Hawks, and Red-tailed Hawks moving west.

Cherokee
An early October flight of 49 Red-tailed Hawks was seen at Cherokee (Kleen, 1980a).

Columbus Junction
Dumont (1935) reported 400 migrating Broad-winged Hawks in late September along the Iowa River four miles north of Columbus Junction in Louisa County.

Davenport Area
Morrissey (1968) stated that Northern Harriers, Cooper's Hawks, Red-

tailed Hawks, and American Kestrels are the most common migrants in the Davenport area, and Petersen (1969) reported 46 Red-tailed Hawks there in late October.

Des Moines
October migrations of 400 Turkey Vultures and 200 Broad-winged Hawks are seen at Des Moines (Petersen, 1969). In early spring, Brown and Brown (1972) also reported Red-tailed Hawk migrations at Saylorville Dam north of Des Moines.

Dickinson County
An early October flight of 200 to 300 Swainson's Hawks was seen at two farms along the Little Sioux River in Dickinson County in western Iowa (Williams, 1941).

Dows
A mid-September migration of 38 Broad-winged Hawks was reported following the general course of the Iowa River on the edge of Dows in Wright County (Spiker, 1933).

Eastern Iowa
Dumont (1934a) reported that migrating Red-tailed Hawks noted in central and eastern Iowa contain approximately 83.3 percent eastern race birds and the rest of the sample (183 hawks) representative of Harlan's and Krider's races.

Red-tailed Hawks noted in eastern Iowa in spring contain Harlan's, Krider's, and western races of this species (Dumont, 1934b).

Fremont County
Daniel Varland reported a low-density autumn hawk migration (1.6 hawks per hour) at the loess bluffs northwest of Sidney in Fremont County. Red-tailed Hawks appear in largest numbers (Varland, 1982).

Iowa City
Autumn hawk migrations at Iowa City (85.8 hawks per day) occasionally contain as many as 50 Northern Harriers, 300 Broad-winged Hawks, and 40 Red-tailed Hawks (Mumford, 1959a, 1960, 1961a; Nolan, 1958a).

Keosauqua
A mid-September flight of 100 Turkey Vultures was reported at Keosauqua (Kleen, 1980a).

Liscomb
Moderate spring migrations of Red-tailed Hawks and a few other raptors are reported at Liscomb (Proescholdt, 1961).

Marble Rock
Mumford (1961a) reported a late September flight of 100 Broad-winged Hawks at Marble Rock.

Marshall County
Autumn flights of Ospreys and Broad-winged Hawks sometimes are re-

ported in Marshall County, such as at the Grammer Grove Wildlife Area southwest of Liscomb (Dorow and Kurtz, 1975).

Marshalltown

A mid-September 1875 hawk flight of about 1,000 birds (probably Broad-winged Hawks) was reported at Marshalltown (Anonymous, 1875).

Mississippi-Missouri Triangle

Youngworth (1935) reported autumn migrations of Sharp-shinned Hawks, Cooper's Hawks, Red-shouldered Hawks, Broad-winged Hawks, Merlins, and Peregrine Falcons in the triangle formed by the Mississippi and Missouri rivers.

Muscatine

A late April count of 230 Bald Eagles was reported at Muscatine (Kleen, 1979b).

Red Rock Refuge

Autumn hawk migrations at Red Rock Refuge northeast of Pleasantville occasionally contain more than 4,200 Turkey Vultures and a few other raptors in very small numbers (Black, 1969a, 1969b).

Sioux City

A September flight of 150 American Kestrels was reported at Sioux City, and occasionally a few Swainson's Hawks also are noted (Graber, 1962; Youngworth, 1960).

State Highway 30

A late October roadside raptor survey along 77 miles of State Highway 30 between Carroll and DeSoto National Wildlife Refuge produced 0.48 hawks per mile, the bulk of which were Red-tailed Hawks. Ages of these Red-tails were 35.8 percent adult and 64.2 percent immature (Bednarz, 1978).

Story County

Autumn hawk migrations in Story County contain Northern Harriers, Red-tailed Hawks, Rough-legged Hawks, and various other raptors; the same species are also reported during the spring migrations (Birkeland, 1934).

Sweet Marsh

A late April flight of 600 migrating Broad-winged Hawks was seen at Sweet Marsh (Kleen, 1980b).

Upper Missouri River Valley

A September flight of 75 Broad-winged Hawks and several other species was reported in the Upper Missouri River valley (Youngworth, 1959).

Waterloo

Petersen (1966a) reported a late September migration of 5,000 Broad-winged Hawks at Waterloo.

Western Iowa

Analysis of decades of Merlin sightings in western Iowa demonstrated that these falcons can migrate as early as 5 September but typically arrive in early October. Some Merlins winter in the state, but spring migrants generally arrive between 7 April and the end of the month (Youngworth, 1963).

Williamson

A mid-September flight of 1,000 Broad-winged Hawks was reported at Williamson (Petersen, 1970a).

KANSAS

A variety of Kansas hawk migrations, especially Broad-winged Hawks and Swainson's Hawks, are reported in the literature. Some of the most interesting flights are summarized here.

Arkansas River Area

A winter 1957–1962 eagle survey along roads within about 25 miles of the Arkansas River demonstrated that Bald Eagles and Golden Eagles appear in largest numbers in December as a result of late autumn migrations. The six-year survey produced 72 Bald Eagles and 88 Golden Eagles in December (Stephens, 1966).

Eastern Kansas

April roadside raptor surveys (devoted especially to Red-tailed Hawks) produced 0.04 hawks per mile. Observed were 10 Northern Harriers, 95 Red-tailed Hawks, and 28 American Kestrels (Fitch, Stephens, and Bare, 1973).

Hutchinson

Allan D. Cruickshank observed an early October migration of 3,400 Swainson's Hawks heading southward over Hutchison during only 1.5 hours of observation. Additional uncounted hawks continued to migrate past the site well into the afternoon, the birds rising aloft to become mere specks after eleven o'clock in the morning (Cruickshank, 1937).

Kansas Plains

A late September flight of 2,000 Broad-winged Hawks was reported on the Kansas plains while Albrecht (1922) was on a train going across the state.

MISSOURI

Organized interest in Missouri hawk migrations began in 1951 (Moore, 1951), and various short-term observations since then are reported in the literature.

Columbia

Autumn hawk migrations at Columbia (151 hawks per day) contain Turkey Vultures, Broad-winged Hawks, Red-tailed Hawks, and American Kestrels in largest numbers, but various other species also are reported in small numbers (Graber, 1962; Rathert, 1977; Wilhelm, 1951).

A late April flight of 120 Broad-winged Hawks also was reported at Columbia (Kleen, 1978b).

Saint Louis

Occasionally during September as many as 300 migrating Broad-winged Hawks and a few Swainson's Hawks are reported at St. Louis (Kleen, 1974; Mumford, 1959a; Petersen, 1971).

Swan Lake National Wildlife Refuge

A field study of color-marked and/or radio-tagged Bald Eagles observed at this refuge demonstrates that the birds originate on breeding grounds in Ohio, Michigan, Wisconsin, Minnesota, Ontario, and Saskatchewan. Some of the birds migrate through Manitoba and Ontario (Griffin, Southern, and Frenzel, 1980).

Table Rock Park

A mid-October flight of 150 Turkey Vultures was reported at this park (Kleen and Bush, 1973a).

NEBRASKA

Various hawk migration reports are published in the ornithological literature (Cary, 1899; Mathisen and Mathisen, 1959; Rapp, 1954). Johnsgard (1980) reported that September to November are the most important autumn migration months for birds of prey, and March to May are the most important spring raptor migration months.

Adams County

An early October migration of 300 Swainson's Hawks was reported in Adams County and various other raptors also are reported as migrants in small numbers (Turner, 1934b, 1944).

Ainsworth

Glandon (1935) reported a late April concentration of at least 500 Swainson's Hawks five miles southeast of Ainsworth in Brown County. The birds were on the ground.

Albion Area

A late September flight of as many as 800 Swainson's Hawks was seen a few miles southwest of Albion (Mollhoff, 1979).

Blue Springs

An "immense" flight of Swainson's Hawks was reported in early October at Blue Springs in Gage County, and other raptors including Golden Eagles also appear there occasionally in late November (Patton, 1940, 1951).

Central City

An early October flock of 1,000 Swainson's Hawks roosted in a grove north of Central City in Merrick County, and 225 of the birds were shot by farmers who considered them "chicken hawks" (Swenk, 1939).

Chadron
Gates (1976) reported a premigratory concentration of 60 Turkey Vultures in early September at Chadron.

Cunningham Lake
An early October flight of 250 Swainson's Hawks was reported at Cunningham Lake in Douglas County (Otto and Otto, 1984).

Ewing Area
Manning (1979) reported a late September flight of 350 Swainson's Hawks near Ewing.

Fairbury
Callaway and Callaway (1940) reported an early October group of more than 200 Swainson's Hawks roosting in trees at Fairbury, but mentioned that the birds were disturbed by gunners.

Fort Niobrara National Wildlife Refuge
A late September migration of 500 Swainson's Hawks was reported at this north-central Nebraska wildlife refuge (Menzel, 1974).

Hastings
Several hundred Swainson's Hawks were reported in early October in a field between Hastings and Fremont (Jones, 1939).

Hyannis
Williams (1981) reported that 200 Swainson's Hawks were seen in a field at Hyannis in late September.

Indian Cave State Park
Three Peregrine Falcons were seen in early April at Indian Cave State Park east of Schubert near the Missouri River (O'Connor, 1977).

Lincoln
Occasionally during autumn as many as 250 Swainson's Hawks, and a few other raptors in small numbers, are reported at Lincoln (Cink, 1972; DiSilvestro, 1976; Fichter, 1939; Whelan, 1934).

Minden Area
Several Gyrfalcons were reported in the Minden area between mid-November and mid-December (Ohlander, 1976, 1980).

Missouri River Valley
Turkey Vultures migrate along the Missouri River valley during spring and autumn, especially in Douglas and Jefferson counties (Rapp, 1954).

Neligh
A late August flight of at least 400 Swainson's Hawks was reported along the Elkhorn River southwest of Neligh (Cary, 1899).

Omaha
An autumn hawk watch bordering the Missouri River produced Sharp-shinned Hawks, Broad-winged Hawks, Red-tailed Hawks, and various other raptors in limited numbers (Bray, 1984).

O'Neill
A flight of 300 Swainson's Hawks was reported in late September at O'Neill (Williams, 1980).

Panhandle Area
A 1957 roadside raptor survey covering 11 counties and 14,000 square miles in the Panhandle area of Nebraska, and 17,807 miles of travel, produced 1.2 hawks per hour of observation. Northern Harriers, Rough-legged Hawks, Golden Eagles, American Kestrels, and Peregrine Falcons were very well represented but various other birds of prey also were seen. September was the peak autumn migration period for Swainson's Hawks, Red-tailed Hawks, Ferruginous Hawks, American Kestrels, and Peregrine Falcons, whereas October was the peak migration period for Northern Harriers. In spring, April was the peak migration period for Northern Harriers and American Kestrels. Rough-legged Hawks and Golden Eagles appeared in largest numbers in winter (Mathisen and Mathisen, 1959).

Valentine
A late September flight of 200 Turkey Vultures and 2,000 Swainson's Hawks and Rough-legged Hawks was reported at the Valentine Fish Hatchery where the birds roosted for the night and departed the next day (Menzel, 1974).

Webster County
Turner (1934a, 1947) reported that autumn hawk migrations in Webster County sometimes contained up to 40 Swainson's Hawks and a few other raptors, and further stated that Northern Harriers, Swainson's Hawks, Red-tailed Hawks, and American Kestrels are common autumn migrants.

York County
Occasionally as many as 15 Prairie Falcons are reported in December in York County (Morris, 1978).

NORTH DAKOTA

A spring hawk watch conducted by David and Sharon Lambeth at Grand Forks produced 11.2 hawks per hour. Sharp-shinned Hawks, Broad-winged Hawks, and Red-tailed Hawks appeared in largest numbers, but various other raptors also were seen in small numbers (Peacock and Myers, 1981).

OKLAHOMA

Various reports of hawk migrations in Oklahoma are published indicating that clear migration periods exist for Ospreys, Sharp-shinned Hawks, Broad-winged Hawks, Swainson's Hawks, Ferruginous Hawks, and Rough-legged Hawks. The main migration periods are March to May and September to November (Gould, 1921; Sutton, 1967; Wood and Schnell, 1984).

Ada

An early October migration of at least 300 Swainson's Hawks was reported eight miles northeast of Ada (Mays, 1969).

Anadarko

Thousands of migrating Swainson's Hawks were seen at Anadarko during October 1967 (Williams, 1968).

Banner

An early October concentration of between 100 and 150 Northern Harriers was reported in a field near Banner (Gould, 1921).

Claremore

In early October a flight of 200 Swainson's Hawks was reported at Claremore (Williams, 1970).

Oklahoma City

Occasionally during autumn as many as 132 Mississippi Kites and 500 Swainson's Hawks are reported at Oklahoma City (Williams, 1968, 1974), and a Gyrfalcon also was reported there in December 1982 (Grzybowski, 1983).

Stillwater

Occasional autumn hawk migrations at Stillwater include as many as 250 Mississippi Kites, 100 Swainson's Hawks, and a few other species in small numbers (Eubanks, 1971; Platt, 1974; Williams, 1970).

Tillman County

Williams (1974) reported a flight of 165 Swainson's Hawks in early October in this county.

Wichita Mountains Wildlife Refuge

An early October migration of 2,000 Swainson's Hawks was reported at this refuge (Baumgartner, 1955).

SOUTH DAKOTA

Various observations of hawk migrations are reported in South Dakota including a Red-tailed Hawk, banded in the state, that migrated 1,700 miles to central Mexico (Adolphson, 1974).

Clay County

An autumn hawk watch along Clay Creek ditch and the Vermillion River in Clay County produced only meager hawk migrations—0.04 hawks per hour of which American Kestrels were most frequently noted (Bartelt and Orde, 1976).

Northeastern South Dakota

September hawk migrations in northeastern South Dakota consist of Northern Harriers, Red-tailed Hawks, and American Kestrels, plus a few other species in limited numbers.

April hawk flights consist mostly of Turkey Vultures and American Kestrels, although several other species occasionally are observed (Elliott, 1967).

Harris (1973) reported that the autumn 1972 hawk migration in this part of South Dakota was outstanding for Northern Harriers and Red-tailed Hawks, but even rarer species such as Northern Goshawks, Prairie Falcons, and Gyrfalcons were reported in larger-than-normal numbers.

Sioux Falls

Several Broad-winged Hawks were seen in late April 1953 in Woodlawn Cemetery in Sioux Falls (Chapman, 1953).

CHAPTER 10

Western States Hawk Migrations

Until recently little organized interest in hawk migrations existed in the 12 western states—Alaska, Arizona, California, Colorado, Idaho, Montana, Nevada, New Mexico, Oregon, Utah, Washington, and Wyoming. Recently, however, some efforts began to identify western hawk watching lookouts and conduct roadside raptor surveys in order to secure basic information against which future western states hawk migration studies can be compared and evaluated (Hoffman, 1981a–1984b; MacRae, 1983, 1984; Spofford, 1978, 1979, 1980; Tilly, 1981, 1983a, 1983b, 1984).

ALASKA

Various authors discussed hawk migrations in general terms in Alaska, often with particular attention to Bald Eagle migrations, but our knowledge of raptor movements in this state remains meager (Gabrielson and Lincoln, 1959; Heintzelman, 1979a; Robards and Taylor, no date). In general, Mindell and Mindell (1984) state that harriers and accipiters tend to use coastal migration routes, whereas buteos and Golden Eagles prefer inter-mountain routes.

Aleutian Islands
A variety of rare North American, or vagrant Eurasian, raptors are reported in the Aleutian Islands. Included are White-tailed Eagles, Stellar's Sea Eagles, a Common Buzzard, and Eurasian Kestrels (American Ornithologists' Union, 1983; Gibson, 1979, 1981, 1983, 1984; Roberson, 1980).

Chilkat River

Up to 3,000 Bald Eagles gather along the Chilkat River near Haines from October through December to feed on salmon (Heintzelman, 1979a: 237). In 1982, Alaska established the Chilkat Bald Eagle Preserve to protect the essential habitat used by the birds (Rymer, 1985).

Copper River Delta

Autumn hawk migrations at the Copper River delta include as many as 1,000 Bald Eagles and small numbers of other raptors including Ospreys, Sharp-shinned Hawks, American Kestrels, and Peregrine Falcons (Gibson, 1969, 1970, 1971).

Denali Highway

Smith (1965, 1966) reported that various raptors occasionally migrate along the Denali Highway during autumn including Northern Harriers, Bald Eagles, Red-tailed Hawks, Rough-legged Hawks, American Kestrels, and Gyrfalcons.

Fairbanks

Occasionally during autumn Northern Goshawks and American Kestrels are reported as migrants in the Fairbanks area (Gibson, 1969, 1972).

Glenn Highway

Red-tailed Hawks are regular late September and early October migrants along the Glenn Highway between King and Sheep mountains in south-central Alaska (Mindell and Mindell, 1984).

Matanuska River

A mid-October low-density migration (9 hawks per day) is reported along the Matanuska River north of Anchorage. Sharp-shinned Hawks, Red-tailed Hawks, Rough-legged Hawks, and Golden Eagles are observed (Smith, 1965).

North Gulf Coast Area

Spring hawk migrations in the North Gulf Coast area include Ospreys, Northern Harriers, Sharp-shinned Hawks, Rough-legged Hawks, Peregrine Falcons, and several other raptors in limited numbers (Gibson and Byrd, 1975).

Prince William Sound

A mid-April influx of Bald Eagles to this site near Anchorage is reported (Heintzelman, 1979a).

Seymour Eagle Management Area

This section of Admiralty Island supports two Bald Eagle nests per mile and doubtless has some migratory movements of these birds (Heintzelman, 1979a).

Sitkagi Beach

The spring hawk migrations reported at Sitkagi Beach, west of Yakutat Bay along the Gulf of Alaska, consists of Ospreys, Northern Harriers, Sharp-shinned Hawks, Rough-legged Hawks, Merlins, and Peregrine Falcons, plus limited numbers of various other species (Mindell and Mindell, 1984).

Autumn hawk migrations at this site (1.9 hawks per hour) contain Ospreys, Northern Harriers, Sharp-shinned Hawks, American Kestrels, and Merlins, plus a few other raptors in small numbers (Tilly, 1984).

Stikine River
A mid-April Bald Eagle migration is reported in this area (Heintzelman, 1979a).

ARIZONA

The basic migratory status of Arizona's birds of prey is defined by Phillips, Marshall, and Monson (1964) and updated by Monson and Phillips (1981).

Various roadside raptor surveys also are conducted in the state and may provide a more productive index of hawk migrations than the use of lookouts. Autumn roadside raptor surveys produce a variety of species of which Black Vultures, Turkey Vultures, Northern Harriers, Swainson's Hawks, Red-tailed Hawks, and American Kestrels are the most numerous, but other species also are noted in limited numbers (Spofford, 1978, 1979).

Spring hawk migrations, as measured by roadside raptor surveys, produce Turkey Vultures, Swainson's Hawks, Red-tailed Hawks, and American Kestrels in largest numbers, but other species are seen as well in small numbers (Spofford, 1980).

Castle Dome Mountains
An early November flight of 41 migrating Turkey Vultures was reported over the desert southwest of the Castle Dome Mountains in Yuma County (Monson, 1956)

Dragoon Mountains
Spring hawk migrations studied by Steve Hoffman at the southern end of the Dragoon Mountains produced a passage rate of 5 hawks per hour. Turkey Vultures, Sharp-shinned Hawks, Cooper's Hawks, Red-tailed Hawks, Golden Eagles, and American Kestrels occurred in the largest numbers, but other species also are seen in limited numbers (Hoffman, 1981a, 1982a).

Autumn hawk migrations (1.6 hawks per hour) contain mostly Red-tailed Hawks, but several other raptors also are seen in lesser numbers (Hoffman, 1981b).

Graham (Pinaleno) Mountains
The autumn hawk migrations at the Graham Mountains in southeastern Arizona (6.5 hawks per hour) consist of Turkey Vultures and Red-tailed Hawks, plus a few other species in limited numbers (Hoffman, 1981b).

Huachuca Mountains
Spring hawk migrations (1.6 hawks per hour) at these mountains contain

Turkey Vultures in largest numbers and occasional views of various other species (Hoffman, 1982a).

Organ Pipe Cactus National Monument

An early October migration of 400 Turkey Vultures was reported at this site (Snider, 1969).

Phoenix Area

In the Phoenix area Parker (1974) reported a late September flight of 68 Swainson's Hawks.

Roll

An early October migration of 200 Turkey Vultures was seen at Roll (Mills, Alden, and Hubbard, 1975).

Signal Peak

Autumn hawk flights (2.3 hawks per hour) reported at Signal Peak contain Turkey Vultures and much smaller numbers of other raptors including Cooper's Hawks and Golden Eagles (Hoffman, 1982b).

Wellton

A late September flight of 20 American Kestrels was reported at Wellton (Mills, Alden, and Hubbard, 1975).

Willcox

Occasionally as many as 250 Swainson's Hawks are reported in late September at Willcox (Mills, Alden, and Hubbard, 1975).

CALIFORNIA

Hawk migrations in California received limited study (Small, 1974) with the exception of Binford's (1979) detailed autumn study at Point Diablo near San Francisco. Wilbur (1978b) also demonstrated that California Condors are separated into two subpopulations which engage in limited intrastate migratory movements.

Chico

A late September migration of 1,000 Turkey Vultures was reported at Chico (DeSante and Remsen, 1973).

Edminston Pump Station

A late October flight of 1,000 Swainson's Hawks was reported at the Edminston Pump Station at the southern end of the San Joaquin Valley (McCaskie, 1980a).

Farallon Islands

A concentration of 12 Rough-legged Hawks was reported in late October 1973 on the Farallon Islands 30 miles west of San Francisco (Remsen and Gaines, 1974).

Grapevine Area

A mid-October migration of 170 Swainson's Hawks was seen in the Grapevine area of Kern County (McCaskie, 1983).

Gray Lodge
Occasionally during October as many as 100 migrating Turkey Vultures are reported at Gray Lodge (Stallcup, DeSante, and Greenberg, 1975).

Imperial Valley
Small (1956) reported 1,000 migrating Turkey Vultures in late February in the Imperial Valley.

Lancaster
At least 2,000 migrating Turkey Vultures were reported in early March at Lancaster (McCaskie, 1980b).

O'Neals Area
Occasionally as many as 3,000 migrating Turkey Vultures are seen in the O'Neals area (Pray, 1953).

Pacific Grove
A late November migration of 17 Broad-winged Hawks was reported at Pacific Grove—a remarkable date and number of birds for California (LeValley and Roberson, 1983).

Point Diablo
Point Diablo is located on the headlands part of the Golden Gate National Recreation Area in Marion County overlooking the mouth of San Francisco Bay. Binford's (1979) autumn hawk watch there produced good hawk flights (33.1 hawks per hour). The species noted in largest numbers include Turkey Vultures, Sharp-shinned Hawks, Cooper's Hawks, Red-tailed Hawks, and American Kestrels, although a variety of other raptors are noted in lesser numbers. Presumably hawks seen at Point Diablo arrive from the northwestern North American coastal belt and some also from the Great Basin and Great Plains.

Spring hawk flights at this location contain various species of which Turkey Vultures, Sharp-shinned Hawks, Cooper's Hawks, and Red-tailed Hawks are the most numerous, but other raptors also appear in small numbers (Remsen and Gaines, 1973).

Point Reyes Peninsula
Autumn hawk migrations on the Point Reyes peninsula (34.6 hawks per day) sometimes contain Sharp-shinned Hawks, Red-tailed Hawks, and Rough-legged Hawks (Chandik and Baldridge, 1968; Laymon and Shuford, 1980; Mans, 1963; Remsen and Gaines, 1974).

Riverdale
A mid-October migration of 260 Swainson's Hawks was reported at Riverdale (Laymon and Shuford, 1980).

Sacramento Area
Occasionally during October as many as 500 Turkey Vultures are reported migrating past the Sacramento area (Chandik and Baldridge, 1969; Cogswell and Pray, 1955).

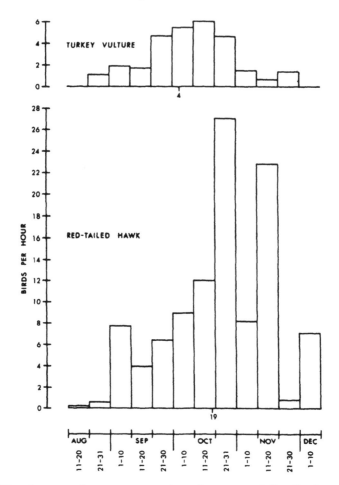

The charts on these two pages show the temporal distribution of migrating raptors observed at Point Diablo, Calif., during 262.6 hours of observation in the autumns of 1972–1977. *Reprinted from Binford (1979).*

Salmon Mountains
A late September flight of 600 Turkey Vultures was reported in the Salmon Mountains (Winter and Laymon, 1979).

Silverdale
In early October a migration of 300 Turkey Vultures was reported at Silverdale (Hunn and Mattocks, 1981).

Springville
Chase and Paxton (1966) reported a late September migration of 3,000 Turkey Vultures at Springville in the Central Valley of Tulare County.

Temecula
In late October a flight of 200 Swainson's Hawks was seen at Temecula in Riverside County (McCaskie, 1980a).
Tracy Area
In early September a flight of 100 Swainson's Hawks was reported 10 miles south of Tracy (Pray, 1953).
Woodland Area
Occasionally during autumn as many as 100 Turkey Vultures and 100

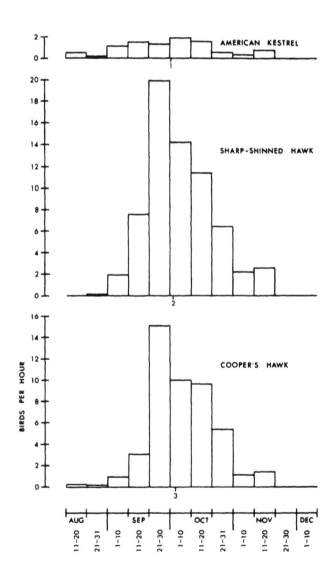

Swainson's Hawks are observed in the Woodland area of Yolo County (Chandik and Baldridge, 1968, 1969; Cogswell, 1956).

Yolo County

During September as many as 200 Turkey Vultures are reported as migrants in Yolo County (Cutler and Pugh, 1960).

COLORADO

Bailey and Niedrach (1965) provided a broad summary of the state's hawk migrations, and several recent hawk watches at Boulder and Dakota Ridge further identified spring migration flight-lines.

Boulder

Freeman Hall's spring hawk watch at three Boulder locations (Bluebell, Flagstaff Mountain, and Table Mountain) indicate a rate of passage of migrants of 16.7 hawks per hour. The species reported in largest numbers are Turkey Vultures, Sharp-shinned Hawks, Cooper's Hawks, Red-tailed Hawks, Golden Eagles, and American Kestrels, but other raptors also are seen in more limited numbers (Hoffman, 1981c).

Dakota Ridge

Spring hawk migrations at Dakota Ridge (17.25 hawks per hour) include Sharp-shinned Hawks, Cooper's Hawks, Golden Eagles, and American Kestrels as well as a few other species in limited numbers (Hoffman, 1982a).

IDAHO

Hawk migrations in Idaho received limited study (Burleigh, 1972; Melquist, Johnson, and Carrier, 1978; Steenhof, Kochert, and Moritsch, 1984).

Northern Idaho

Recoveries of Ospreys banded as nestlings in northern Idaho and eastern Washington demonstrate that these birds leave the natal areas during September and early October, move south rapidly, and winter along the Pacific coasts of El Salvador, Honduras, and Costa Rica—an area not used by Ospreys from the Midwest or East Coast populations. Adult Ospreys then return to Idaho in late March and early April, whereas younger birds remain on the wintering grounds until they are two years old (Melquist, Johnson, and Carrier, 1978).

Snake River Birds of Prey Natural Area

The migrations of Red-tailed Hawks, Golden Eagles, and Prairie Falcons banded and color-marked as nestlings in the Snake River Birds of Prey Natural Area near Boise demonstrate considerable difference in dispersal patterns.

Red-tailed Hawks migrate southwestward—some well into the Pacific coastal lowlands of Mexico and Central America, one hawk moving 2,628 miles to Guatemala. These hawks begin their migrations in late autumn.

Although 78 percent of the Golden Eagles remain within 62.5 miles of the natal locations, those that migrate longer distances disperse in all directions. One eagle flew 878 miles to eastern New Mexico, but most go to Great Basin shrub-steppe habitats because of feeding ecology requirements.

Prairie Falcons migrate eastward following prey movements. More than one-half move east of the Continental Divide, the maximum distance covered being 1,356 miles (Steenhof, Kochert, and Moritsch, 1984).

MONTANA

Little is known about hawk migrations in Montana except for the pioneering field studies summarized here.

Bridger Mountains

Autumn hawk migrations in the Bridger Mountains northeast of Bozeman (6.7 hawks per hour) consist of various species the bulk of which are Sharp-shinned Hawks, Cooper's Hawks, Red-tailed Hawks, and Golden Eagles, although other raptors also are seen in limited numbers (Tilly, 1980, 1981b, 1983b).

Glacier National Park

Annually between late September and mid-December hundreds of Bald Eagles migrate into Glacier National Park to feed on Kokanee Salmon (Heintzelman, 1979a: 249–250).

Mission Valley

Autumn hawk migrations in the Mission Valley east of Arlee (8 hawks per hour) contain Sharp-shinned Hawks and Golden Eagles in largest numbers, but various other species also are noted in limited numbers (Tilly, 1983b).

Shields Valley

An autumn roadside raptor count in the Shields Valley north of Wilsall produced 33.7 hawks per day. The most numerous species are Bald Eagles and Golden Eagles, but a variety of other birds of prey also are seen in small numbers (Tilly, 1983a)

NEVADA

Information on Nevada hawk migrations is very limited, but several recent field studies provide some important insights into the magnitude of hawk flights in this state.

Goshute Mountains
Autumn hawk migrations (15.8 hawks per hour) reported at the Goshute
Mountains in Nevada, 25 miles south-southwest of Wendover, Utah, con-
tain various species of which the most numerous are Turkey Vultures,
Northern Harriers, Sharp-shinned Hawks, Cooper's Hawks, Red-tailed
Hawks, Golden Eagles, and American Kestrels (Hoffman, 1981b, 1982b,
1984b).

Kingsley Lookout
The autumn hawk migrations reported at the Kingsley Lookout, a northeastern Nevada site at the southern end of a narrow, isolated ridge 20 miles
south-southwest of the Goshute Mountains, exhibit a passage rate of 19.1
hawks per hour. Sharp-shinned Hawks, Cooper's Hawks, and American
Kestrels are seen in largest numbers (Hoffman, 1981b).

Potosi Mountain
Potosi Mountain, the most southerly peak in the Spring Mountains eight
miles north-northwest of Goodsprings, is located on a fine Sharp-shinned
Hawk and Cooper's Hawk flight line. Highest counts of accipiters occur on
days when cold fronts pass the site, even when no obvious changes in
surface weather variables are noted. Apparently, increased hawk counts
result from more birds moving within visible range, more birds using
deflective updrafts instead of thermals, and reluctance of accipiters to cross
deserts, thus resulting in the hawks moving across boreal forests along
ridgetops (Millsap and Zook, 1983).

NEW MEXICO

It is only since Hubbard (1977) conducted a pioneering hawk watch in 1976
at Sierra Grande that organized interest in hawk migrations developed in
New Mexico. Several additional sites now are used as lookouts resulting in
some basic knowledge of New Mexico hawk migrations.

Bosque del Apache National Wildlife Refuge
An early April concentration of 40 Ferruginous Hawks was reported at
Bosque del Apache National Wildlife Refuge (Monson, 1973).

Clines Corner
A late September flight of 100 Turkey Vultures was reported at Clines
Corner (Monson, 1958).

Fort Bayard
At least 100 Turkey Vultures were seen in mid-September at Fort Bayard in
Grant County (Monson, 1972).

Manzanos Mountains
Autumn hawk counts made by David Ponton and Kris Frame at the Manzanos Mountains—a large range 35 miles south-southwest of Albuquerque,

produced a passage rate of 9.1 hawks per hour. The most commonly observed species are Turkey Vultures, Sharp-shinned Hawks, Cooper's Hawks, Red-tailed Hawks, Golden Eagles, and American Kestrels (Hoffman, 1981b, 1982b, 1983b, 1984b).

Spring hawk flights (10.3 hawks per hour) contain Turkey Vultures, Sharp-shinned Hawks, and Cooper's Hawks in largest numbers (Hoffman, 1983a).

Organs
Steve Hoffman's October hawk watch at San Augustine Peak in the Organs Range east of Las Cruces (3.2 hawks per hour) produced Sharp-shinned Hawks in largest numbers, although various other raptors also are reported in limited numbers (Hoffman, 1981b, 1982b).

San Augustine Plains
A late September flight of 300 Turkey Vultures was reported on the San Augustine plains in Catron County (Monson, 1958).

Sandia Mountains
Spring hawk migrations (6.2 hawks per hour) at Sandia Crest at the southern end of the Sandia Mountains east of Albuquerque contain various species of which Turkey Vultures, Sharp-shinned Hawks, Cooper's Hawks, Golden Eagles, and American Kestrels are the most numerous (Hoffman, 1984a).

Sierra Grande
This northeastern New Mexico site is a volcanic shield cone reaching an elevation of 8,732 feet. John Hubbard's autumn hawk watch at this site produced 90.5 hawks per day the bulk of which are Turkey Vultures, Sharp-shinned Hawks, Red-tailed Hawks, and American Kestrels, although other raptors also appear in limited numbers (Hoffman, 1981b; Hubbard, 1977).

OREGON

Some hawk migration information is available from Oregon (Gabrielson and Jewett, 1970; Leopold, 1942; Littlefield, Thompson, and Ehlers, 1984). More field study, however, is needed in this area.

Ashland
An early October flight of 70 Turkey Vultures was reported at Ashland (Hunn and Mattocks, 1979).

Baskett Slough National Wildlife Refuge
An early November migration of 40 Red-tailed Hawks and 12 Rough-legged Hawks was reported at this refuge (Hunn and Mattocks, 1979).

Brookings Area
As many as 80 Turkey Vultures are seen in early September in the Brookings area (Crowell and Nehls, 1969).

Corvallis

An early September migration of 100 Turkey Vultures was seen at Corvallis (Crowell and Nehls, 1973).

Malheur National Wildlife Refuge

The numbers of breeding and migrating Swainson's Hawks at this refuge declined sharply since the late 1950s according to comparisons of historic and recent raptor survey data (Littlefield, Thompson, and Ehlers, 1984). Leopold (1942), for example, reported 37 Swainson's Hawks on a 67 kilometer roadside raptor survey in the Blitzen Valley in August 1941 (0.55 hawks per kilometer), whereas August raptor surveys in the same area between 1975–1983 average only 0.056 Swainson's Hawks per kilometer (Littlefield, Thompson, and Ehlers, 1984). There is no satisfactory explanation for the lower hawk densities.

Mount McLoughlin

An early September hawk watch by Craig Roberts at the 8,000 foot level of Mount McLoughlin south of Crater Lake National Park produced 8 hawks per day. Species observed included Northern Harrier, Sharp-shinned Hawk, Cooper's Hawk, Red-tailed Hawk, American Kestrel, and Prairie Falcon (Crowell and Nehls, 1975).

Mount Scott

Pettingill (1981) states that good late August and September hawk flights pass Mount Scott in Crater Lake National Park near Medford.

Rogue River Valley

During autumn, various hawk migrations are reported in the Rogue River Valley in southern Oregon. Some flights contain as many as 160 Turkey Vultures and 250 Red-tailed Hawks (Crowell and Nehls, 1968, 1973).

UTAH

More field study is devoted to Utah's hawk migrations than in most other western states. Thus the basic framework of autumn hawk migrations is documented in the state.

Bountiful Peak

Autumn hawk migrations (3.6 hawks per hour) at Bountiful Peak, along the Wasatch Plateau, include in largest numbers Northern Harriers, Sharp-shinned Hawks, Cooper's Hawks, Red-tailed Hawks, and American Kestrels, but other raptors also appeared in small numbers (Hoffman, 1982b).

Brighton

This ridge lookout, east of Salt Lake City, is part of the Wasatch Range. The autumn hawk migrations there (7.1 hawks per hour) consist of Sharp-shinned Hawks, Cooper's Hawks, Red-tailed Hawks, Golden Eagles, and American Kestrels in largest numbers, plus small numbers of various other species (Hoffman, 1981b, 1982b).

Brown's Peak
The autumn hawk flights (21 hawks per hour) reported at Brown's Peak consist of Sharp-shinned Hawks, Red-tailed Hawks, and American Kestrels in larger numbers, but a few other species also are noted (Hoffman, 1982b).

Camel Pass
Autumn hawk migrations at Camel Pass (5.2 hawks per hour) are formed by Sharp-shinned Hawks, Red-tailed Hawks, and Golden Eagles, plus a few sightings of other raptors (Hoffman, 1982b).

Clarkston Mountains
The autumn migrations of hawks reported at this site exhibited a passage rate of 12.6 hawks per hour. The most numerous species are Northern Harriers, Sharp-shinned Hawks, Cooper's Hawks, Red-tailed Hawks, and American Kestrels, although other raptors also appear in smaller numbers (Hoffman, 1982b, 1983b).

Cutler Dam
Cutler Dam, in northern Utah, produces autumn hawk flights (7.5 hawks per hour) containing Northern Harriers, Sharp-shinned Hawks, Cooper's Hawks, Red-tailed Hawks, and American Kestrels, in addition to a few other birds of prey in small numbers (Hoffman, 1984b).

Della Ridge
Spring hawk flights at Della Ridge (14 hawks per hour) include Sharp-shinned Hawks, Cooper's Hawks, and Red-tailed Hawks as well as other raptors in small numbers (Hoffman, 1982a).

Diamond Peak
Diamond Peak autumn hawk migrations (41 hawks per hour) contain Turkey Vultures, Sharp-shinned Hawks, and Red-tailed Hawks in the largest numbers (Hoffman, 1982b).

Gunsight Peak
Fair autumn hawk flights are reported at Gunsight Peak west of Clarkston (Heintzelman, 1979a: 212–213).

Hyde Park Knoll
Fair autumn and spring hawk migrations are reported at Hyde Park Knoll on the western slope of the Bear River Range near Hyde Park (Heintzelman, 1979a: 213).

Johnson's Pass
Spring hawk migrations (4.3 hawks per hour) reported at this site include Northern Harriers and Red-tailed Hawks plus various other species in small numbers (Hoffman, 1982a).

Lakeside Mountains
Spring hawk migrations at the Lakeside Mountains (6.9 hawks per hour) contain various species of which the most numerous are Turkey Vultures,

Bald Eagles, Northern Harriers, Cooper's Hawks, Red-tailed Hawks, Rough-legged Hawks, and Golden Eagles (Hoffman, 1982a, 1983a).

Manti Mountain

The autumn hawk flights reported at Manti Mountain (33.6 hawks per hour) contain Red-tailed Hawks and Golden Eagles in largest numbers, but other raptors also are seen there (Hoffman, 1982b).

Needles

In autumn, the hawk migrations at Needles (5.5 hawks per hour) include in largest numbers Sharp-shinned Hawks, Cooper's Hawks, Red-tailed Hawks, and American Kestrels (Hoffman, 1982b).

North Oquirrh

A spring hawk watch at this location (2.7 hawks per hour) produced a Northern Harrier, Golden Eagles, and various unidentified birds (Hoffman, 1982a).

Ophir

The spring hawk migrations reported at Ophir (7.6 hawks per hour) include a variety of species of which the most numerous are Bald Eagles, Red-tailed Hawks, Rough-legged Hawks, Golden Eagles, and American Kestrels (Hoffman, 1982a, 1983a).

Peoa

A spring hawk watch at Peoa produced 12.8 hawks per hour with the bulk of the birds being Red-tailed Hawks and Golden Eagles, although other raptors were seen in small numbers (Hoffman, 1982a).

Pioneer Trail State Park

An autumn hawk watch at this park produced 3.6 hawks per hour. Sharp-shinned Hawks, Cooper's Hawks, and Red-tailed Hawks are among the raptors observed (Hoffman, 1982b).

Promontory Point

Fair autumn accipiter migrations are reported at Promontory Point in the Southern Promontory Mountains near Brigham City (Heintzelman, 1979a: 213–214).

Rock Canyon

Hoffman (1982b) reported 27 hawks per hour at Rock Canyon in autumn. Sharp-shinned Hawks appear in largest numbers, but other species also are noted.

Rockport

Spring hawk migrations at Rockport (9.8 hawks per hour) include a variety of species of which the most numerous are Turkey Vultures, Bald Eagles, Red-tailed Hawks, and Golden Eagles (Hoffman, 1983a, 1984a).

Scott's Peak

This peak near Salt Lake City produces autumn hawk flights at the rate of 7.4 hawks per hour. Northern Harriers, Sharp-shinned Hawks, Cooper's

Hawks, Red-tailed Hawks, Golden Eagles, and American Kestrels are the most numerous species reported (Hoffman, 1983b, 1984b).

Sheep Ridge
An autumn hawk watch at Sheep Ridge produced 2 hawks per hour with a Sharp-shinned Hawk, Red-tailed Hawk, and Golden Eagle in the flight (Hoffman, 1982b).

Slate Canyon
The spring hawk migrations reported at Slate Canyon at the base of the Wasatch Front in Provo exhibit a passage rate of 11 hawks per hour. Bald Eagles, Red-tailed Hawks, and Golden Eagles are the species noted in largest numbers, but other raptors also are seen (Hoffman, 1984a).

Tintics
Autumn hawk flights in the Tintics, mountains in central Utah southwest of Provo, are modest (5.5 hawks per hour). Cooper's Hawks, Red-tailed Hawks, Golden Eagles, and American Kestrels appear in largest numbers (Hoffman, 1983b).

Uinta National Forest
Autumn hawk migrations at the Uinta National Forest at the Wasatch Front forming the eastern border of the Great Basin near Provo (8.5 hawks per hour) contain mostly Sharp-shinned Hawks, Cooper's Hawks, Red-tailed Hawks, Golden Eagles, and American Kestrels (Mosher, et al., 1978).

Utah State University
Mike Tove watched autumn hawk flights from the campus of Utah State University at Logan (89.5 hawks per hour) and reported Turkey Vultures, Sharp-shinned Hawks, Red-tailed Hawks, and American Kestrels in largest numbers (Hoffman, 1983b).

Utah Valley Hawk Migration Lookouts
Fair autumn hawk migrations are reported at a knoll and a scenic roadside lookout near Orem (Heintzelman, 1979a: 214).

Wellesville Mountain Hawk Lookout
The Wellesville Mountains are located four miles west of Mendon and produce autumn and spring hawk flights (Heintzelman, 1979a: 214–215).

The autumn hawk migrations (19.5 hawks per hour) contain Northern Harriers, Sharp-shinned Hawks, Cooper's Hawks, Red-tailed Hawks, Golden Eagles, and American Kestrels in largest numbers (Hoffman, 1981b, 1982b, 1983b, 1984b).

Spring hawk migrations (22 hawks per hour) contain Golden Eagles in largest numbers; other species also are reported, although the coverage is very fragmentary (Hoffman, 1982a).

Willard
Spring hawk flights at Willard (10.5 hawks per hour) include Sharp-shinned Hawks and American Kestrels in largest numbers, but several other raptors also are observed (Hoffman, 1982a).

WASHINGTON

Current knowledge of hawk migrations in Washington is fragmentary for many parts of the state, although some more detailed studies were made of the migrations and wintering activities of eagles along rivers.

Dungeness

An early September flight of 50 Turkey Vultures was reported at Dungeness (Hunn and Mattocks, 1979).

Glacier Pass

Richard Knight reported some September migrations of Northern Harriers, American Kestrels, and other species at Glacier Pass in the Brush Creek drainage of Okanogan County (Tilly, 1981b).

Harts Pass

Richard Knight's observations of Northern Harriers, Red-tailed Hawks, and American Kestrels in late September at Harts Pass in Whatcom County suggest a flight-line there (Tilly, 1981b).

Methow Pass

Brief observations by Richard Knight at Methow Pass in Skagit County in late September suggest a flight-line for Red-tailed Hawks, American Kestrels, and Merlins (Tilly, 1981b).

North Cascades National Park Service Complex

A late October and mid-March migration of Bald Eagles is reported at the Skagit River section of the North Cascades area (Heintzelman, 1979a).

Olympic National Park

A Bald Eagle migration into Olympic National Park is reported in mid-November to late February (Heintzelman, 1979a).

Panhandle Lake

A Broad-winged Hawk was reported on 5 August 1976 at Panhandle Lake west-southwest of Newport—the first state record (Larrison, 1977).

Rocky Pass

Richard Knight's September hawk watch at Rocky Pass, Three Fool's drainage in the Cascade Mountains, suggests a migration flight-line for Northern Harriers, Red-tailed Hawks, Golden Eagles, and American Kestrels (Tilly, 1981b).

Rufus Woods Lake

Rufus Woods Lake, between Chief Joseph and Grand Coulee dams, experiences a Bald Eagle migration toward the end of October and in April with peak eagle numbers appearing in February. Some Golden Eagles also migrate to the lake and winter there (Knight, et al., 1979).

San Juan Island National Historic Park

From mid-November through late February, Bald Eagles migrate to San Juan Island to winter there (Heintzelman, 1979a: 257–258).

Seattle

Occasionally during late September and early October as many as 40

Turkey Vultures are reported migrating past Seattle (Crowell and Nehls, 1972; Eddy, 1948).

Skagit River Bald Eagle Natural Area

Between late October and mid-March, Bald Eagles migrate to the Skagit River between Rockport and Marblemount where they receive protection in the Skagit River Bald Eagle Natural Area (Heintzelman, 1979a: 258).

WYOMING

Little hawk migration information is available from Wyoming, but good autumn hawk flights are reported at Indian Medicine Wheel near Lovell (Heintzelman, 1979a: 228–229).

CHAPTER 11

Central American Hawk Migrations

Some hawks from North America migrate across Central America to reach their wintering grounds there or in South America. Thus Central America plays an important role in the migrations of many hawks from North America—especially Broad-winged Hawks and Swainson's Hawks.

BELIZE

Detailed information on hawk migrations in Belize is unavailable, but Russell (1964: 44–54) summarized the status of birds of prey in that country and indicated that Ospreys, American Swallow-tailed Kites, Northern Harriers, Broad-winged Hawks, American Kestrels, Merlins, and Peregrine Falcons are migrants from North America.

A definite migration of falcons was reported between 7 April and 6 May 1958 on seven days at Half Moon Cay by J. Verner, who observed a maximum of 20 Merlins and 4 Peregrine Falcons on 16 April (Russell, 1964: 53–54).

COSTA RICA

Slud (1964) described the status of birds of prey in Costa Rica. Skutch (1945: 80–89), however, described spring Broad-winged Hawk and Swainson's Hawk migrations in the country in much more detail. He stated that enormous flights of Swainson's Hawks migrate northward in March and April across the Pacific drainage of southern Costa Rica, whereas much smaller numbers of migrating Broad-winged Hawks also appear during the same months. Autumn migrations of Broad-winged Hawks and Swainson's Hawks, however, almost never appear along the Pacific drainage but may pass over the Caribbean lowlands in autumn.

EL SALVADOR

Dickey and Van Rossem (1938: 111) stated that Broad-winged Hawks were common autumn and spring migrants in El Salvador. The first birds they noted were scattered individuals among a flight of Turkey Vultures, Swainson's Hawks, and Red-tailed Hawks over Divisadero on 12 October 1925. Broad-winged Hawks occurred in other hawk flights until 6 November, and other individuals of that species wintered on Mount Cacaguatique.

GUATEMALA

A general statement of the status of raptors in Guatemala was provided by Land (1970: 57–80), who reported that Turkey Vultures, Ospreys, American Swallow-tailed Kites, Mississippi Kites, Plumbeous Kites, Northern Harriers, Sharp-shinned Hawks, Cooper's Hawks, Broad-winged Hawks, Swainson's Hawks, Zone-tailed Hawks, Red-tailed Hawks, American Kestrels, Merlins, and Peregrine Falcons migrate into or through Guatemala.

Land (1970: 66–67) stated that Swainson's Hawks appear in large flocks in March–April and October, especially on the Pacific slope, but that these hawks pass through the country rapidly. On 23 March 1966, a migration of 800 Turkey Vultures, 4,000 Swainson's Hawks, and 6 Red-tailed Hawks was seen near Panajachel at Lake Atitlan at 5,000 feet elevation, and Guatemalan Indians reported observing large hawk flights for centuries (Hundley, 1967: 29).

Broad-winged Hawks also migrate through Guatemala, especially in large flocks on the Caribbean slope along the foot of the mountains (Land, 1970: 67). At Tikal, the Broad-winged Hawk is an uncommon but regular migrant reported on 8 October, 4 December, and 28 March (Smithe, 1966: 31).

A Red-tailed Hawk, banded as a nestling at the Snake River Birds of Prey Natural Area in southwestern Idaho, was recovered in southern Guatemala suggesting the Pacific lowlands of Central America and Mexico serve as wintering grounds for this species (Steenhof, Kochert, and Moritsch, 1984: 359, 361).

HONDURAS

In his distributional survey of the birds of Honduras, Monroe (1968: 65–91) stated that migrant raptors include Turkey Vultures, Ospreys, Black-shouldered Kites, American Swallow-tailed Kites, Mississippi Kites, Plumbeous Kites, Northern Harriers, Sharp-shinned Hawks, Cooper's Hawks, Broad-winged Hawks, Swainson's Hawks, Red-tailed Hawks, American Kestrels, Merlins, and Peregrine Falcons.

Broad-winged Hawks migrate across Honduras during October, sometimes in mixed flocks with Swainson's Hawks. As many as 6,000 Broad-winged Hawks were seen on 6 October 1962 along the Pacific slope. Swainson's Hawks also are reported migrating in early to mid-October in Honduras mostly on the Pacific slope, with a few birds also reported on the Caribbean slope. Curiously there are few spring migration sightings in Honduras for either Broad-winged Hawks or Swainson's Hawks.

MEXICO

A general statement of Mexican hawk migrations was provided by Friedmann, Griscom, and Moore (1950: 46–68), as did Edwards (1972) and Peterson and Chalif (1973). These sources, and the more detailed reports summarized here, demonstrate that Mexico is a major migration route for several species including Turkey Vultures, Broad-winged Hawks, and Swainson's Hawks.

Pacific Coastal Lowlands
Recoveries of Red-tailed Hawks, banded as nestlings in southwestern Idaho, in the Pacific coastal lowlands of Mexico suggest that these Mexican and Central American areas serve as wintering grounds for some Red-tailed Hawks reared in Idaho (Steenhof, Kochert, and Moritsch, 1984: 361, 366).

Tamaulipas
A summary of raptor migrants from North America that occurred in southwestern Tamaulipas was presented by Sutton and Pettingill (1942: 8–10). They included Ospreys, Northern Harriers, Sharp-shinned Hawks, Cooper's Hawks, Swainson's Hawks, Red-tailed Hawks, and American Kestrels on their list. Large flocks of migrating Swainson's Hawks were seen from 25 March to 3 April 1941, and a few additional individuals appeared until 3 May 1941.

Veracruz
An excellent summary of North American raptor migrants in the state of Veracruz was provided by Loetscher (1955: 24–26). He included Ospreys, American Swallow-tailed Kites, Mississippi Kites, Bald Eagles, Northern Harriers, Sharp-shinned Hawks, Cooper's Hawks, Red-shouldered Hawks, Broad-winged Hawks, Swainson's Hawks, Red-tailed Hawks, American Kestrels, Merlins, Peregrine Falcons, and Prairie Falcons.

On 6 and 10 April 1939, Eugene Eisenmann saw groups of up to 50 Turkey Vultures migrating northward through southern Veracruz (Bussjaeger, et al., 1967: 426). On 27 March 1966, along Highway 125 some 30 miles south of Tecolutla in Veracruz, a three-mile-long concentration of migrating Turkey Vultures estimated to contain several thousand birds also was observed drifting north. Some hawks were included among the migrating vultures. Additional small vulture flocks were seen on 28 March 1966 west of Veracruz (Bussjaeger, et al., 1967: 425–426).

Another large migration of approximately 1,600 Turkey Vultures and 1,600 Swainson's Hawks was seen on 22 March 1970 along Highway 180 west of Cardel, Veracruz. More migrating raptors appeared along the coastal plain toward Tamarindo to the south on 23 March 1970. Thousands of migrating Turkey Vultures also were seen on 26 March 1970 between Vega de Alatorre and Tecolutla (Purdue, et al., 1972: 92–93).

The most detailed and comprehensive field study of spring hawk migrations in Veracruz was conducted by Thiollay (1980: 13–20) for 23 days between 6 April and 6 May 1978 along 12.5 miles of road north of Palma Sola to the valley of Plan de Las Hayas. The effort produced counts of 333 Black Vultures, 25,820 Turkey Vultures, 380 Ospreys, 2 American Swallow-tailed Kites, 14 Snail Kites, 12,432 Mississippi Kites, 170 Northern Harriers, 1,396 Sharp-shinned Hawks, 45 Cooper's Hawks, 29 Red-shouldered Hawks, 202,147 Broad-winged Hawks, 16,684 Swainson's Hawks, 6 Red-tailed Hawks, 2,597 American Kestrels, 7 Merlins, and 48 Peregrine Falcons—in all 262,110 raptors.

Tilly (1985) also studied spring hawk migrations between mid-March and early April along Mexico's east coast in Veracruz. He reported tens of thousands of migrant Turkey Vultures and Broad-winged Hawks, plus limited numbers of other raptors, migrating along the Sierra Madre Oriental and the coast at various sites between Chachalacas and Tampico.

NICARAGUA

No information is available. See comments by Monroe (1968) on hawk flights in the Caribbean drainage area of Honduras near San Francisco.

PANAMA

A good general outline of the migratory status of North American hawks in Panama was provided by Wetmore (1965), whereas Ridgely (1981) presented an excellent and more recent list of these migrants. Included are Black Vultures, Turkey Vultures, Ospreys, American Swallow-tailed Kites, Mississippi Kites, Plumbeous Kites, Northern Harriers, Cooper's Hawks, Broad-winged Hawks, Swainson's Hawks, Zone-tailed Hawks, Red-tailed Hawks, American Kestrels, Merlins, and Peregrine Falcons.

The migratory status of Black Vultures in Panama is subject to further study, but observations made during November 1962 at Panama City and its vicinity by Eugene Eisenmann revealed groups of these birds (often in association with *Buteo* hawks) flying eastward toward South America in an unidirectional movement (Eisenmann, 1963: 244–249). Similar unidirectional Black Vulture movements in Costa Rica between September and December 1964 also suggest a migratory movement, thus helping to support Eisenmann's tentative conclusion of a Black Vulture migration in Panama (Skutch, 1969: 726–729).

Turkey Vultures, on the other hand, are abundant migrants in Panama from late February to early April and in October and November (Ridgely, 1981: 56–57). Chapman (1933: 30–34) was one of the first ornithologists to provide fairly detailed counts of spring Turkey Vulture migrations at Barro Colorado Island in the Canal Zone. He observed dozens to hundreds of migrating vultures moving south to southwest as part of their northward migrations because of the odd shape of the land at that site. On 28 October 1967, a count of more than 100,000 migrating Turkey Vultures was made at Almirante on the Caribbean coast (Kleen, 1969: 172). Another mid-October 1968 migration of these birds at Barro Colorado Island contained at least 1,500 birds according to Charles Leck (Heintzelman, 1975: 135–137). In another detailed study of these migrations at Panama City by N. G. Smith (1980: 51–65), a migratory flight of 7,450 Turkey Vultures was seen within two hours on 19 October 1977 and annual autumn Turkey Vulture counts for 1970–1976 ranged between about 31,500 birds in 1974 to about 230,000 birds in 1973.

Mississippi and/or Plumbeous Kites also migrate through Panama (Ridgely, 1981: 61–62). On 15 October 1965, for example, a flight of approximately 1,000 kites (either Mississippi or Plumbeous) was seen near Almirante on the Caribbean coast (Hicks, Rogers, and Child, 1966).

Unquestionably it is the annual migrations of Broad-winged Hawks across Panama that are the largest and most spectacular. Between 1970 and 1976, autumn counts of migrating Broad-wings over Panama City ranged from 42,209 birds in 1974 to 395,003 birds in 1972 (N. G. Smith, 1980: 61). During October of 1963–1965 hundreds to thousands of migrating Broad-wings were seen at Almirante (Hicks, Rogers, and Child, 1966), and during late October and early November 1967 thousands of these hawks also were seen migrating over Almirante (Kleen, 1969: 172). Charles Leck also counted thousands of migrating Broad-wings in October 1968 over Barro Colorado Island in the Canal Zone (Heintzelman, 1975: 135–137) and Smith (1973) photographed and counted 41,333 Broad-winged Hawks on 11 October 1972 over Ancon Hill in the Canal Zone. Further studies of Broad-winged Hawk and Swainson's Hawk migrations over the Isthmus of Panama indicate that the visible flights represent only part of the complete migrational movement of these birds. An unknown (but very considerable) additional portion of the migration passes at very high altitudes, sometimes above tropical storm clouds or along thermal streets and are invisible to observers on the ground (N. G. Smith, 1980, 1985). Not all migrating Broad-winged Hawks, however, avoid tropical storms by flying above them. Sometimes considerable numbers of hawks are forced to land due to odd or strong wind currents. On 17 April 1963, for example, at a ranch in northwestern Panama, approximately 200 hawks of this species landed during a driving rain and remained grounded for two days until the rain stopped and favorable flight conditions prevailed again (Loftin, 1967: 29). N. G. Smith (1985: 287–288) also reported examples of Broad-winged Hawks and Swainson's Hawks being forced to land due to unfavorable weather conditions. Indeed, some of the hawks actually crashed to the ground and were injured.

Swainson's Hawks also migrate across Panama in large numbers from late February to early April and from late September to early November (Ridgely, 1981: 65–66). Indeed, between 1970 and 1982, autumn counts of Swainson's Hawks in the Isthmus of Panama ranged between 54,403 birds in 1974 and 344,409 birds in 1972 (N. G. Smith, 1980: 61, 1985: 273).

C H A P T E R   12

West Indies Hawk Migrations

Knowledge of hawk migrations in the West Indies is fragmentary, inadequate, and in need of organized field study. Ironically it was in the West Indies that the first observations of New World hawk migrations were made in the early 1500s by Oviedo (Baughman, 1947: 304). Bond (1980) listed the raptor fauna of the West Indies and reported that North American raptor migrants include Ospreys, American Swallow-tailed Kites, Northern Harriers, Sharp-shinned Hawks, perhaps the Broad-winged Hawk, and Red-tailed Hawks, American Kestrels, Merlins, and Peregrine Falcons.

BAHAMA ISLANDS

These islands are not located on a major hawk migration route, but a few species—Ospreys, Northern Harriers, Sharp-shinned Hawks, American Kestrels, Merlins, and Peregrine Falcons—apparently migrate across the islands during autumn in limited numbers (Bond, 1956; Brudenell-Bruce, 1975: 37–40). Several other species are vagrants or occasional winter visitors—Black Vultures, Turkey Vultures, Sharp-shinned Hawks, Red-tailed Hawks, and American Kestrels (Brudenell-Bruce, 1975: 37–40; Jackson, 1983: 17).

BARBADOS

Bond (1956) reported that Northern Harriers are migrants on Barbados.

CUBA

The first observations of hawk migrations crossing Cuba come from the writings of Oviedo in the early 1500s who saw hawks coming from the direction of North America, cross Cuba, and continue southward across the West Indies toward Panama (Baughman, 1947: 304). It is not possible, however, to identify which species of raptors he observed. Nevertheless migrant hawks from North America reported on Cuba by Bond (1956, 1980: 53–64) include Black Vultures, Ospreys, American Swallow-tailed Kites, Northern Harriers, American Kestrels, Merlins, and Peregrine Falcons.

An endemic, nonmigratory subspecies of the Broad-winged Hawk *(Buteo platypterus cubanensis)* lives on Cuba, but at least one specimen of the North American subspecies of the Broad-wing *(B. p. platypterus)* was collected on the island (James Bond, letter of 16 April 1973; Heintzelman, 1975: 299).

GUADELOUPE

Northern Harriers are reported as migrants in Guadeloupe (Bond, 1956).

HISPANIOLA

Migrant hawks from North America reported on Hispaniola include Ospreys, Northern Harriers, and Peregrine Falcons. American Kestrels and Merlins also may reach the island, whereas the endemic Ridgway's Hawk replaces the Broad-winged Hawk on the island (Bond, 1956, 1980).

JAMAICA

Bond (1956, 1980) stated that North American migrant raptors reported in Jamaica include Black Vultures, Ospreys, American Swallow-tailed Kites, Northern Harriers, and Peregrine Falcons. The Merlin also may migrate across Jamaica.

MARTINIQUE

Observers on Martinique report migrant Ospreys from North America (Bond, 1956, 1980). Sightings of Northern Harriers are questionable (Bond, 1956), but a specimen of the Eurasian Kestrel was secured on the island on 9 December 1959 (Pinchon and Vaurie, 1961: 92–93).

PUERTO RICO

North American migrant hawks reported on Puerto Rico include Ospreys, Northern Harriers, Merlins, and Peregrine Falcons (Bond, 1956; Raffaele, 1983: 97–101). A rare, dark, endemic subspecies of the Broad-winged

Hawk *(Buteo platypterus brunnescens)* also lives on Puerto Rico (Friedman, 1950; Bond, 1956), but observations of Broad-winged Hawks on the island during the spring of 1964 and 1965 were claimed to be migrating North American Broad-wings *(B. p. platypterus)* despite serious doubts about the validity of that subspecific identification (Heintzelman, 1975: 295). A vagrant Bald Eagle also appeared on the island on 8 October 1975 (Raffaele, 1983: 195).

TRINIDAD AND TOBAGO

North American migrant raptors on Trinidad include Ospreys, perhaps Broad-winged Hawks, and Peregrine Falcons (ffrench, 1973).

On Tobago, migrant Ospreys and Peregrine Falcons from North America also are reported, whereas the migratory status of Broad-winged Hawks on the island is unclear. A flock of 41 Broad-winged Hawks was seen on 2 May 1972 over Little Tobago (ffrench, 1973: 105), and a kettle of approximately 500 Broad-wings was seen on 17 March 1977 over northeastern Tobago (Rowlett, 1980: 54). Unfortunately an endemic race of the Broad-winged Hawk *(Buteo platypterus antillarum)* is a common Tobago hawk (Bond, 1970: 4) so that it is difficult to know whether large kettles of these birds are local birds or North American migrants. Nevertheless, I agree with Rowlett (1980: 54) that large numbers of Broad-winged Hawks seen over Tobago presumably are migrants, the fate of which remains unknown.

VIRGIN ISLANDS

Several North American raptor migrants are reported in the Virgin Islands—Turkey Vultures, Ospreys, Northern Harriers, Merlins, and Peregrine Falcons. A vagrant Bald Eagle also was seen on 21 February 1977 on St. John (Raffaele, 1983: 97–101, 195).

PART
THREE

Hawk Migrations and
Weather Conditions

C H A P T E R 13

Hawk Migrations and General Weather Systems

An understanding of North American hawk migrations is intimately related to the interaction of general weather systems with hawk movements. Two methods are used to consider these interactions: (1) traditional comparisons of hawk count data from one or more sites with movements of weather systems plotted on weather maps to determine the systems responsible for triggering migrations and/or the use of sophisticated statistical analyses to detect the relationships of general weather systems with hawk movements, or (2) traditional comparisons and/or more sophisticated statistical analyses of specific local weather variables with hawk count data from one or more sites in an attempt to isolate and identify key variables responsible for producing hawk flights at various locations as discussed in the next chapter.

AUTUMN MIGRATIONS

In North America, ornithologists recognized the influence of general weather systems in triggering autumn hawk migrations as early as the end of the last century. Since then the volume of literature reporting similar relationships increased steadily.

Trowbridge (1895) was the first person to study the correlations of autumn hawk migrations with weather systems from a coastal Connecticut location. He determined that hawk flights increased greatly when air temperatures lowered and northwest winds developed—conditions resulting from the passage of a cold front. Ferry (1896) also noticed similar correlations of weather to hawk flights in Illinois. Well into this century, Poole (1934) and Broun (1935) discovered these same correlations at Hawk Mountain, Pennsylvania. At Cape May Point in New Jersey the largest hawk flights were seen after cold fronts passed and northwest winds prevailed (Allen and Peterson, 1936), and a similar situation was documented at Cape Charles, Virginia (Rusling, 1936). By 1949, Bagg (1950) demonstrated that the passage of cold fronts correlated well with hawk flights in the Connecticut Valley of Massachusetts. Tanner (1950), in Tennessee, likewise noted the impact of the passage of cold fronts upon the production of hawk flights. The field studies of Robbins (1950) in Maryland and DeGarmo (1953) in West Virginia tended to provide further confirmation to the earlier studies elsewhere. Meanwhile, hawk flights in Texas also developed after the passage of cold fronts (Fox, 1956).

The detailed field studies of weather and hawk migrations at Cedar Grove, Wisconsin, provided important additional confirmation to the concept that large hawk flights developed soon after the passage of cold fronts (Mueller and Berger, 1961). At Fire Island, New York, much the same correlation of weather conditions and hawk flights was reported by Darrow (1963), Hofslund (1966) noted similar correlations at Duluth, Minnesota, and in Kentucky the situation also was similar (Stamm, 1969). Further studies in Maryland (Lee and Sykes, 1975) and also in Pennsylvania (Heintzelman, 1975), at Cape May Point, New Jersey (Dunne and Clark, 1977), and more recently along coastal South Carolina (Laurie, McCord, and Jenkins, 1981), in Florida (Duncan and Duncan, 1975; Darrow, 1983), in Nevada (Millsap and Zook, 1983), and in Indiana (Brock, 1984) all produced correlations of large hawk flights with the passage of cold fronts. A low-pressure area coupled with a cold front also seemed to be an important additional factor in relationship to large hawk flights at Hawk Mountain, Pennsylvania (Broun, 1951, 1963). Haugh (1972) also confirmed the importance of a low-pressure area and the passage of a cold front in producing large hawk migrations in autumn. It appears, therefore, that an area of low barometric pressure north of a site triggers autumn hawk migrations, and the passage of a cold front aids the migrants by producing northwesterly winds that create favorable deflective updrafts along mountain ridges upon which hawks soar and/or otherwise tends to concentrate migrants at lower flight altitudes, thereby resulting in more hawks being seen under those conditions (Haugh, 1972; Millsap and Zook, 1983).

The fact that not all major autumn hawk flights conform to the classic weather pattern described by Broun (1951, 1963) and others suggests that other factors also influence hawk flights. Robbins (1956: 211) summarized the situation pointing out that New England's largest autumn hawk flights tended to develop from one to three days after cold fronts passed the region. He stated further, however, that some very large hawk flights in the

Middle Atlantic states exhibited weaker correlations with the passage of weather systems. He concluded that weather systems seemed to be more important in initiating autumn hawk migrations than in halting them once started—a conclusion also reached by Haugh (1972).

Autumn hawk migrations in western North America also are related to the movements of general weather systems and various local weather variables (Hoffman, 1981a: 2–3), but the general weather condition that seems to produce the largest hawk flights is an approaching northwest cold front rather than the passage of the front. The passage of a weak cold front seems not to decrease western hawk migrations (Hoffman, 1982b: 11). In southern Nevada, however, Millsap and Zook (1983) conducted a very sophisticated statistical analysis of the effects of weather systems on autumn Sharp-shinned Hawk and Cooper's Hawk migrations. They determined that the largest hawk flights were seen after cold fronts passed because the

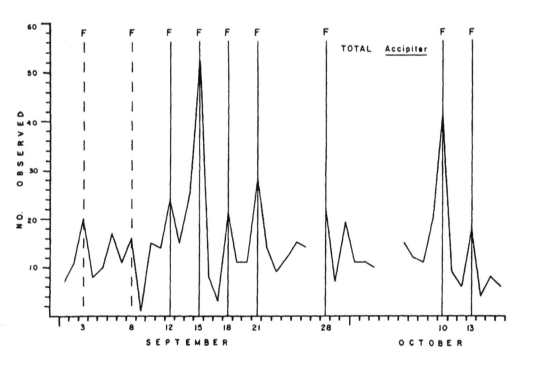

Correlation of autumn cold front passage (F) with total accipiter (Sharp-shinned Hawk + Cooper's Hawk + unidentified accipiters) counts for complete observation days in southern Nevada. *Reprinted from Millsap and Zook (1983).*

hawks flew at lower altitudes and were more visible to ground observers, the birds tended to use deflective updrafts more than thermals during those periods, and the birds shifted their flight-lines over forested ridgetops rather than over open deserts. It appears, therefore, that much more detailed field study is needed in western North America to evaluate the effects of weather systems on autumn hawk migrations there.

SPRING MIGRATIONS

Studies of spring hawk migrations in North America are limited, but several people noticed relationships between general weather systems and northward migratory hawk movements. Listman, Wolf, and Bieber (1949) stated that warm southwest winds (warm fronts) coupled with a low pressure area to the west or north produced large hawk flights at Braddock Bay Park near Rochester, New York, along the south shore of Lake Ontario—a conclusion confirmed by Moon and Moon (1982e). In Kentucky, Able (1965) also reported a relationship between warm fronts and spring hawk flights, and similar correlations occurred in Connecticut (Hopkins and Mersereau, 1972) and in eastern Pennsylvania (Kranick, 1982).

Haugh (1972), however, provided the most detailed analysis of the relationship of general weather systems with hawk migrations in spring and concluded that the most favorable weather factors associated with spring flights were opposite to those reported during autumn. Thus the key spring weather factors were an approaching low pressure area to the west of a site, rising air temperatures, warm southerly winds, and decreasing pressure at the migration observation site—in other words, a warm front.

CHAPTER 14

Hawk Migrations and Local Weather Variables

Ornithologists as long ago as the last century recognized that certain local weather variables, such as wind direction, influenced autumn hawk migrations (Dunn, 1895; Ferry, 1896; Trowbridge, 1895). It is now recognized that numerous local weather variables actually influence visible hawk migrations at a site either individually or in various combinations (Broun, 1935, 1939, 1949a; Allen and Peterson, 1936; Haugh, 1972; Heintzelman, 1975). The following are particularly important: (1) wind direction, (2) wind velocity, (3) amount of cloud cover, (4) air temperature, (5) barometric pressure, and (6) visibility.

WIND DIRECTION

Wind direction is one of the most obvious local weather variables that influences hawk migrations greatly during autumn and spring, but in different ways during each season.

Autumn Migrations

Almost from the beginning of systematic autumn hawk migration study in North America, ornithologists noticed that wind direction was a key local weather variable in influencing hawk migrations. Generally, northwest winds prevailed when large hawk flights were observed. Dunn (1895) noticed that northwest winds produced hawk flights along coastal Rhode Island, for example, and Ferry (1896) also observed that northwest winds occurred when he saw Illinois hawk flights. Trowbridge (1895), in his classic study of coastal Connecticut hawk migrations, also noted the obvious influence of northwest winds upon production of large hawk flights there. At Fishers Island, New York, northwest winds also were key local weather factors when large hawk flights appeared (Ferguson and Ferguson, 1922) and a similar situation developed at Cape May Point, New Jersey (Stone, 1922; Allen and Peterson, 1936; Dunne and Clark, 1977). At Hawk Mountain, Pennsylvania, Poole (1934) and Broun (1935, 1939, 1949a) also noted that northwest winds tended to accompany many of the largest hawk flights there. Robbins (1956) also reported that many of Maryland's large hawk flights occurred on northwest winds. So, too, were many of West Virginia's large hawk flights (DeGarmo, 1953). At Cedar Grove, Wisconsin, in excess of 92 percent of the hawk flights reported at that site also occurred on northwest or westerly winds (Mueller and Berger, 1961: 189). Hawk migrations in South Dakota also tended to occur in largest numbers when northwest winds prevailed (Bartelt and Orde, 1976), as did Illinois hawk flights (Johnson, 1977). West winds, though, produce some large Minnesota hawk flights (Hofslund, 1954a, 1954b; Rosenfield and Evans, 1980).

Not all large hawk flights, however, correlate well with northwest winds. Bailey (1912) reported hawk migrations on easterly winds, and Broun (1949a) also observed some large Broad-winged Hawk flights on easterly winds at Hawk Mountain, Pennsylvania. In Massachusetts, some large hawk flights also were seen when northeast winds prevailed (Robinson, 1981). South winds also produce some large hawk flights in Maryland (Hackman, 1954).

Autumn hawk migrations at Bake Oven Knob, Pennsylvania, were studied in 1957 and annually from 1961 through 1984 and produced a collective total of 330,400 raptors tabulated during 1,546 days (9,444 hours) of observation (Heintzelman, 1963a through 1984e). Analysis of the effects of wind direction upon counts of selected species at the Knob produced various results.

There were 80 days (5.1 percent) of the coverage period when 20 or more Ospreys were counted. Northwest winds developed on 23 (28.7 percent) of those 80 days, southwest winds occurred on 19 (23.7 percent) days,and 14 days (17.5 percent) had west winds. Collectively there were 56 days (70 percent) of the 80-day coverage with some westerly component winds. Three or more Bald Eagles also appeared on each of 42 days (2.7 percent) of the coverage period with northwest winds prevailing on 29 (69 percent) of those days, west winds on six (14.2 percent) of the days, and southwest winds on three (7.1 percent) of the days. Collectively westerly wind components occurred on 38 (90.4 percent) of the 42-day sample.

Sharp-shinned Hawk flights, numbering at least 150 birds per day, occurred on 130 days (8.4 percent) of the coverage period. Northwest winds occurred on 58 days (44.6 percent), west winds prevailed on 26 days (20 percent), and southwest winds occurred on 23 days (17.6 percent)

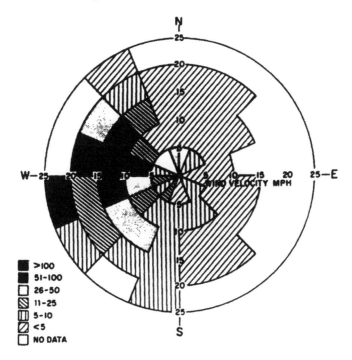

Correlation of Sharp-shinned Hawk migration with wind direction and velocity at Cedar Grove, Wis. The wind data are the vector means of the twelve hourly readings taken between 0600 and 1700 at Milwaukee, Wis. Radial lines enclose 22.5° increments of mean wind vector-directions (e.g., all mean winds between west and west-northwest). Concentric circles represent 5-mile-per-hour increments of the vector-mean velocity. The intensity of shaded segments indicates the mean number of Sharp-shinned Hawks observed per day under the indicated wind direction and velocity conditions. *Redrawn from Mueller and Berger (1967b).*

resulting in an overall westerly component on 107 (82.3 percent) of the 130-day sample.

There were 50 days (3.2 percent) of the coverage period on which at least 1,000 Broad-winged Hawks were counted. Of those 50 days, northwest winds occurred on 24 days (48 percent) and southwest winds occurred on 10 days (20 percent), whereas east winds prevailed on seven days (14 percent) and southeast winds prevailed on six days (12 percent). Westerly component days numbered 35 days (70 percent) and easterly wind component days numbered 14 days (28 percent). One day (2 percent) had calm winds.

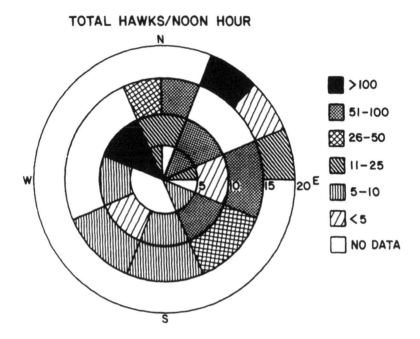

Correlation of autumn hawk migrations with wind direction and velocity at Fort Johnson, S.C. in 1980. *Reprinted from Laurie, Mc-Cord, and Jenkins (1981).*

There also were 53 days (3.4 percent) out of the 1,546 days of coverage at Bake Oven Knob on which at least 200 Red-tailed Hawks were counted. Northwest winds occurred on 40 days (75.4 percent) and west winds were recorded on four days (7.5 percent). Indeed, 45 days (84.9 percent) had westerly wind components of some sort, thus indicating the importance of that compass area for Red-tailed Hawks.

At least three Golden Eagles were counted at the Knob on each of 57 days (3.6 percent) of the 1,546 day coverage period. Northwest winds occurred on 40 (70.1 percent) of the 57 days, west winds on five days (8.7 percent), and southeast winds on six days (10.5 percent). There also were 47 days (82.4 percent) with winds from some sort of westerly component.

In addition to traditional methods of considering relationships of wind direction to hawk migrations, several sophisticated statistical analyses of these variations also were conducted for several North American locations. Haugh (1972), for example, analyzed autumn hawk count data gathered from 1954–1968 at Hawk Mountain, Pennsylvania. He determined that Red-tailed Hawk and American Kestrel migrations exhibited highly significant preferences for winds from northerly and westerly directions, whereas the preference of Red-shouldered Hawks for northerly winds was significant. Broad-winged Hawks preferred both northwest and northeast winds, but a few very large flights on northeast winds tended to distort the analysis and further analysis failed to demonstrate a statistically significant difference between northwest or northeast winds. Sharp-shinned Hawks preferred westerly component winds regardless of southerly or northerly aspects. Northern Harriers exhibited no significant preferences for any wind direction.

Another statistical analysis of autumn Sharp-shinned Hawk, Broad-winged Hawk, and Red-tailed Hawk data at Dan's Rock and High Rock, Maryland, for the period 1975–1979, demonstrated no significant interaction for wind direction for any of the three species at High Rock. At Dan's Rock, however, Sharp-shinned Hawks and Broad-winged Hawks showed some correlation with winds from directions other than south to west, whereas Red-tailed Hawks exhibited migrations on winds from the south to west (Titus and Mosher, 1982).

Spring Migrations

Various observers in North America also noticed the influence of wind direction on spring hawk migrations. Magee (1922) reported that south winds accompanied hawk flights at Whitefish Point, Michigan, and a similar correlation was reported in Iowa (Proescholdt, 1961) and Michigan (Sheldon, 1965). At Braddock Bay Park near Rochester, New York, southwest winds were noticed to correlate well with hawk flights along the south shoreline of Lake Ontario (Listman, Wolf, and Bieber, 1949; Moon and Moon, 1982e).

Haugh (1972) also made a more sophisticated statistical analysis of wind direction and spring hawk flights along the southeastern shoreline of Lake Ontario at Derby Hill, New York, and discovered a statistically signficant preference for southwest and southeast winds in migrations of Northern Harriers, Sharp-shinned Hawks, Red-shouldered Hawks, Broad-winged Hawks, Red-tailed Hawks, and American Kestrels.

WIND VELOCITY

There are many general references to the correlation of wind velocity with North American hawk migrations, but few detailed statistical analyses of this local weather variable. Even from the available anecdotal literature, however, it is clear that wind velocity is an important variable (in combination with wind direction) in the production of many large hawk flights.

Autumn Migrations

Trowbridge (1895) was one of the first to recognize that wind velocity, as well as wind direction, is an important component of autumn hawk flights along coastal Connecticut. He stated that strong northwest winds correlated with high hawk counts, as did Ferry (1896) in Illinois. Broun (1935, 1939,

1949a, 1951, 1963) also confirmed that brisk wind velocity was an important local weather variable in hawk flights at Hawk Mountain, Pennsylvania, and further pointed out that Broad-winged Hawks leave the Kittatinny Ridge and spread over broad migratory fronts during periods of light or variable winds (Broun, 1946: 3, 1960: 7; Edge, 1940a: 1, 1940b: 1). Similar effects of wind velocity upon hawk migrations also apply to the flights observed at Bake Oven Knob, Pennsylvania (Heintzelman, 1975: 187–189). Broun and Goodwin (1943) also studied flight speeds of migrating hawks in relation to wind velocity and other factors at Hawk Mountain.

At Cedar Grove, Wisconsin, the most desirable wind velocities for migrating Broad-winged Hawks were 10 to 15 miles per hour, whereas winds from 15 to 25 miles per hour produced the largest numbers of other hawk species. Wind velocities above 30 miles per hour, however, inhibited hawk migrations (Mueller and Berger, 1961: 187).

Spring Migrations

At Derby Hill, New York, along the southeastern shoreline of Lake Ontario, large hawk flights generally occurred on days when the wind velocity was between 10 and 25 miles per hour, but hawk flights generally did not occur if winds increased in velocity above 35 miles per hour (Haugh and Cade, 1966: 101). Moon and Moon (1982e: 2) also stated that the best hawk flights observed at Braddock Bay Park along the south shoreline of Lake Ontario near Rochester, New York, occurred when strong southwest winds prevailed.

CLOUD COVER

A variety of observations of hawks flying into, through, out of, or above opaque clouds are published or otherwise available. Sightings include Turkey Vultures thermal soaring into cumulus clouds and disappearing from view in New Jersey (Heintzelman and MacClay, 1974: 849), a California Condor rising to about 5,000 feet altitude and disappearing into opaque stratus clouds and another condor entering a cumulus cloud at 15,000 feet altitude in California (Borneman, 1976: 636), and several Bald Eagles entering and disappearing into cumulus clouds at altitudes of 3,000 to-ʲ 4,000 feet in Washington (Serveen, 1976: 387).

There also are many observations of Broad-winged Hawks soaring into opaque clouds, or unexpectedly appearing out of the bottom of complete cloud cover. Forbes and Forbes (1927: 101–102) watched hawks (probably Broad-wings) gain altitude and disappear into cumulus clouds over Mt. Monadnock, New Hampshire, and Broun (1961: 8–9) also observed Broad-winged Hawks flying "exceedingly high and scarcely discernible as they moved through thin clouds" at Hawk Mountain, Pennsylvania.

At Bake Oven Knob, Pennsylvania, I occasionally watched Broad-winged Hawks rising into opaque cloud cover, or losing altitude and appearing out of the bottom of clouds. On 14 September 1965, for example, many hawks appeared at an extremely high altitude with some birds partly obscured by cloud mist. On 15 September 1967 a group of 81 hawks flew out of an opaque cloud layer, and on 19 September 1969 a mid-afternoon flight of 2,099 hawks glided out of dense opaque clouds and passed overhead. The next day more hawks flew out of the tops of cumulus clouds. On 19 September 1972 Herbert Douglas watched many Broad-winged Hawks appearing out of dark clouds, and on 20 September 1978 Maurice Broun and I watched these hawks and Ospreys rise in thermals into the bottom of clouds. Barry Reed and I also watched Sharp-shinned Hawks and Broad-winged Hawks flying through dense mist and fog on 13 September 1979, and another 50 Broad-wings rose into complete opaque cloud cover on 9 September 1982.

In Connecticut, in September, a motor-glider pilot also followed and photographed a kettle of Broad-winged Hawks into opaque clouds where the birds were lost to view. In one instance, however, the hawks later were seen emerging from the cloud about 800 feet higher than the cloud base (Welch, 1975: 17–18). In Texas, an autumn flight of 500 Broad-winged Hawks also appeared out of a low cloudbank near sunset (Donohue, 1979b: 24).

In New England, however, most Broad-winged Hawks generally changed from thermal soaring to downward glides just below the cloud base (Hopkins, Mersereau, and Welch, 1975: 19), and in Panama migrating Broad-winged Hawks are known to avoid tropical storms by gliding over the face of advancing storms at altitudes of more than 21,000 feet (N. G. Smith, 1980: 56). At times some migrant Swainson's Hawks may even reach altitudes of 29,500 feet (N. G. Smith, 1985: 286).

On 12 November 1972, at Bake Oven Knob, Pennsylvania, I also watched a Red-tailed Hawk circle high overhead and move in and out of opaque clouds.

The effect of cloud cover upon hawk migrations in North America is disputed by various ornithologists. Haugh and Cade (1966: 103), and Haugh (1972: 3), for example, stated that clouds sometimes may affect hawk migrations by reducing the formation of thermals or updrafts but otherwise doubted the importance of cloud cover to migrant hawks. Griffin (1973: 118), on the other hand, stated that current knowledge of oriented bird migration in opaque cloud layers was limited and that few direct observations existed of the phenomena. In addition, I pointed out that large numbers of migrating Broad-winged Hawks flying through (or above) opaque clouds could produce a serious bias in hawk counts (Heintzelman, 1975: 232). Presumably migrations of less thermal-dependent hawks are less affected by local cloud cover variations.

AIR TEMPERATURE

Relatively few studies of the effects of air temperature upon hawk migrations exist. Those that do provide conflicting conclusions concerning the importance of this local weather variable.

Autumn Migrations

Bagg (1950: 75–80) conducted a detailed analysis of air temperatures and September 1949 Broad-winged Hawk flights in the Connecticut Valley of Massachusetts. He concluded that a major drop in temperature was required to trigger a major hawk flight in New England early in September, but later in the Broad-wing migration season lesser temperature drops were capable of triggering remaining Broad-winged Hawk migrations.

At Cedar Grove, Wisconsin, "no reasonable correlation" existed between short-term changes in air temperature and hawk migrations, although a decrease in minimum temperature seemed to produce the best correlations (Mueller and Berger, 1961: 187–188). There also was no clear correlation between air temperature and hawk flights at Hawk Mountain, Pennsylvania, although there seemed to be some correlation between American Kestrel flights and colder air temperatures (Haugh, 1972: 30–31). At Bake Oven Knob, Pennsylvania, some of the larger Sharp-shinned Hawk flights occurred when air temperatures ranged between about 7 and 20°C., many larger Broad-winged Hawk migrations occurred on 10–20°C. days, and larger Red-tailed Hawk flights were seen on 0–15°C. days (Heintzelman, 1975: 189–193).

Spring Migrations

Spring hawk migrations at Derby Hill, New York, exhibited a statistically significant correlation with warm air temperatures. On days with southerly winds, however, the correlation with temperature was unimportant, but when winds with northerly components developed, hawks were more stimulated to engage in spring migrations as a result of temperature than otherwise might be expected (Haugh, 1972: 24).

BAROMETRIC PRESSURE

The importance of barometric pressure in triggering North American hawk migrations was confirmed by various studies.

Autumn Migrations

In his classic analysis of weather variables that correlated well with large hawk migrations in autumn at Hawk Mountain, Pennsylvania, Broun (1951b, 1963a) demonstrated that an intense area of low barometric pressure crossing upstate New York and New England and followed by a cold front triggered major hawk flights. Haugh's (1972: 30–31) analysis of barometric pressure at Hawk Mountain, however, produced a strong correlation with rising pressure and high hawk counts when northerly winds occurred. The migrations passing Hawk Mountain, however, probably were triggered several days earlier in New England under the influence of a low pressure center there.

At Duluth, Minnesota, the autumn hawk migrations seen at the Hawk Ridge Nature Reserve also tended to occur in largest numbers one or two days after a cold front passed and when the barometer was rising (Hofslund, 1966: 81). There also seemed to be a trend at Cedar Grove, Wisconsin, whereby hawk migrations occurred when barometric pressure was rising (Mueller and Berger, 1961: 189).

Spring Migration

At Derby Hill, New York, hawk flights were significantly correlated with falling barometric pressure coupled with the approach of a low pressure area and a warm front (Haugh and Cade, 1966: 95; Haugh, 1972: 24).

VISIBILITY

No detailed analysis is available regarding the effects of visiblity on hawk migrations. Nevertheless, severe conditions of local atmospheric haze and/ or air pollution might seriously affect the accuracy and completeness of hawk counts reported at a site because the hawks might not be identified correctly or seen at all.

OTHER WEATHER VARIABLES

Several additional local weather variables also can affect hawk migrations and are discussed briefly in the hawk migration literature.

Rain

The effect of rain upon hawk migrations varies depending upon the extent of the precipitation. Robbins (1956: 211) stated that migrating hawks sometimes will fly around scattered showers, and at Cedar Grove, Wisconsin, a few autumn hawk flights were seen on days with light rain or brief showers but prolonged heavy rains generally terminated migrations (Mueller and Berger, 1961: 188–189).

At Bake Oven Knob, Pennsylvania, I noted a similar effect of rain upon autumn hawk migrations. On 2 October 1977, for example, heavy rain and dense fog occurred at the Knob and hawks were not migrating. By late in the morning, however, the rain stopped and some of the fog lifted as northwest winds prevailed. Almost at once a large Sharp-shinned Hawk flight developed and continued during the rest of the afternoon—1,148 Sharp-shins being counted by Barry Reed and Jan Sosik. The flight continued into 3 October 1977, despite scattered rain showers most of that morning, as many Sharp-shins flew through the rain in large numbers. Indeed, Charles and Betty Wonderly and I counted 1,022 hawks of that species during six hours of observation that day. Occasionally I also observed small numbers of migrating Broad-winged Hawks and other hawks flying through brief, light showers.

In Panama, migrating Broad-winged Hawks also are known to glide up the face of advancing tropical storms in order to avoid them (N. G. Smith, 1980: 56). On the other hand, if it is raining in the morning in Panama and thermals do not form, Turkey Vultures, Broad-winged Hawks, and Swainson's Hawks do not attempt to fly until the rain stops. Indeed, the birds may remain perched for several days while waiting for suitable flight conditions to develop (N. G. Smith, 1985: 286).

Lake and Sea Breezes

Occasionally breezes blowing inland from Lakes Erie and Ontario form a frontal boundary resulting in updrafts where warmer land air flows meet the cooler lake breezes. Red-shouldered Hawks, Red-tailed Hawks, and other hawks sometimes use these lake breezes to aid their migrations. Similar sea breezes can develop along coast lines and also may be used by migrating hawks (Haugh, 1972: 37).

Shoreline Updrafts

During March and April, cold air over Lake Ontario occasionally is blown inland resulting in a rapid drop in air temperature over the land. When warmer inland breezes meet the colder lake breezes, the warmer air is deflected upward providing migrating hawks with suitable air currents and narrow flight-lines along the lake shoreline (Haugh, 1972: 36–37).

Squall Lines

Squall lines are severe wind gusts associated with thunderstorms forming ahead of, or along, cold fronts. Some squall lines can be 100 miles long, others much shorter. Frequently an updraft zone develops ahead of a squall line and migrating hawks sometimes use these air currents to aid their migrations. On 21 April 1966, at Derby Hill, New York, a group of some 4,300 hawks migrated along an updraft zone ahead of a severe squall line (Haugh, 1972: 38).

Thermal Streets

During the hottest periods of the day in Panama, migrating Broad-winged Hawks and Swainson's Hawks sometimes use lenticular thermal lift zones, or thermal streets, to aid them in their migrations (N. G. Smith, 1985: 286). Haugh (1972) also noted some limited use of thermal streets by migrant hawks in North America. The use of thermal streets by migrating birds of prey may be of more importance in the neotropics than in more temperate areas.

Lee Waves

Haugh (1972) mentioned lee waves but was not certain what role they played as aids to migrating hawks in North America. Along the mountain passes in Panama, however, lee waves sometimes are used by migrating hawks when thermal activity is inadequate or unavailable to provide suitable flight conditions (N. G. Smith, 1985: 283).

C H A P T E R 15

Hawk Migrations and Deflective Updrafts

Some of eastern North America's finest autumn hawk migrations are seen from lookouts located on long, relatively unbroken, mountain ridges oriented in a northeast-to-southwest direction because of deflective updrafts that occur there. Thousands of hawks annually pass Bake Oven Knob and Hawk Mountain, Pennsylvania, and Raccoon Ridge, New Jersey—three notable points on the crest of the famous Kittatinny Ridge (Broun, 1949a; Heintzelman, 1975). Other ridges, shorter in length and/or oriented in different directions such as those in Vermont, also have deflective updrafts but tend to produce lesser numbers of migrant hawks (Will, 1980). Autumn hawk migrations in western North America also use deflective updrafts along some north-to-south oriented mountain ridges—especially those that are single, narrow, and steep with wide valleys on each side (Hoffman, 1981a). Hilly areas and river bluffs also have deflective updrafts used by some migrant raptors.

Deflective Updraft Formation

A key reason why autumn hawk migrations occur along mountain ridges, bluffs, or other natural barriers is because of the ability of those geographic features to produce deflective updrafts. Thus it is necessary to consider how such updrafts are formed at those places.

Mountain Ridges

In autumn, for example, prevailing westerly surface winds strike the sides of ridges and are deflected upward, creating powerful updrafts. Since most

231

hawks and eagles are soaring birds, they are well adapted to using updrafts as aids to their migratory efforts while at the same time conserving energy by doing so. Hence mountain ridges—particularly those such as the Kittatinny Ridge, which extend virtually unbroken for many miles—provide natural migration flight-lines along which soaring hawks and eagles can travel relatively effortlessly.

Broun (1935, 1939, 1949a) and Poole (1934) early recognized the basic importance of updrafts along mountains as a significant influence upon migrating hawks. Later studies of autumn hawk flights elsewhere along the Kittatinny Ridge further confirmed the basic importance of deflective updrafts as aids to migrating hawks (Heintzelman, 1975).

Deflective updrafts also are produced along mountain ridges oriented in directions other than northeast-to-southwest, but it is still the surface winds striking the sides of those mountains that produce the updrafts. Brockway Mountain in Michigan, for example, is oriented in an east-to-west direction yet hawk migrations also use deflective updrafts there (Isaacs and Hennigar, 1980). The shorter, broken ridges in Vermont also produce deflective updrafts used by migrant hawks despite being heterogeneous rather than being oriented in well-defined ridge systems (Will, 1980). Much the same situation pertains to deflective updrafts generated along the many broken mountains in West Virginia (DeGarmo, 1953). Similar updrafts also are generated along western North American mountain ridges (Hoffman, 1981a).

Downwind from mountain ridges lee waves also occur as a result of surface winds striking the windward side of a ridge, flowing over the top of the ridge, then extending in wave-like formations on the lee side. At times migrating hawks doubtless use lee waves, but their roles as factors in hawk migrations is not yet studied or fully understood (Haugh, 1972: 39).

Hilly Terrain

Hilly terrain in Saskatchewan also is followed by some migrating Bald Eagles in spring because of the deflective updrafts generated at such places (Gerrard and Hatch, 1983).

River Bluffs

There are various locations in North America where the updrafts generated along river bluffs are used by migrating raptors in autumn. In Saskatchewan, for example, migrating Bald Eagles in spring also use river bluff updrafts (Gerrard and Gerrard, 1982), and autumn raptor migrations also use updrafts along Mississippi River bluffs in Minnesota (Reese, 1973). More than a thousand migrating Bald Eagles also use deflective updrafts along high Mississippi River bluffs at the Eagle Valley Nature Preserve near Cassville, Wisconsin (Ingram, Brophy, and Sherman, 1982), while farther south in Kentucky the updrafts along Mississippi River bluffs again are used by migrant Red-tailed Hawks in autumn (Mengel, 1965).

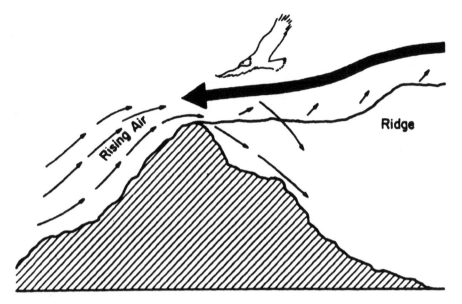

Mechanics of updraft formation along mountain ridges. Surface winds strike the sides of mountains and are deflected upward, thereby creating favorable air currents upon which migrating hawks soar. *Redrawn from Heintzelman (1972d).*

SOARING FLIGHT

Birds of prey that use deflective updrafts along mountain ridges or other geographic barriers make extensive use of soaring flight of one type or another. Soaring flight, however, is divided into two types: (1) dynamic and (2) static. The former is used primarily by albatrosses and other seabirds but not by raptors. Static soaring, on the other hand, is the major method of flight employed by many hawks during migration (and during other periods of their annual cycles). Cone (1961) classified static soaring into three basic types based upon the manner in which vertical air currents are produced: (1) deflective updrafts, (2) thermals, and (3) combinations of deflective updrafts and thermals. This chapter discusses deflective updrafts and combinations of those updrafts and thermals, whereas thermal soaring is discussed in the next chapter.

Deflective Updraft Soaring

At places like Bake Oven Knob and Hawk Mountain in Pennsylvania, deflective updrafts are used extensively by all the migrant hawks seen there (Broun and Goodwin, 1943). Depending upon the velocity of the prevailing winds on a particular day, the strength of the deflective updrafts varies as does the manner in which hawks hold their wings to secure maximum use of those air currents. When relatively weak updrafts occur hawks may extend their wings fully or nearly so, but when stronger deflective updrafts are being generated hawks fold their primaries back to streamline their overall shape to produce less drag. Occasionally, when extremely strong deflective updrafts occur, a hawk folds its wings almost against its body and becomes a projectile hurtling through the air. Ospreys are particularly adept at streamlining their shape to accommodate various deflective up-draft conditions, resulting in their characteristic crooked-wing appearance. All migrant raptors, however, employ similar tactics, and one sometimes sees a Sharp-shinned Hawk or American Kestrel with its wings folded almost against its body while soaring on very strong deflective updrafts. The buteos also adjust their wing shapes as wind conditions require, but the overall appearance of their adjustments may be somewhat less dramatic than some of the other hawks.

Deflective Updraft and Thermal Soaring

On days when light surface winds occur along mountain ridges the result-ing deflective updrafts may be relatively weak. Migrant hawks continue to use those updrafts but also tend to use thermals they encounter to gain lift and altitude. Thus combinations of deflective updrafts and thermals some-times are used as necessary or as opportunities occur (Broun and Goodwin, 1943). Buteos such as Broad-winged Hawks and Red-tailed Hawks are very prone to the use of combinations of such air currents, but Sharp-shinned Hawks and many other species also use combinations as well.

COMPARATIVE SPECIES
USE OF DEFLECTIVE UPDRAFTS

To some extent a trend exists concerning the comparative species use of deflective updrafts along mountain ridges as opposed to coastal or other areas.

At Hawk Mountain, Pennsylvania, Broun (1949a: 151) reported that four *Buteo* species formed about 68.11 percent of the autumn hawk flights, three *Accipiter* species formed another 26.29 percent of the flights, and three falcon species formed 1.09 percent of the flights. Collectively two *Buteo* and one *Accipiter* species represented 90.43 percent of the flights: Broad-winged Hawks (41.86 percent), Red-tailed Hawks (24.96 percent), and Sharp-shinned Hawks (23.61 percent).

At Bake Oven Knob, Pennsylvania (Heintzelman, 1963a—1984e; Heintzelman and MacClay, 1972–1979; Heintzelman and Reed, 1982), four *Buteo* species represented 69.3 percent of the autumn hawk flights, three *Accipiter* species another 23 percent of the flights, and falcons a

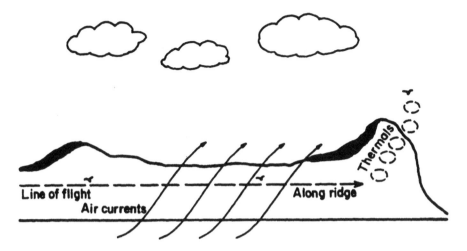

A mountain ridge showing surface winds being deflected upward after striking the sides of the mountain. Migrating hawks soar on these updrafts but also may use thermals as shown at the right. *Modified and redrawn after Simpson (1954).*

further 1.45 percent of the flights. Collectively Sharp-shinned Hawks, Broad-winged Hawks, and Red-tailed Hawks represented 89.86 percent of the Bake Oven Knob flights: Sharp-shinned Hawks (21.8 percent), Broad-winged Hawks (53.2 percent), and Red-tailed Hawks (14.7 percent).

Autumn hawk flights reported along coastal New Jersey at Cape May Point (Allen and Peterson, 1936; Choate *in* Heintzelman, 1970e; Choate, 1972; Choate and Tilly, 1973; Clark, 1972, 1973, 1975b, 1976; Dunne, 1978b, 1979b, 1980b; .Dunne, el al; 1981; Dunne and Clark, 1977; LeGrand, 1984; LeGrand, et al; 1983), however, are composed of 9.8 percent buteos, 58.2 percent accipiters, and 25.9 percent falcons. Sharp-shinned Hawks, Broad-winged Hawks, and Red-tailed Hawks collectively represent 65.62 percent of the Cape May Point flights: Sharp-shinned Hawks (56.1 percent), Broad-winged Hawks (6.48 percent), and Red-tailed Hawks (2.97 percent). Apparently buteos are much more dependent upon the deflective updrafts encountered along mountain ridges than are accipiters, whereas falcons clearly prefer coastal migration routes but not necessarily because of differences in air currents there and at inland ridge sites.

C H A P T E R   16

Hawk Migrations and Thermals

The use of thermals by hawks, particularly Broad-winged Hawks and Swainson's Hawks, is of major importance when considering hawk migrations. It is essential to the success of these birds reaching their winter ranges in the neotropics and their return to the North American breeding grounds.

THERMAL FORMATION

Thermals are formed when sunlight strikes a land surface with features such as rocks, sand, gravel, quarries, forests, and towns on which there is a discontinuity in the rate at which air is heated. Thus the ground, whose temperature rises more rapidly than air temperature, provides radiating surfaces which are in contact with the air and results in localized increases in air temperature (Foster, 1955). Large differences in temperature are not required for thermal production—fractions of a degree centigrade over horizontal distances of a half mile are sufficient (Eastwood, 1967: 184). Eventually a layer of warming air forms a large bubble, breaks away from the thermal-producing source, and rises into the atmosphere until the dew point is reached where water vapor in the bubble forms a cumulus cloud (Foster, 1955).

THERMAL STRUCTURE

There are two important theories of thermal structure: (1) vortex rings within bubbles, and (2) continuous columns of rising warm air.
Vortex Rings Within Bubbles
The most widely accepted explanation of thermal structure is that the air mass is an invisible bubble, containing a vortex ring of warm air, rising into

236

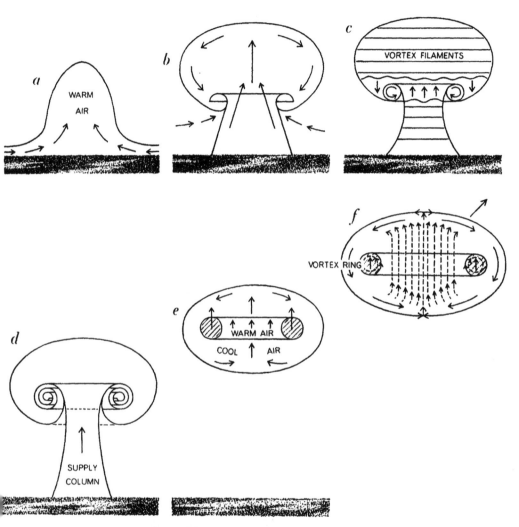

Mechanics of thermal bubble formation. Warm air on the ground rises (a) and assumes a bubble-like shape (b). Turbulence at the surface of the bubble forms twisting filaments that rapidly coil into a toroidal vortex ring similar to a ring of smoke (c, d). The warm air mass is buoyant and is pinched off from the ground by cool air flowing inward (e). The thermal bubble or shell then floats away (f). The bubble then entrains some cool air from the outside. This air circulates upward through the center of the vortex ring then downward around the outside of the ring. *Reprinted from "The Soaring Flight of Birds" by Clarence D. Cone, Jr. Copyright © 1962 by Scientific American, Inc. All rights reserved.*

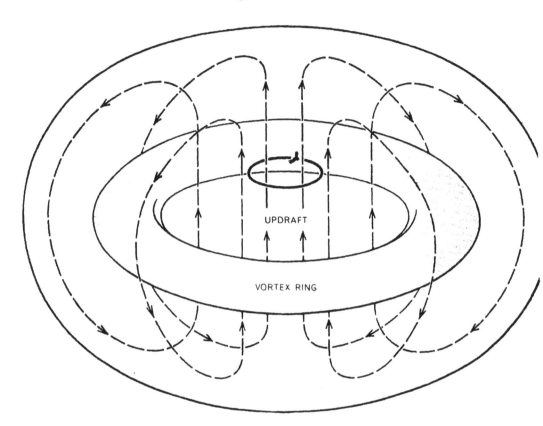

A thermal bubble or shell contains a vortex ring of warm air around which a current of cooler air circulates in a streamlined pattern (broken lines). A soaring hawk flies in a circle with the radius giving the hawk an aerodynamic sinking velocity equal to the updraft velocity. The bird thus is held in an equilibrium position inside the thermal bubble which, in turn, rises aloft and can be subject to wind drift. *Reprinted from "The Soaring Flight of Birds" by Clarence D. Cone, Jr. © 1962 by Scientific American, Inc. All rights reserved.*

the atmosphere from its point of origin (Cone, 1961, 1962a, 1962b). Observations of kettles of Broad-winged Hawks milling around in a thermal tend to assume the general shape of a bubble, thus supporting this theory.

Columns of Warm Air

In some of the older hawk migration literature thermals are referred to as a continuous column of warm air rising from its source to the dew point, where cumulus cloud development signals the point of decay. Generally this continuous column theory is not accepted currently.

Dust Devils

Dust devils are odd variations of standard thermal structure. They are whirling, rising columns of warm air whose presence can be seen when dust particles are lifted to high altitudes within the air mass. Black Vultures and other raptors sometimes appear to locate dust devils visually, fly into them, and use them to ride to great altitudes without expending much energy (Kessler, 1975: 113–114). Dust devils, however, probably play a relatively minor role as aids to hawk migrations in North America.

RAPTOR USE OF THERMALS

Several important factors relate to raptor use of thermals to aid them in their migrations: (1) species preference for thermal soaring, (2) seasonal considerations, and (3) behavior in thermals.

Species Preference for Thermal Soaring

Among migrant raptors in North America, all species use thermal soaring as a migration aid to some degree (Broun and Goodwin, 1943). Two species are especially dependent upon this flight style, however, and deserve particular attention—Broad-winged Hawks and Swainson's Hawks. Both species move to and from their North American breeding grounds in large flocks. In temperate North America, however, thermals and deflective updrafts are used as opportunities occur, but as soon as the hawks arrive in the neotropics thermal soaring becomes the dominant flight style for both species (N. G. Smith, 1980: 54). Other migrating hawks also use thermals to some extent but generally not as their dominant flight styles and rarely in large flocks. Nevertheless, individual birds, or occasionally small numbers of the same species, engage in thermal soaring.

There is some evidence to suggest that spring hawk flights occur at much higher altitudes than are reported for autumn hawk migrations at inland ridge sites such as Raccoon Ridge, New Jersey. The spring migrants also seemed to be less dependent upon deflective updrafts generated along the ridge, but more dependent upon thermal soaring (Dunne, 1977). During the spring 1977 hawk migration at Raccoon Ridge, Dunne (1978a: 45) reported that 38 percent of the ten most common species were observed using thermal soaring during the period each bird was watched passing the observation site: Ospreys (34 percent), Northern Harriers (33 percent), Sharp-shinned Hawks (40 percent), Cooper's Hawks (62 percent), Northern Goshawks (33 percent), Red-shouldered Hawks (47 percent), Broad-winged Hawks (41 percent), Red-tailed Hawks (23 percent), Rough-legged Hawks (40 percent), and American Kestrels (38 percent). Similar data are needed from many other sites during autumn and spring before an overall statement of percentage of thermal dependence can be made.

Seasonal Considerations

The seasonal restrictions on hawk migrations in North America are crucial when considering raptor use of thermals. Each species has a definite range of autumn and spring dates between which migrations occur. Fairly well-defined peak migratory periods exist between the extreme dates. It is during those peak periods that the bulk of the individuals of a particular species are seen in North America (Broun, 1949a; Haugh, 1972; Heintzelman, 1979a; Mueller and Berger, 1961; Rosenfield and Evans, 1980; Bednarz, 1978; Dumont, 1934; Binford, 1979; Haugh and Cade, 1966).

It is general knowledge, therefore, that mid-September is the peak migratory period for Broad-winged Hawks and Swainson's Hawks in the more northern latitudes, late September to early October is the peak period in more southerly North American latitudes, and late October and early November in Panama. The northward spring migration peak in Panama is in mid-March, and the birds tend to arrive in temperate North America by mid- to late April. Broad-winged Hawks average approximately 28 days and Swainson's Hawks average 70 days engaged in each seasonal migration (N. G. Smith, 1980).

Behavior in Thermals

The manner in which both Broad-winged Hawks and Swainson's Hawks use thermals during their migratory periods is spectacular and more characteristic of those two species than of other migrant North American raptors, although they also sometimes use thermal soaring. Observations made from a motor-glider in New England indicated that some Broad-winged Hawk kettles in thermals are formed, used, and vacated more or less as a group as opposed to a "self-perpetuating" kettle in which some hawks enter a thermal at a lower altitude as other hawks at a higher altitude glide from the top of the thermal (Hopkins, Mitchell, and Welch, 1978).

Not infrequently hundreds, at times even thousands, of these hawks form kettles (nonsocial flocks) in thermals. The birds mill around within the kettle, and as the thermal carries them aloft, there is no apparent organization among the birds. Some photographs of Broad-winged Hawks in kettles, however, seem to suggest a distribution of hawks in a pattern similar to parts of the vortex ring structure discussed by Cone (1962b) as part of the internal structure of thermal bubbles.

As the kettle of hawks within a thermal approaches the dew point, however, the thermal cools and expands, energy within the thermal escapes, and the hawks receive no further lift. They then begin gliding downward, either singly or a few at a time, forming a line or stream of birds. Eventually some or all of the hawks detect another thermal, perhaps visually (Hopkins, Mersereau, and Welch, 1975: 4) as suggested by observations in which hawks were observed changing direction to join other hawks in a kettle, and repeat the process.

It is this process of thermal soaring that successfully carries Broad-winged Hawks and Swainson's Hawks from North America, through Central America, to their South American winter ranges (Heintzelman, 1975; N. G. Smith, 1980).

Occasionally other species of hawks also are seen within kettles of Broad-winged Hawks or Swainson's Hawks, and sometimes limited numbers of Red-tailed Hawks or other species also form very small kettles, but they are not the ordinary flight styles used by those birds.

CHAPTER 17

Altitudes, Daily Rhythms, and Noon Lulls in Hawk Flights

Among the important and fascinating aspects of North American hawk migrations are (1) altitudes at which the birds fly, (2) daily rhythms exhibited by some of the species, and (3) noon lulls that occur in some flights of certain species.

ALTITUDES OF FLIGHTS

Measurements of the altitudes at which migrant hawks can fly in North America and the neotropics are of academic interest, important to ornithologists and hawk watchers making hawk counts at migration lookouts because birds passing too high to be seen from the ground could produce a serious bias in the counts, and have important implications concerning the evolution of the endemic insular subspecies of the Broad-winged Hawk. Thus a review of information on altitudes of hawk flights is desirable.

As a general rule, it appears that many migrating hawks in North America (at least in autumn) utilize an altitude zone between approximately 1,950 and 2,925 feet as measured by fixed-beam radar (Kerlinger and Gauthreaux, 1983: 14; Kerlinger, 1983: 7). These altitudes generally are consistent with those reported by observers on a blimp over northern New Jersey (Stearns, 1948a, 1948b, 1949), and also generally are consistent with some measurements made in New England by observers on a motor-glider (Hopkins, Mitchell, and Welch, 1978). The tracking radar studies near Albany, New York, reported by Kerlinger (1980), however, suggested that some hawks migrate at different altitudes—Sharp-shinned Hawks lower than Broad-winged Hawks, and Red-tailed Hawks higher than Broad-winged Hawks.

Vultures and Condors

Few measurements are available of the altitudes at which North American vultures fly. In New England, in autumn, several observations of Turkey Vultures were made by hawk watchers on a motor-glider at altitudes ranging between 1,000 and 2,200 feet (Hopkins, et al., 1979). A Turkey Vulture at Cape May Point, New Jersey, however, was recorded via radar at 3,200 feet (Kerlinger and Gauthreaux, 1983: 14). Air Force pilots in Panama also often reported Turkey Vultures at altitudes between 11,700 and 13,000 feet, and occasionally these birds reached 20,800 feet when they glided over the top of tropical storms (N. G. Smith, 1980: 56).

At Hollister, California, two California Condors also were reported from a jet airplane at an altitude estimated at about 15,000 feet (Greenberg and Stallcup, 1974: 847).

Ospreys and Harriers

Flight altitudes of Ospreys were measured by several people at several locations in North America. In New England in autumn, these birds were seen from a motor-glider between 1,000 and 4,400 feet (Hopkins, Mitchell, and Welch, 1978; Hopkins, et al., 1979) whereas an Osprey at 3,200 feet was recorded via radar at Cape May Point, New Jersey (Kerlinger and Gauthreaux, 1983: 14).

A Northern Harrier at Cape May Point, New Jersey, was tracked by radar at 3,000 feet (Kerlinger and Gauthreaux, 1983: 14).

Eagles

In New England, autumn flight altitudes of two Bald Eagles were 1,300 feet as observed from a motor-glider (Hopkins, et al., 1979). I have no measurements of flight altitudes of Golden Eagles, but at Bake Oven Knob, Pennsylvania, I estimated that these birds sometimes reached at least 2,500 feet.

Accipiters

Autumn flight altitudes are available only for Sharp-shinned Hawks which were determined in New England by observers on a motor-glider and ranged between 1,200 and 3,700 feet (Hopkins, Mitchell, and Welch, 1978; Hopkins, et al., 1979). In autumn at Albany, New York, a tracking radar measured Sharp-shinned Hawks at an average altitude of 1,104 feet with a range between 417 and 1,658 feet (Kerlinger, 1980: 36). In autumn at Cape May Point, New Jersey, radar tracks of these hawks indicated an altitude of 3,200 feet (Kerlinger and Gauthreaux, 1983: 14).

Buteos

Autumn flight altitudes of Red-shouldered Hawks in New England were determined by observers on a motor-glider and ranged between 2,800 and 3,500 feet (Hopkins, Mitchell, and Welch, 1978).

Numerous measurements of migrating Broad-winged Hawk altitudes are reported in the literature. In New Jersey, autumn flight altitudes were determined from a blimp and ranged between 1,500 and 2,900 feet (Stearns, 1948b). In New England, altitudes as measured from a motor-glider ranged between 800 and 4,700 feet with many birds noted between 2,000 and 3,000 feet (Hopkins, Mitchell, and Welch, 1978; Hopkins, et al., 1979). Forbes and Forbes (1927: 101–102) also watched hawks (probably Broad-wings) soar into clouds estimated at an altitude of 7,000 feet over Mt. Monadnock. Tracking radar determination of Broad-wing altitudes in autumn at Albany, New York, ranged between 447 and 1,949 feet with an average of 1,155 feet (Kerlinger, 1980: 36). In Kentucky, in autumn, approximately 3,000 Broad-winged Hawks in two large kettles were seen from an airplane at 4,500 feet and later watched rising into opaque clouds at 6,000 feet altitude (Mayfield, 1980: 23).

Flights of migrating Broad-winged Hawks and Swainson's Hawks in Panama also were seen by pilots at altitudes between 11,700 and 13,000 feet, and occasionally as high as 20,800 feet (N. G. Smith, 1980: 56).

Tracking radar determination of Red-tailed Hawk altitudes at Albany, New York, ranged between 409 and 2,050 feet with an average of 1,252 feet (Kerlinger, 1980: 36).

Falcons

Measurements of the autumn migration altitudes of American Kestrels are available from several locations in North America. In autumn in New England, for example, these falcons were seen from a motor-glider between 2,100 and 3,500 feet and averaged 2,825 feet (Hopkins, Mitchell, and Welch, 1978; Hopkins, et al., 1979). An American Kestrel at Cape May Point, New Jersey, was recorded via tracking radar at 2,500 feet (Kerlinger and Gauthreaux, 1983: 14).

A Merlin also was seen from a motor-glider at 1,500 feet altitude in New England (Hopkins, et al., 1979), and a Peregrine Falcon was recorded via tracking radar at 2,100 feet altitude at Cape May Point, New Jersey (Kerlinger and Gauthreaux, 1983: 14).

DAILY RHYTHMS

Hawk watchers at various hawk lookouts in the East long have discussed the daily rhythm patterns of migrating diurnal birds of prey. Mueller and Berger (1973) correctly pointed out, however, that relatively little information is published on the subject.

At Cape May Point, New Jersey, Allen and Peterson (1936) reported that most Sharp-shinned Hawks appeared in the morning, whereas Merlins tended to appear in the afternoon. At Hawk Mountain, Pennsylvania, Broun (1939, 1949a) reported that the majority of Sharp-shinned Hawks seen during September appeared during the afternoon. Since many ideas about daily rhythms of migrant hawks are based upon subjective impressions instead of statistical data, it is worthwhile to examine trends in the daily rhythms of selected raptors more carefully.

Ospreys and Harriers

Daily rhythms of autumn migrant Ospreys at Cedar Grove, Wisconsin, exhibited approximately equal daily activity patterns with no obvious peak period during the day (Mueller and Berger, 1973: 592–593). At Bake Oven Knob, Pennsylvania, analysis of daily rhythm patterns of 986 Ospreys observed in autumn from 1979–1984 yielded no sharply defined activity peak although 162 birds (16.4 percent) of the sample appeared between 1400 and 1500 hours, EST. Between 1000 and 1600 hours, however, 796 Ospreys (80.7 percent of the sample) were seen indicating only slight hourly variation during that period.

Northern Harriers showed a peak daily rhythm period during 0800 and 0900 hours, CST, at Cedar Grove, Wisconsin (Mueller and Berger, 1973: 592–593).

Eagles

A long-standing idea, reported by Broun (1949a: 164) at Hawk Mountain, Pennsylvania, suggested that eagles in autumn are particularly likely to appear late in the afternoon, although statistical data were not provided to support the idea.

At Bake Oven Knob, Pennsylvania, however, my analysis of 326 autumn Bald Eagle sightings made between 1961 and 1981 demonstrated that these eagles reached a peak in their migratory activity between noon and 1300 hours, EST, with the greatest possibility of seeing Bald Eagles between 1000 and 1600 hours (Heintzelman, 1982b: 65–67). With the addition of another 66 Bald Eagles seen at the Knob from 1982–1984 (Heintzelman, 1982i, 1983c, 1984e), raising the sample size to 392 birds, the daily rhythm pattern of these birds was not altered in any important way. Based upon the new sample of 392 birds, only 20 eagles (5.1 percent) were seen after 1600 hours whereas 74 birds (18.8 percent) appeared between 1200 and 1300 hours, and 347 eagles (88.5 percent) passed the Knob between 1000 and 1600 hours.

At Cedar Grove, Wisconsin, autumn Bald Eagle migrations showed a peak in daily rhythm between 1100 and 1200 hours, CST, but the sample contained only 13 birds (Mueller and Berger, 1973: 592).

The Bald Eagle daily rhythm noted at Bake Oven Knob is similar to that reported for the species during spring migration near Saskatoon, Saskatchewan. The peak in the daily rhythm there also occurred between 1200 and 1300 hours with 24 percent of the sample sighted then. Between 1000 and 1600 hours, sightings of migrant eagles formed 82 percent of the sample. After 1600 hours, only 12 percent of the sample was observed (Gerrard and Gerrard, 1982: 98).

Winter activity rhythms of Bald Eagles along the Missouri River in South Dakota exhibited the most movement or activity during the first six

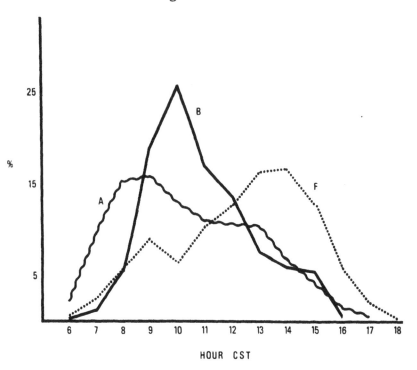

The daily rhythm of migrating hawks at Cedar Grove, Wis. The graph shows the hourly distribution of migrant *Accipiter* species (wavy line, n = 4,602), *Buteo* species (solid line, n = 2,317) less the Broad-winged Hawk, and *Falco* species (dotted line, n = 987). The value for each hour interval (e.g., 0700–0759) represents the percentage of the total number of hawks observed during the four-year period 1958–1961. *Reprinted from Mueller and Berger (1973).*

hours after sunrise after which activity declined significantly (Steenhof, 1983).

Analysis of 450 autumn Golden Eagle sightings at Bake Oven Knob, Pennsylvania, between 1961 and 1981 demonstrated a peak in migratory activity or daily rhythm between 1300 and 1400 hours, EST, with the bulk of these birds passing the site between 1000 and 1600 hours (Heintzelman, 1982b: 65–67). Adding an additional 108 Golden Eagle sightings gathered between 1982–1984 increased the sample size to 558 birds, but failed to change the basic daily rhythm of the species. Between 1000 and 1600 hours, EST, 473 eagles (84.7 percent of the sample) appeared, whereas only 25 eagles (4.4 percent) occurred after 1600 hours.

Accipiters

The daily rhythms of accipiters migrating past Cedar Grove, Wisconsin, in autumn tended to peak in early morning. Sharp-shinned Hawks had an activity peak between 0900 and 1000, CST, for example, whereas Cooper's Hawks peaked between 1000 and 1100, CST (Mueller and Berger, 1973: 592–593). A similar daily rhythm pattern exists at Bake Oven Knob, Pennsylvania, for accipiters migrating past that site in autumn (Heintzelman, 1975: 214).

Buteos

The daily rhythm of buteos migrating in autumn past Cedar Grove, Wisconsin, exhibited a late morning activity peak with the exception of Broad-winged Hawks which showed midmorning and late afternoon activity peaks (Mueller and Berger, 1973: 591–596). At Bake Oven Knob, Pennsylvania, however, buteos showed an afternoon peak in daily rhythm (Heintzelman, 1975: 214).

Falcons

The daily rhythms of autumn falcon migrations at Cedar Grove, Wisconsin, showed an early afternoon peak. American Kestrels at that site showed a peak activity rhythm between 1300 and 1400 hours, CST (Mueller and Berger, 1973).

Daily rhythms of migrant Merlins in autumn at Bake Oven Knob, Pennsylvania, between 1962 and 1984 are based upon analysis of a sample of 217 birds. The peak activity occurred between 1400 and 1500 hours, EST, when 14.7 percent of the sample appeared whereas 87.5 percent of the sample passed the Knob between 0800 and 1500 hours. A rapid decrease in Merlin migratory activity occurred after 1500 hours—only 11.5 percent of the sample being seen then.

These daily rhythm patterns are somewhat similar to those exhibited by autumn Merlin flights at Cape May Point, New Jersey, and Cape Charles, Virginia (Dunne, 1979b: 20), although Merlin activity at the Knob developed more rapidly during the morning. At all three locations, however, a rapid decrease in migratory activity occurred in late afternoon.

The peak in daily rhythm of autumn Merlin migrations at Cedar Grove, Wisconsin, also occurred between 1400 and 1500 hours, CST, with 20 percent of the sample reported then (Mueller and Berger, 1973: 592).

Daily rhythm in autumn Peregrine Falcon migrations at Bake Oven Knob, Pennsylvania, between 1962 and 1984, is based upon analysis of a sample of 183 birds. Three activity periods or peaks occurred—one between 1000 and 1100 hours, EST, another between 1200 and 1300 hours, and a third larger peak between 1400 and 1500 hours. Between 0900 and 1600 hours, 171 falcons (93.4 percent of the sample) appeared, whereas the 89 falcons seen collectively in the three peak periods accounted for 48.6 percent of the entire sample.

At Cedar Grove, Wisconsin, autumn Peregrine Falcon migrations showed a peak in daily rhythm between 1300 and 1400 hours, CST, whereas 70 percent of the sample appeared between 1000 and 1500 hours

(Mueller and Berger, 1973: 592). Along the Texas coast, however, autumn Peregrine Falcon migrations showed a daily rhythm with an activity peak between 0900 and 1000 hours, and 79 percent of the entire sample (249 birds) was seen after 0800 hours and prior to 1100 hours (Hunt, Rogers, and Slowe, 1975: 113).

NOON LULLS IN MIGRATIONS

A particularly curious feature of autumn hawk migrations is the lull which often develops in some hawk flights during the middle of the day. These noon lulls can vary by an hour or more on either side of noon, and some flights do not contain them. However, it is notable that flights without lulls usually occur on days when heavy cloud cover prevails.

Noon lulls are well known to Pennsylvania hawk watchers. Broun (1949a: 152), for example, mentioned their regular occurrence during September and October hawk flights at Hawk Mountain. Heintzelman and Armentano (1964) also reported similar lulls in many Bake Oven Knob, Pennsylvania, hawk flights, and data from various other lookouts in eastern North America also show noon lulls in some flights.

Species for which noon lulls in some daily flights are especially easy to detect are Sharp-shinned Hawks, Broad-winged Hawks, and Red-tailed Hawks. Lulls also may occur in flights of other species, but they may be more difficult to detect because of smaller numbers of birds in those samples.

Descriptions of Noon Lulls

To illustrate some typical noon lulls in autumn hawk flights, graphs of selected Sharp-shinned Hawk, Broad-winged Hawk, and Red-tailed Hawk data for single days are provided with the hawk count data grouped into frequency distributions based upon one-hour intervals.

Four examples of Sharp-shinned Hawk flights illustrate various noon lull patterns—some lulls more pronounced than others. Nevertheless, a noon lull occurred in each flight.

A curious sidelight of the noon lull phenomenon in Sharp-shinned Hawk flights at Hawk Mountain, Pennsylvania, *during September*, is that the bulk of these birds *"always"* appear during the afternoon (Broun, 1949a: 157), whereas this pattern does not occur at nearby Bake Oven Knob, Pennsylvania. Analysis of September Sharp-shinned Hawk flights at the Knob between 1964 and 1984, based upon 396 observation days and a sample of 14,922 Sharp-shinned Hawks, demonstrated that 7,917 birds (53 percent) were seen prior to noon and 7,005 birds (47 percent) were seen during the afternoon. At Cape May Point, New Jersey, these hawks fly largely during the morning (Broun, 1949a: 157), and at Cedar Grove, Wisconsin, autumn Sharp-shin flights also peaked in the morning (Mueller and Berger, 1973).

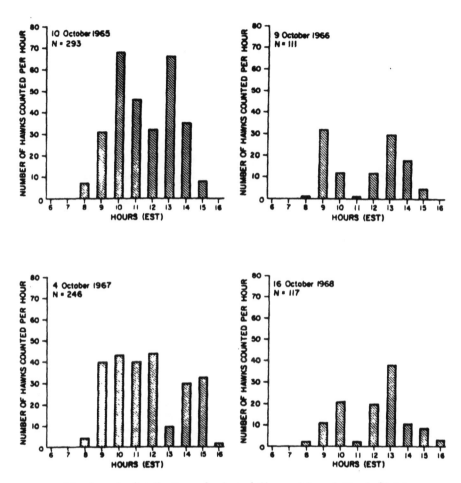

The hourly distribution of selected Sharp-shinned Hawk flights
at Bake Oven Knob, Pa. Noon lulls are more evident in some
flights than in others. *Reprinted from Heintzelman (1975).*

Noon lulls frequently appear in Broad-winged Hawk flights, too, appar-
ently as a result of the dependence of these birds upon thermal soaring.
Some typical examples of lulls in Broad-wing flights at Bake Oven Knob,
Pennsylvania, are illustrated in the accompanying graphs.

The Red-tailed Hawk is one of the largest soaring hawks passing Bake
Oven Knob, Pennsylvania during autumn. These birds make extensive use
of deflective updrafts along the Kittatinny Ridge and other ridges. Occa-
sionally in autumn a few Red-tails also make limited use of thermal soaring
(Broun, 1949a: 158–159), but never to the extent that Broad-winged
Hawks do. In spring, however, Red-tails were reported to use thermal

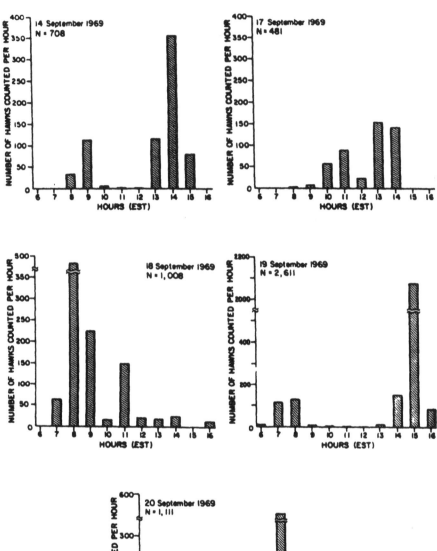

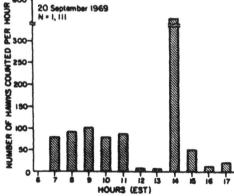

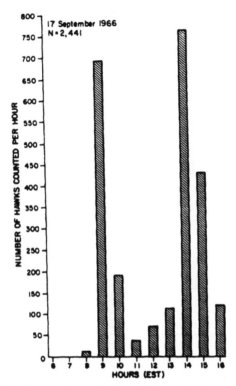

The hourly distribution of selected Broad-winged Hawk flights (Figures on opposite page and above) at Bake Oven Knob, Pa. Conspicuous noon lulls are evident in each of these flights. *Reprinted from Heintzelman (1975).*

soaring in as much as 23 percent of their migration (Dunne, 1978a: 45). Despite their relatively limited use of thermals, Red-tails often exhibit noon lulls in their migrations—especially on days with blue skies and limited or no cloud cover. Four typical variations in noon lulls in Red-tailed Hawk flights at Bake Oven Knob, Pennsylvania, are illustrated in the accompanying graphs.

Possible Causes of Noon Lulls

A variety of possible causes of noon lulls in hawk flights were suggested by ornithologists in the past, some much more likely to be correct than others (Broun, 1949a: 152; Heintzelman, 1975: 223–234). Least likely as a major cause of noon lulls in hawk flights are landing and feeding, interspecific behavior of migrants in flight, and low-altitude parallel-to-lookout flights, although each occasionally may cause minor lulls in hawk migrations. More likely as occasional causes of noon lulls in hawk flights are high-altitude

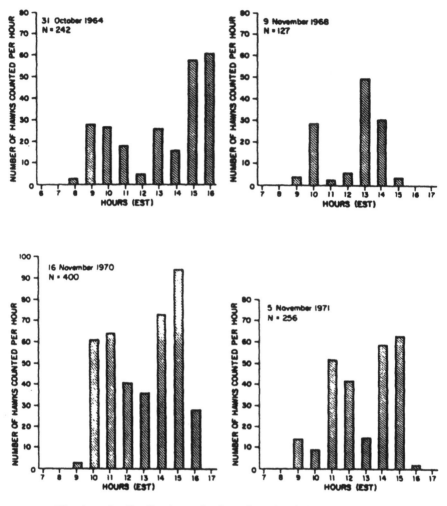

The hourly distribution of selected Red-tailed Hawk flights at
Bake Oven Knob, Pa. Noon lulls are more evident in some flights
than in others. *Reprinted from Heintzelman (1975).*

parallel-to-lookout flights, and diagonal broad-front migrations combined
with ridge-hopping.

Sometimes migrant hawks, including Sharp-shinned Hawks, Broad-
winged Hawks, and Red-tailed Hawks fly at extremely high altitudes, following a course parallel to a mountain such as the Kittatinny Ridge in
Pennsylvania but far out over the valley north or south of the ridge. Noon
lulls could develop in the visual migration if the birds increased the altitude
and/or distance at which they flew from the ridge to the point that observers

on a lookout failed to see some of the birds. On 26 October 1968, for example, because of poor light conditions and the distance of the hawks from Bake Oven Knob, Pennsylvania, a 20X telescope was used to identify the 161 Red-tailed Hawks which passed the Knob; some birds could have been missed had only binoculars been used.

Diagonal broad-front hawk migrations combined with ridge-hopping also can produce noon lulls in some flights. As hawks diagonally approach a mountain they may be seen at one site but not at another. In New Jersey, for example, Edwards (1939), Eynon (1941), and Lang (1943) were well aware of such cross-ridge migrations, sometimes at high altitudes. Similar flights sometimes are seen in Pennsylvania in autumn. On 18 September 1965, for example, Alan and Paul Grout were stationed on the Kittatinny Ridge at Bear Rocks less than two miles southwest of Bake Oven Knob, Pennsylvania. They noticed 335 Broad-winged Hawks flying toward them from a northerly direction. These hawks were not seen upridge at Bake Oven Knob. Apparently the hawks were migrating over a broad front. Upon reaching the mountain, they changed direction and followed it as a diversion-line or leading-line for an unknown distance—an example of a diagonal broad-front flight coupled with ridge-hopping. An example of a similar flight occurred on 14 September 1963, between Bake Oven Knob and Hawk Mountain (Heintzelman and Armentano, 1964: 8–10). Nagy (1970: 9) also reported a flight of Broad-winged Hawks at Hawk Mountain, Pennsylvania, that approached, and crossed, the ridge at right angles under conditions of strong northerly winds. Kettles of hawks were seen in all directions at the limit of his vision. Evidently this also was a broad-front migration. On 18 November 1983, at Bake Oven Knob, Pennsylvania, I also watched a Gyrfalcon engaged in cross-ridge migrations (Heintzelman, 1983c: 5).

The most likely and reasonable explanation for the cause of noon lulls in some Sharp-shinned Hawk, Broad-winged Hawk, Swainson's Hawk, and perhaps Red-tailed Hawk flights is the use of thermal soaring by these birds which carries them to high altitudes at midday beyond the limit of vision of observers on the ground (Heintzelman, 1975: 230–234). Stone (1937) and Kerlinger and Gauthreaux (1983: 14) reached similar conclusions, and the measurements of altitude of flight presented earlier in this chapter further support the validity of the conclusion as do the observations of hawks entering into, and flying within, opaque clouds or even flying above tropical storms in Panama.

PART FOUR

Migration Routes, Geography, and Hawk Counts

C H A P T E R   18

Age Classes and Differential Migrations

Age classes and differential hawk migrations in North America sometimes are related and thus should be discussed together.

AGE CLASSES

A variety of age class data are reported in the North American hawk migration literature based upon field observations or bird-banding data. A review of this information is useful, therefore, as a prelude to a discussion of differential hawk migrations.

Ospreys and Northern Harriers

Between 1982–1984, at Bake Oven Knob, Pennsylvania, ages were determined for 285 autumn Osprey migrants of which 275 (96.5 percent) were adults and 10 (3.5 percent) were juvenile (Heintzelman, 1982i, 1983c, 1984e), clearly indicating that adult birds preferred the inland mountain migration route. First year Ospreys fledged in Maryland and Virginia, however, used a narrow coastal migration route southward (Kennedy, 1973).

Ages of Ospreys migrating northward in spring from their neotropical wintering grounds, however, are at least two years old (Gillespie, 1960; Henny and Van Velzen, 1972).

At Hawk Mountain, Pennsylvania, the autumn Northern Harrier migration tended to contain between 33.9 and 48.3 percent immatures, but in 1942 immatures formed 77 percent of the sample (Broun, 1949a: 165). Between 1982 and 1984 at Bake Oven Knob, Pennsylvania, ages were determined for 298 autumn harrier migrants of which 110 (37 percent) were adult males, 54 (18 percent) were adult females, and 134 (45 percent) were immatures (Heintzelman, 1982i, 1983c, 1984e). At another Kittatinny Ridge hawk banding station in New Jersey, upridge from Bake Oven Knob, Pennsylvania, a sample of 38 autumn harrier migrants consisted of 5 (13 percent) adult males, 7 (18 percent) adult females, and 26 (64 percent) immatures (Bildstein, et al., 1984: 145).

Along coastal Connecticut, some 95–98 percent of Northern Harriers seen in autumn are immatures (Currie, 1978). At Cape May Point, New Jersey, however, a sample of 415 harriers contained 31 (7 percent) adult males, 29 (7 percent) adult females, and 355 (86 percent) immatures. Along the Great Lakes, 433 harriers at Hawk Cliff, Ontario, contained 11 (3 percent) adult males, 10 (2 percent) adult females, and 411 (95 percent) immatures whereas 370 harriers studied at Duluth, Minnesota, contained 7 (2 percent) adult males, 24 (6 percent) adult females, and 339 (92 percent) immatures (Bildstein, et al., 1984: 145).

Eagles

Age class data for migrant Bald Eagles are available from various sites. At Hawk Mountain, Pennsylvania, Broun (1961: 9) reported that 36.5 percent of these eagles counted there between 1931 and 1941 were immatures, whereas 23.1 percent were immatures between 1954 and 1960, which apparently reflected the impact of DDT upon eagle reproduction in the later years. However, a sample of 383 Bald Eagles seen at Hawk Mountain since 1967 (as reported in the Hawk Mountain newsletters) contained 279 (72.8 percent) adults and 104 (27.2 percent) immatures.

Upridge at nearby Bake Oven Knob, Pennsylvania, 177 Bald Eagles were seen between 1961 and 1971 with 144 (81.4 percent) adults and 33 (18.6 percent) immatures (Heintzelman and MacClay, 1972: 20; Heintzelman, 1975: 354). Between 1972 and 1984, at the Knob, another 215 Bald Eagles were seen of which 135 (62.8 percent) were adult and 80 (37.2 percent) were immature (Heintzelman, 1982i, 1983c, 1984e; Heintzelman and MacClay, 1973, 1975, 1976a, 1979; Heintzelman and Reed, 1982)—age classes that closely match the 1931–1941 pre-DDT era Hawk Mountain data but differ somewhat from recent Hawk Mountain Bald Eagle data. The autumn 1935 Bald Eagle migration at Cape May Point, New Jersey, contained 60 birds of which 50 (83.3 percent) were immature and 10 (16.7 percent) adult (Allen and Peterson, 1936: 403).

Spring Bald Eagle migrations near Saskatoon, Saskatchewan, as studied in 1981, produced a sample of 257 birds of which 192 (74.7 percent) were adult and 65 (25.3 percent) were immature (Gerrard and Gerrard, 1982: 98).

Age classes of Golden Eagles also are available from several locations. At Hawk Mountain, Pennsylvania, between 1935 and 1941, there were 434 eagles counted of which 159 (36.7 percent) were immatures and 275 (63.3 percent) were adult, whereas 290 Golden Eagles appeared between 1954 and 1960 of which 216 (74.5 percent) were adult and 74 (25.5 percent) were immature (Broun, 1961: 9). Golden Eagle age classes reported since 1967 in the Hawk Mountain newsletter represent 521 birds of which 278 (53.3 percent) were adult and 243 (46.7 percent) were immature.

At nearby Bake Oven Knob, Pennsylvania, also on the Kittatinny Ridge, between 1961 and 1971, a sample of 193 autumn Golden Eagle migrants contained 122 (63.2 percent) adults and 71 (36.8 percent) immatures (Heintzelman and MacClay, 1972; Heintzelman, 1975: 354), whereas between 1972 and 1984 another eagle sample of 353 birds contained 191 (54.1 percent) adult and 162 (45.9 percent) immature birds (Heintzelman, 1982i, 1983c, 1984e; Heintzelman and MacClay, 1973, 1975, 1976a, 1979; Heintzelman and Reed, 1982)—age classes very similar to those recently reported at Hawk Mountain.

Spring Golden Eagle migrations seen between 1978 and 1980 at Derby Hill, New York, produced 43 eagles of which 17 (39.5 percent) were adults and 26 (60.5 percent) were subadults and/or immatures (Muir, 1978, 1979, 1980).

Accipiters

Age class data for migrant Sharp-shinned Hawks are reported from various locations. At Hawk Mountain, Pennsylvania, 50 percent of these hawks were immatures in 1935, whereas in 1936 only 12 percent were immature (Broun, 1949a: 157). Upridge at nearby Bake Oven Knob, my data from 1982–1984 produced a sample of 1,325 Sharp-shins of which 715 (54 percent) were adult and 610 (46 percent) were immature, but with considerable yearly variation (Heintzelman, 1982i, 1983c, 1984e).

Along coastal New Jersey, at Cape May Point, the autumn age class of Sharp-shinned Hawks is almost entirely immature as demonstrated in 1935 when 8,180 (99.6 percent) of 8,206 birds were immatures (Allen and Peterson, 1936: 402). Farther north, along coastal Connecticut, the autumn migrations of Sharp-shinned Hawks contain mostly immatures (Trowbridge, 1895; Currie, 1978).

Age classes of autumn Sharp-shinned Hawk migrants at Duluth, Minnesota, between 1973 and 1978, based on a sample of 11,515 trapped birds, consisted of 3,091 (26.8 percent) adults and 8,424 (73.2 percent) immatures (Rosenfield and Evans, 1980). At Hawk Cliff, Ontario, between 1971 and 1981, an autumn sample of 15,509 Sharp-shinned Hawks was caught of which 3,296 (21.3 percent) were AHY (after hatching year) and 12,213 (78.7 percent) were HY (hatching year) birds (Duncan, 1982b).

Cooper's Hawks migrating past Hawk Mountain, Pennsylvania, during autumn are immatures in the majority (Broun, 1949a: 158), and immatures also predominate in autumn at Cape May Point, New Jersey (Allen and

Peterson, 1936: 402). Between 1971 and 1980 at Hawk Cliff, Ontario, a sample of 960 autumn Cooper's Hawk migrants contained 220 (23 percent) AHY birds and 740 (77 percent) HY birds (Duncan, 1981).

Northern Goshawks exhibit cyclical peaks every three-to-five years (Heintzelman, 1982h). The age classes of autumn migrants of this species vary considerably depending upon where in a cycle the data are taken and other ecological factors. At Bake Oven Knob, Pennsylvania, the autumn 1972 Northern Goshawk flight contained 79.3 percent adults and 20.7 percent immatures based on a sample of 343 birds (Heintzelman and MacClay, 1973: 3), whereas the autumn 1973 echo-flight produced 64 (64.6 percent) adults and 35 (35.4 percent) immatures based on a sample of 99 birds (Heintzelman and MacClay, 1975: 17). Between 1982 and 1984 at the Knob, ages of 155 Northern Goshawks separated into 98 (63.2 percent) adults and 57 (36.8 percent) immatures (Heintzelman, 1982i, 1983c, 1984e).

At Duluth, Minnesota, the autumn 1972 Northern Goshawk invasion produced 5,352 birds of which about 15 percent were immatures (Hofslund, 1973). At Cedar Grove, Wisconsin, autumn Northern Goshawk flights contained predominately juveniles in non-invasion years, whereas adults predominated during the 1962–1963 and 1972–1973 invasions (Mueller, Berger, and Allez, 1977).

Buteos

Autumn Red-shouldered Hawk migrations at Hawk Mountain, Pennsylvania, contain largely adults (Broun, 1949a: 160). Upridge at Bake Oven Knob, between 1982 and 1984, a sample of 240 autumn Red-shoulder migrants also contained 198 (82.5 percent) adults and 42 (17.5 percent) immatures (Heintzelman, 1982i, 1983c, 1984e).

Age classes of autumn Broad-winged Hawks in North America are reported from various locations. Along the Atlantic coast, where buteos are not the predominate migrants, immatures form most of the flights in Connecticut (Currie, 1978), 90 percent of the flights at Cape May Point, New Jersey (Darrow, 1983: 38), and most of the autumn flights and winter population in the Florida Keys (Darrow, 1983: 38; Tabb, 1979: 60). In northwest Florida, approximately 83.3 percent of autumn Broad-wing migrants also are immatures (Duncan and Duncan, 1975: 2). Inland along the Kittatinny Ridge at Bake Oven Knob, Pennsylvania, a sample of 831 migrant Broad-wings was secured in 1983–1984 of which 678 (81.6 percent) were adults and 153 (18.4 percent) were immatures (Heintzelman, 1983c, 1984e).

Of the Swainson's Hawks reported in autumn and winter in the Florida Keys, most also tend to be immatures (Darrow, 1983).

Red-tailed Hawk age class data also are reported from various locations in North America. At Hawk Mountain, Pennsylvania, a sample of 16,620 autumn Red-tail migrants contained 14,838 (89.3 percent) adults and 1,782 (10.7 percent) immatures (Broun, 1949a: 159). Upridge at Bake Oven Knob, Pennsylvania, between 1982 and 1984, a sample of 2,467 autumn Red-tail migrants contained 2,259 (91.6 percent) adults and 208 (8.4 percent) immatures (Heintzelman, 1982i, 1983c, 1984e).

At Hawk Cliff, Ontario, between 1971 and 1982, an autumn sample of 3,012 Red-tailed Hawks trapped and banded contained 2,466 (81.9 percent) HY and 546 (18.1 percent) AHY birds (Duncan, 1983). Another sample of 28 migrant Red-tails seen on 29 October 1977 in western Iowa contained 10 (35.7 percent) adults and 18 (64.3 percent) immatures (Bednarz, 1978: 141–142), whereas a sample of 19 Red-tails seen on 22 September 1976 at Sierra Grande, New Mexico, contained 16 (84.2 percent) adults and 3 (15.8 percent) immatures (Hubbard, 1977: 7).

Falcons

At Cape May Point, New Jersey, approximately 95 percent of the Peregrine Falcons and other hawks seen in autumn are immatures (Clark, 1976: 6). Autumn age classes of Peregrines observed in 1970 and 1971 at Assateague Island in Maryland and Virginia also were 85 percent immatures in 1970 and 91 percent immatures in 1971 (Ward and Berry, 1972). At Padre Island along the Texas coast, autumn Peregrine Falcon age records were secured between 1954 and 1973 for a sample of 733 sightings of which 149 (20.3 percent) were adults and 584 (79.7 percent) were immatures (Hunt, Rogers, and Slowe, 1975).

Inland at Bake Oven Knob, Pennsylvania, a sample of 81 autumn Peregrine Falcon migrants observed between 1967 and 1984 contained 73 (90.1 percent) adults and 8 (9.1 percent) immatures.

DIFFERENTIAL MIGRATIONS

When the age classes of various migrant hawks are examined during various periods of a migration season (especially autumn), and/or at various coastal and inland locations, it is clear that striking differences exist in the migratory patterns of some species based upon age classes. Thus two types of differential hawk migrations exist: (1) geographic differential migrations, and (2) seasonal differential migrations.

Geographic Differential Migrations

Age class data for Ospreys seen migrating past several eastern North American locations during autumn clearly demonstrate a geographic differential migration used by this species. Adult Ospreys (96.5 percent) tend to follow the Appalachian ridges southward (Heintzelman, 1982i, 1983c, 1984e), whereas juvenile Ospreys form the bulk of the birds reported in a narrow coastal migration route (Kennedy, 1973).

No clear geographic differential migration pattern exists for autumn

Northern Harrier migrations, although immatures tend to form most of the flights along the Atlantic coast and the Great Lakes (Currie, 1978; Bildstein, et al., 1984), whereas about one-half of the harriers seen along the Appalachian ridges are immatures (Bildstein, et al., 1984; Heintzelman, 1982i, 1983c, 1984e).

Sharp-shinned Hawks also exhibit a marked geographic differential migration. Immatures form most of the autumn migrants along the Atlantic coast (Allen and Peterson, 1936; Currie, 1978; Trowbridge, 1895), whereas along the Appalachian ridges about half of the birds seen are immatures (Broun, 1949a; Heintzelman, 1982i, 1983c, 1984e). Along the Great Lakes, autumn Sharp-shin flights consist of about 75 percent immatures (Duncan, 1982b; Rosenfield and Evans, 1980).

Autumn Broad-winged Hawk migrations tend to result in immatures following the Atlantic coastal route southward (Currie, 1978; Darrow, 1983; Duncan and Duncan, 1975) and adults following the Appalachian ridges (Heintzelman, 1983c, 1984e).

Autumn Red-tailed Hawk migrations also tend to show adults using the Appalachian ridges (Broun, 1949a; Heintzelman, 1982i, 1983c, 1984e) and immatures the predominate age class along the Great Lakes (Duncan, 1983).

Peregrine Falcon migrations also exhibit autumn geographic differential migrations. Immatures form most of the Atlantic and Gulf coast flights in autumn (Clark, 1976; Ward and Berry, 1972; Hunt, Rogers, and Slowe, 1975), whereas mostly adult Peregrines are seen along the Appalachian ridges as at Bake Oven Knob, Pennsylvania, where 90.1 percent of the sightings of this species were adults.

Seasonal Differential Migrations

A seasonal aspect also exists in the autumn migrations of some species of hawks in that one age class (usually immatures) predominates early in the season and is followed later by a shift to another age class (usually adults). This seasonal differential migration pattern exists in the autumn migrations of several species including Northern Harriers, Sharp-shinned Hawks, Cooper's Hawks, and Red-tailed Hawks (Broun, 1949a; Heintzelman, 1982i, 1983c; Allen and Peterson, 1936; Ferguson and Ferguson, 1922; Duncan, 1981, 1982b, 1983). In some instances, such as Sharp-shinned Hawks at Hawk Cliff, Ontario, slight sex-linked differences also exist in the seasonal differential migrations—adult males peaking slightly later than adult females (Duncan, 1982b).

C H A P T E R 19

Hawk Migration Routes and Geography

Geography plays a vital role in the formation of migration routes used by North American hawks. Thus an examination of the various routes and their more important features upon migrating hawks is necessary as is wind in relation to wind-drift or broad-front hawk migrations.

WIND-DRIFT

There are various reports suggesting that autumn concentrations of migrating hawks observed along the Atlantic coastline, inland mountains, and Great Lakes shorelines occur there because hawks are subject to wind-drift. Presumably the hawks are carried to those locations, especially when northwest winds occur, because winds push or shift the birds from courses elsewhere. Lesser numbers of hawks tend to be seen at such locations on easterly winds, however, because fewer birds then are pushed and concentrated there. Hawk flights on easterly winds also appear at much higher altitudes which makes their visual detection from the ground more difficult (Allen and Peterson, 1936; Broun, 1949a: Mueller and Berger, 1967; Trowbridge, 1895).

BROAD-FRONT MIGRATIONS

An alternative concept to wind-drift is that autumn and spring hawk migrations in North America tend to progress over broad fronts, but are subject to considerable influence by local wind conditions and prominent geo-

263

graphic features (Broun, 1940: 1, 1947: 1, 1949a; Edge, 1940a: 1, 1940b: 1; Heintzelman, 1975; Heintzelman and Armentano, 1964; Kellogg, 1983; Nagy, 1979: 25–29). Radar tracking of autumn hawk migrations in Ontario tended to confirm that broad-front movements are more reflective of the actual migrations of hawks there than are concentrations of these birds along prominent geographic features such as a lake shoreline where biased views of the migrations were secured (Richardson, 1975).

DIVERSION-LINES

Diversion-lines or leading-lines are prominent geographic features such as the Atlantic coastline, inland mountain ridges, the Great Lakes shorelines, or other features along which concentrations of migrant hawks appear and follow for limited distances. The exact role of diversion-lines in relation to hawk migrations is disputed.

Murray (1964, 1969) was the first person to develop a formal diversion-line explanation for Sharp-shinned Hawk migrations in autumn along the Atlantic coastline. He stated that these hawks move over a broad front and generally in a southwestward direction, often at such high altitudes that visual detection of the birds from the ground is difficult. He also stated that concentrations of migrant Sharp-shins at prominent geographic features merely represent part of a broader flight and that the birds seen were diverted along a diversion-line. Mueller and Berger (1967), however, refuted Murray's broad-front and diversion-line concept. They suggested that migrant hawks observed along a diversion-line appeared there because of the concentrating influence of wind-drift and that such concentrations were not part of a broad-front migration.

Tracking radar studies of autumn Sharp-shinned Hawks at Cape May Point, New Jersey, by Kerlinger and Gauthreaux (1983: 14), however, support the broad-front and diversion-line concept suggested by Murray (1964, 1969). The radar studies demonstrated that Sharp-shinned Hawks merely flew at higher altitudes on days with calm or easterly winds than on days with westerly winds as earlier visual observations there reported (Allen and Peterson, 1936). Thus the large hawk flights seen at Cape May Point on northwest winds represent biased and distorted reflections of the actual mechanics of hawk migrations. When large numbers of hawks are seen under those conditions the birds merely are using a lower altitude layer (and thus are more visible) than is used when winds from other directions occur.

Clearly, complex relationships exist between broad-front hawk migrations, wind direction, and diversion-lines formed by various prominent geographic features whether coastal, mountain ridges, or lake shorelines. Moreover, the visible hawk migrations reported at various locations may not fully or accurately reflect the full magnitude and scope of the flights passing in a broad front through an area.

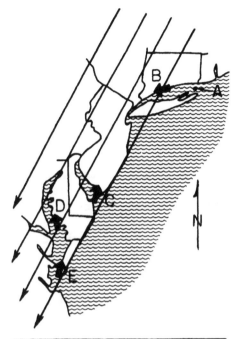

←

Diversion of migrating Sharp-shinned Hawks has occurred at various coastal sites in the northeastern United States (short thick arrows): (A) Fishers Island, N.Y.; (B) New Haven, Conn.; (C) Cape May Point, N.J.; (D) Hooper Island, Md.; and (E) Cape Charles, Va. At Fishers Island, N.Y., hawks continue southwestward toward Long Island. The general direction of the broad-front migration is indicated by the long arrows pointing southwestward. *Reprinted from Murray (1964).*

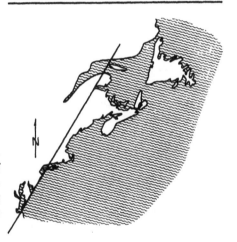

→

The line extending northeast to southwest along the coast indicates the limit of the eastern flank of the *bulk* of the migration of Sharp-shinned Hawks according to the diversion-line theory. *Reprinted from Murray (1964).*

NORTH AMERICAN
HAWK MIGRATION ROUTES

Although Murray's (1964, 1969) broad-front and diversion-line concept seems to best describe most autumn hawk migrations in North America, for practical purposes of viewing visual migrations it is useful to identify and evaluate the many locations and hawk migration flight-lines where concentrations of these birds occur under favorable local weather conditions. The important point to remember when considering these sites is that the visual migrations seen there may represent only an unknown part of the larger broad-front migration passing through the region.

Overwater Migrations

The one important exception to the concept of broad-front hawk migrations is the appearance of migrating hawks over large areas of open water such as the Atlantic and Pacific Oceans and the Great Lakes. Almost certainly no broad-front hawk flights occur over those aquatic areas. Never-theless, the use of overwater migrations is well known and documented, but varies from species to species and genera to genera. Ospreys, Northern Harriers, and Peregrine Falcons, for example, routinely use overwater migrations (Gillespie, 1960; Kennedy, 1973; Enderson, 1965), whereas buteos such as Broad-winged Hawks and Red-tailed Hawks tend to avoid crossing large expanses of water during their migrations (Perkins, 1964; · Pettingill, 1962). Kerlinger (1985) determined that a correlation exists be-tween high aspect-ratio raptors and the use of overwater migrations. Per-egrine Falcons, Ospreys, Northern Harriers, Merlins, Sharp-shinned Hawks, Rough-legged Hawks, and American Kestrels in decreasing order tend to engage in overwater crossings, whereas low aspect-ratio raptors such as Broad-winged Hawks, Red-tailed Hawks, and Turkey Vultures seldom engage in overwater migrations.

A variety of migrant hawks were seen from ships in the Atlantic Ocean off the northeastern coast of the United States, including Ospreys, Sharp-shinned Hawks, American Kestrels, Merlins, and Peregrine Falcons. The Ospreys, Merlins, and Peregrine Falcons probably engage in regular off-shore migrations in that area, whereas the Sharp-shinned Hawks and American Kestrels presumably occur there accidentally (Anderson and Powers, 1979; Kerlinger, Cherry, and Powers, 1983). Several Northern Harriers also were seen 22 miles east of Cape May Point, New Jersey, in mid-October (Dunne, 1980a: 15), and Cooper's Hawks and Merlins were observed in early October some 75 miles off the Carolinas heading toward Charleston, South Carolina (Simpson, 1954: 20). On Bermuda, Ospreys, Northern Harriers, Sharp-shinned Hawks, American Kestrels, Merlins, and Peregrine Falcons also are frequent or regular autumn migrants, whereas Turkey Vultures, American Swallow-tailed Kites, Bald Eagles, Cooper's Hawks, Northern Goshawks, Red-tailed Hawks, Rough-legged Hawks, and Eurasian Kestrels are rare or very rare autumn migrants (Wingate, 1973). An immature Northern Harrier also was seen migrating over the Straits of Florida between Cuba and Florida in late November (Pough, 1939).

In the Gulf of Mexico off the Texas coast, observations made from an oilrig platform during the autumn of 1976 produced no buteos and only 1 Sharp-shinned Hawk, 5 American Kestrels, 1 Merlin, and 1 Peregrine Falcon (Kennedy, 1977b: 10–11).

Various migrating hawks also were reported over the Pacific Ocean off the California coast, including Sharp-shinned Hawks heading *northeast* in early October (Mans, 1963: 64) and a Peregrine Falcon 15 miles offshore in mid-September (Chandik and Baldridge, 1969: 101). A Hawaiian Hawk also came aboard a ship 200 miles offshore from Hilo, Hawaii, and re-mained aboard until the California coast appeared, then flew landward

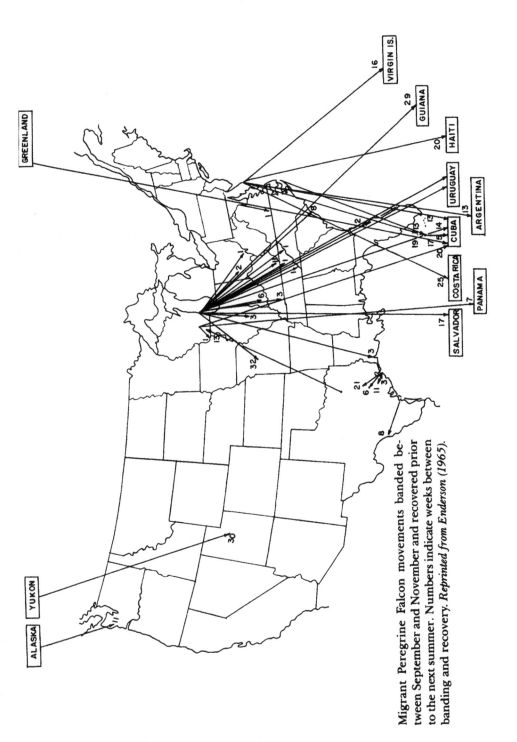

Migrant Peregrine Falcon movements banded between September and November and recovered prior to the next summer. Numbers indicate weeks between banding and recovery. *Reprinted from Enderson (1965).*

267

(Henshaw, 1901). Occasional sightings of Ospreys, Steller's Sea-Eagle, Northern Harriers, Golden Eagles, American Kestrels, and Peregrine Falcons also are reported in the Hawaiian Islands (Berger, 1981; Heintzelman, 1979a: 270–271).

The Great Lakes also represent large expanses of water and major obstacles to be crossed or diverted around by migrating hawks. Buteos generally do not cross the lakes, but follow a migration course along the shorelines during autumn, whereas Ospreys, eagles, accipiters, and falcons sometimes cross the lakes (Perkins, 1964; Pettingill, 1962). The northward spring hawk flights also follow a similar pattern regarding overwater migrations and tend to follow the lake shorelines around the water (Haugh and Cade, 1966; Haugh, 1972; Moon, 1977a–1978e; Moon and Moon, 1979a–1984c).

Atlantic Coastline

The Atlantic coastline of North America serves as a major autumn hawk migration route for thousands of birds of prey, especially accipiters and falcons, from Canada to the Florida Keys and a less important northward spring hawk migration route.

In Atlantic Canada, for example, the coastline, shorelines of bays, long peninsulas, and small islands off the southern tip of the major islands all tend to divert migrating hawks southward or (in the case of islands) act as migration traps. The numbers of migrant hawks reported that far north, however, tends to be less than are seen farther south (Christie, 1980–1983b), but at major autumn migration traps like Brier Island, Nova Scotia, thousands of migrating Sharp-shinned Hawks, Broad-winged Hawks, and other raptors occasionally are seen (Christie, 1983b; Finch, 1969; Vickery, 1978a, 1983). Some low-magnitude northward spring hawk flights also occur along Atlantic Canada coastlines as, for example, at Fundy National Park, New Brunswick (Christie, 1981a) and elsewhere.

Farther south in northern New England, the magnitude of coastal hawk flights in autumn increases. Thus in Maine about 21 hawks per hour pass Mount Agamenticus (Cote, unpublished data, 1982; Currie and Cote, 1982, 1983; MacRae, 1981) whereas at Casco Bay about 37 hawks per hour are reported (Appell, unpublished data; Cote, 1982, 1983; Currie and Cote, 1983; MacRae, 1978; Phinney, 1977). The magnitude of autumn flights along coastal Connecticut, however, is larger. At Lighthouse Point Park, in New Haven, about 40 hawks per hour are reported (Currie, 1979; Currie and Cote, 1982, 1983; Currie and MacRae, 1979, 1980; Mersereau, 1976; Trichka, 1982). Thousands of accipiters and falcons also migrate along Long Island's barrier beaches (Buckley, Paxton, and Cutler, 1978), and at Cape May Point, New Jersey, the autumn passage rate is about 85 hawks per hour (Allen and Peterson, 1936; Choate in Heintzelman, 1970e; Choate, 1972; Choate and Tilly, 1973; Clark, 1972, 1973, 1975b, 1976;

Dunne, 1978b, 1979b, 1980b; Dunne, et al., 1981; Dunne and Clark, 1977; LeGrand, 1984; LeGrand, et al., 1983), whereas the northward spring passage rate there is only 66 hawks per day (Sutton, 1980, 1981; Sutton and Sutton, 1984).

At Kiptopeke, Virginia, the autumn hawk migration passage rate also is about 85 hawks per hour (Dunne, 1978c, 1979b, 1980a, 1981, 1982, 1983; Rusling, 1936; Sutton, 1984), but farther south at Fort Johnson, South Carolina, about 32 hawks per hour are reported (Dunne, 1979b, 1980a, 1982, 1983; Laurie, McCord, and Jenkins, 1981; Sutton, 1984). At Port Canaveral, Florida, mostly accipiters and falcons form the flights which pass at the rate of about 6 hawks per hour (Duncan and Duncan, 1978b), while at Cape Florida, south of Miami, hundreds of accipiters and other hawks sometimes are seen heading south over the upper keys (Atherton and Atherton, 1981, 1983). In the lower keys hundreds of migrating Turkey Vultures, Sharp-shinned Hawks, Broad-winged Hawks, and American Kestrels also are occasionally reported heading southward (Darrow, 1983; Edscorn, 1974, 1976; Robertson, 1970; Stevenson, 1973).

Along much of the Atlantic coastline barrier beach islands are important hawk migration flight-lines in autumn, but in many locations large flights also are reported on the mainland as at Cape May Point, New Jersey, because its location at the end of a peninsula serves as a migration trap. Wind direction at all of the coastal observation sites, however, greatly affects the numbers of migrating hawks seen. In general, northwest winds tend to produce the largest visible migrations.

Gulf Coast

A few autumn hawk migration studies are reported from the Gulf Coast. In the Mobile Bay, Alabama, area the passage rate is about 8 hawks per hour (Parrish and Wischusen, 1980). In the Baton Rouge, Louisiana, area the passage rate is 93 hawks per hour but the figure is greatly skewed by large Broad-winged Hawk counts (Rogers, 1983b, 1984). Along the Texas Coastal Bend area as many as 136,000 migrating Broad-winged Hawks may cross the area (Rowlett, 1978), as do as many as 62,000 Broad-wings in the Corpus Christi area (Webster, 1981). Autumn Peregrine Falcons at Padre Island passed at the rate of 4.7 falcons per day (Hunt, Rogers, and Slowe, 1975).

Appalachian Mountains

In the broadest sense, the Appalachian Mountains form the spine of eastern North America from Quebec southward to northeastern Alabama. The various ridges and mountains that form the chain serve as both minor and major autumn hawk migration routes depending upon the axis of the ridge, its length, and its location.

In Vermont and New Hampshire, autumn hawk flights tend to be dispersed and of relatively low density at least partly because of the numerous isolated mountains lacking very long and well-defined ridge systems in a northeast to southwest direction (Will, 1980). Thus, in New Hampshire, a passage rate of about 25 hawks per hour was reported at the Birchwood Ski Area near Londonderry (Currie and MacRae, 1980), and about 10 hawks per hour passed Ellsworth Hill (Currie and Cote, 1982), 18 hawks per hour passed Little Round Top (Currie and MacRae, 1980), 18 hawks per hour passed Mount Kearsarge (Currie and Cote, 1982, 1983), and at Pack Monadnock about 28 hawks per hour were reported (Currie and Cote, 1982, 1983; Currie and MacRae, 1980). The passage rate of autumn hawk flights in Vermont are of a similar density. At Blue Gate 8 hawks per hour were seen (Norse, 1979b; Pistorius, 1977, 1978; Rowlett, 1977), 4 hawks per hour passed Brandon Gap (Will, 1974, 1975), 16 hawks per hour passed Gile Mountain (Norse, 1979a, 1980b, 1981b, 1982b, 1983b; Pistorius, 1978; Will, 1974, 1976), at Glebe Mountain 33 hawks per hour were seen (Will, 1974, 1976; Rowlett, 1977), 15 hawks per hour passed Grafton (Norse, 1979b, 1980b, 1981b, 1982b, 1983b; Pistorius, 1977, 1978; Rowlett, 1977; Will, 1976), and 20 hawks per hour passed Putney Mountain (Norse, 1977, 1978, 1979b, 1980b, 1981b, 1982b, 1983b; Pistorius, 1977, 1978; Rowlett, 1977; Will, 1974, 1976).

In Massachusetts, autumn hawk flights at Mount Tom (1978–1983) passed at the rate of 94 hawks per hour (Currie, 1979; Currie and Cote, 1981, 1982, 1983; Currie and MacRae, 1979, 1980; Fischer, 1979), whereas 70 hawks per hour passed Mount Wachusett in autumn (Currie, 1979; Currie and Cote, 1981, 1982, 1983; Currie and MacRae, 1979). Farther south in New York, autumn hawk migrations at Franklin Mountain passed at the rate of 18 hawks per hour (Kerlinger and Bennett, 1977; Nagy, 1978b), 15 hawks per hour at the Helderberg Mountains (Nagy, 1978b, 1979b), 31 hawks per hour at Hook Mountain (Benz, 1980b, 1981b, 1982b, 1983b, 1984b; Heintzelman, 1975; Mills and Mills, 1971; Nagy, 1976, 1977b, 1978b, 1979b; Single, 1974, 1975; Single and Thomas, 1975; Thomas, 1973), 22 hawks per hour at Mount Peter (Bailey, 1967, 1969; Cinquina, 1975, 1976, 1977; Cinquina and Martin, 1979, 1980, 1981, 1982, 1983, 1984; Martin, 1978; Rasmussen, 1974; Rogers, 1971, 1972), and 16 hawks per hour at the Near Trapps near the northern end of the Shawangunk (Kittatinny) Mountains (Benz, 1981b, 1983b, 1984b; Carroll, 1980; Nagy, 1976, 1977b, 1978b, 1979b).

In New Jersey, autumn hawk migrations at the Montclair Hawk Lookout Sanctuary in the Watchung Mountains had a passage rate of 206 hawks per day (Bihun, 1967–1983; Breck, 1962, 1963; Breck and Breck, 1964–1966; Redmond and Breck, 1961), while along the Kittatinny Ridge 42 hawks per hour passed Raccoon Ridge (Benz, 1980b, 1981b, 1982b, 1983b; Cant, 1946; Heintzelman, 1975; Tilly, 1972a, 1972b, 1972c, 1973, 1979; Wolfarth, 1975), and 27 hawks per hour appeared at Sunrise Mountain upridge from Raccoon Ridge (Benz, 1980b, 1981b, 1982b, 1983b, 1984b; Nagy, 1976, 1977b, 1978b, 1979b; Tetlow, 1973 field notes).

Rates of passage of autumn hawk migrations along the Kittatinny Ridge in eastern Pennsylvania vary from site to site in a northeast to southwest direction. At Little Gap, 37 hawks per hour were reported (Benz, 1980b, 1981b, 1982b, 1983b, 1984b; Nagy, 1976, 1977b, 1978b, 1979b), 35 hawks per hour passed Bake Oven Knob (Heintzelman, 1963a–1968, 1969b, 1970d, 1982c, 1982i, 1983c, 1984e; Heintzelman and Armentano, 1964; Heintzelman and MacClay, 1972, 1973, 1975, 1976a, 1979; Heintzelman and Reed, 1982), 35 hawks per hour were seen at Bear Rocks (Benz, 1980b, 1981b, 1982b, 1983b, 1984b; Heintzelman, 1975; Nagy, 1977b, 1978b, 1979b), 29 hawks per hour passed Hawk Mountain (Broun, 1949a; Hawk Mountain newsletters), and 35 hawks per hour were seen at Waggoner's Gap (Benz, 1980b, 1981b, 1982b, 1983b, 1984b; Heintzelman, 1975; Kotz, 1973; Nagy, 1976, 1977b, 1978b).

Careful studies of ages and times of sightings of Bald Eagles and Golden Eagles observed along the Kittatinny Ridge in autumn in Pennsylvania, however, demonstrated that more eagles use sections of the ridge than appear at any single ridge lookout. There also are year-to-year variations in utilization of sections of the ridge. Apparently eagles approach the ridge, use it for limited distances, then leave at random locations (Frey, 1940; Heintzelman, 1975: 271, 1982g).

Several studies also are available regarding the influence of intersecting diversion-lines such as rivers upon Ospreys and other hawks migrating along the Kittatinny Ridge. Available data, however, suggest that rivers cutting through the ridge do not serve as important diversion-lines and redirect the flight-lines of most of the hawks passing such intersections, although some birds change course and follow rivers upriver or downriver (Bergey, 1975; Heintzelman, 1973, 1983a).

Elsewhere in Pennsylvania, 18 hawks per hour in autumn passed the Cornwall Fire Tower (George and Schreffler, 1984) south of the Kittatinny Ridge, 16 hawks per hour passed Militia Hill near Philadelphia (Benz, 1980b, 1981b, 1982b, 1983b; Nagy, 1978b, 1979b), and 13 hawks per hour in autumn passed The Pulpit on Tuscarora Mountain north of the Kittatinny Ridge (Benz, 1980b, 1981b, 1982b, 1983b, 1984b; Nagy, 1976, 1977b, 1978b, 1979b; Heintzelman, 1975).

Autumn hawk flights in Virginia passed Harvey's Knob Overlook at the rate of 25 hawks per hour (Finucane, 1976b, 1977b, 1978b, 1979b, 1980b, 1981b; Puckette, 1982b, 1983b, 1984b), 34 hawks per hour appeared at Purgatory Mountain Overlook (Finucane, 1976b, 1977b, 1978b, 1979b, 1980b, 1981b), and 55 hawks per hour were reported at Rockfish Gap (Finucane, 1976b, 1977b, 1978b, 1979b, 1980b, 1981b; Puckette, 1982b, 1983b, 1984b). Autumn flights in West Virginia, where many hawk lookouts are located, included 51 hawks per hour seen at Bear Rocks (De-Garmo, 1953; Finucane, 1979b, 1980b, 1981b; Heimerdinger, 1974; Puckette, 1982b, 1983b, 1984b), and 68 hawks per hour at Hanging Rocks Fire Tower (Finucane, 1979b; Hurley, 1970, 1975; Puckette, 1982b, 1983b, 1984b).

In Tennessee, 44 hawks per hour were seen in autumn at the Rogersville-Kyles Ford Fire Tower (Finucane, 1959, 1960, 1961a, 1961b, 1963, 1964, 1965, 1966, 1967, 1968, 1969, 1971, 1973, 1974, 1975a, 1976a, 1977a, 1978c; Fowler, 1982; Turner, 1981), whereas 10 hawks per hour passed Signal Point (Finucane, 1969, 1976a, 1977a, 1978c, 1980c; Fowler, 1982; Puckette, 1983b, 1984b; Turner, 1981).

Autumn hawk flights in North Carolina produced 16 hawks per hour at East Fork Mountain (Finucane, 1980b, 1981b) and 25 hawks per hour at Pilot Mountain (Finucane, 1979b; Puckette, 1983b), while in Georgia at Lookout Mountain 92 hawks per hour were reported (Finucane, 1975b).

Occasional large flights of Broad-winged Hawks accounted for many of the unusually high passage rates at many of the more southern hawk watching lookouts.

Great Lakes

Large autumn hawk flights occur at many locations along the northern or western shorelines of the Great Lakes. At Hawk Cliff, Ontario, along Lake Erie, major flights appeared (Field and Field, 1979, 1980a, 1980b; Field and Rayner, 1974, 1976, 1977, 1978; Fowler and Fowler, 1981, 1982; Haugh, 1972) and farther west at Holiday Beach Provincial Park 832 hawks per day were reported (Goodwin, 1975, 1977a, 1979a, 1980, 1982a; Kleiman, 1981; Weir, 1983a). At Cedar Grove, Wisconsin, considerable autumn hawk flights also are reported (Berger, 1954; Mueller and Berger, 1961), and at Duluth, Minnesota, at the Hawk Ridge Nature Reserve 82 hawks per hour were seen (Kohlbry, 1980a; Peacock and Myers, 1982b; Sundquist, 1973).

Northward spring hawk migrations also are reported at many locations along the southern shorelines of the Great Lakes, including several key locations. At Whitefish Point, Michigan, 47 hawks per hour are reported (Baumgartner, 1979, 1980, 1981, 1982, 1983), 63 hawks per hour passed Braddock Bay Park, New York (Dodge, 1981; Listman, Wolf, and Bieber, 1949; Moon, 1977a–1983; Moon and Moon, 1979a–1984c; Wolf, 1984), and 543 hawks per day passed Derby Hill, New York (Anonymous, 1984; Baker, 1983; Haugh, 1972; Muir, 1978–1982; Peakall, 1962; Smith, 1973; Smith and Muir, 1978).

Great Plains

Autumn hawk migrations tend to scatter as they migrate southward across the central states and plains with birds appearing at many locations rather than at a few well-defined flight-lines annually. Thus in early October,

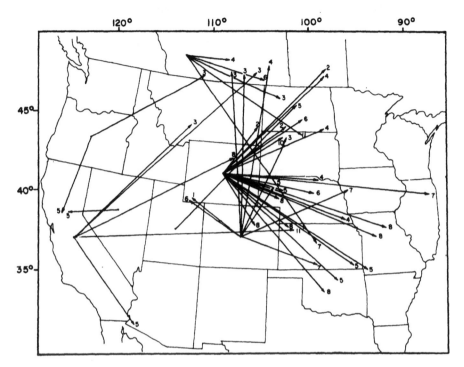

Recoveries of Prairie Falcons banded as nestlings and recovered between 1930 and 1961. Numbers indicate months between banding and recovery. *Reprinted from Enderson (1964).*

3,400 Swainson's Hawks were seen within 1.5 hours over Hutchinson, Kansas (Cruickshank, 1937: 385), for example, and 2,000 Broad-winged Hawks appeared over the Kansas plains in late September (Albrecht, 1922: 5).

In addition to north-to-south hawk migrations, Prairie Falcons engage in west-to-east October migrations from open, unforested Rocky Mountain valleys to the Great Plains where their primary food supply of larks concentrates in winter, then return in March to the mountain valleys (Enderson, 1964).

Rocky Mountains

Field studies in western North America demonstrate that hawk migrations pass along various ridges in the Rocky Mountains (Hoffman, 1981a, 1982b). Autumn hawk migrations in Montana at the Bridger Mountains produced 7 hawks per hour (Tilly, 1981b, 1983b), whereas in Nevada at the Goshute Mountains 16 hawks per hour were reported (Hoffman, 1981b, 1982b, 1984b). In New Mexico, at the Manzanos Mountains 9 hawks per hour were reported (Hoffman, 1981b, 1982b, 1983b, 1984b), and in Utah at Brighton on the Wasatch Range autumn hawk flights passed at the rate of 7 birds per hour (Hoffman, 1981b, 1982b). Utah's Wellesville Mountains produced autumn hawk flights at 20 hawks per hour (Hoffman, 1981b, 1982b, 1983b, 1984b). A spring hawk migration at Dakota Ridge in Colorado produced a passage of 17 hawks per hour (Hoffman, 1982a).

Pacific Coast

Very limited systematic hawk migration data exist for the Pacific coast of North America. In Alaska, at Sitkagi Beach, about 2 hawks per hour in autumn were reported (Tilly, 1984), and in autumn at Point Diablo, California, 33 hawks per hour were reported (Binford, 1979). In addition, many fragmentary sightings of migrating hawks appeared in the literature as discussed earlier in this book.

ABANDONMENT OF FLIGHT-LINES

Almost nothing is known about the possibility that migrant hawks may permanently abandon established migration flight-lines. Nevertheless, several published comments hint at such a possibility. On Long Island, New York, for example, Elliott (1960: 155) stated that ornithologists today have a unique opportunity to study "changes in migration routes due to human occupancy of the coastline." Elkins (1974: 103) also stated that the spring flight-line through the Sudbury Valley in Massachusetts used 30 years earlier by accipiters and buteos disappeared, with the decline of Red-shouldered Hawks being an important factor in the change. At Gulf Breeze, Florida, a shift from Red-shoulderd Hawks to Broad-winged Hawks as the most abundant autumn migrants also occurred between 1943 and 1965 (Duncan and Duncan, 1975: 2).

While different factors may be responsible for the permanent abandonment or changes in flight-lines at each of these locations, habitat change or loss was the factor mentioned at each site. Long-term data bases and carefully documented related ecological changes in sites are needed to further evaluate possible permanent abandonment of hawk migration flight-lines. It is possible, however, that certain ecological factors such as changes in vegetation may be much more important factors in relation to flight-lines used by migrating hawks than currently is appreciated. Haugh (1972: 15) pointed out, for example, that habitat variations in the area around Hawk Cliff, Ontario, produced three separate hawk migration flight-lines—accipiters followed a wooded ravine, buteos used varying paths, and falcons flew over open fields and along the cliff shoreline. In southern Nevada, autumn Sharp-shinned Hawk migrations also followed a forested ridge but tended to avoid crossing inhospitable deserts (Millsap and Zook, 1983). Presumably one ecological factor responsible for the

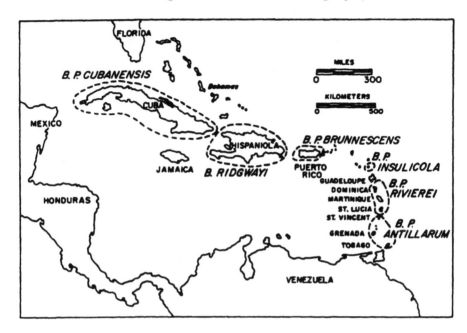

Geographic distribution of the insular subspecies of the Broad-winged Hawk in the West Indies. The distribution of Ridgway's Hawk also is shown. *Reprinted from Heintzelman (1975).*

limited numbers of American Kestrels migrating along the Kittatinny Ridge in New Jersey and Pennsylvania is the forested mountain ridge rather than open field-like habitat generally required by the falcon.

Leck (1980) also pointed out that some migratory birds sometimes develop new breeding centers along traditional migration routes, or even in their winter range, and that such populations can produce new subspecies or species. Perhaps birds from some of those new populations also might select new migration flight-lines, or cause abandonment of established flight-lines, if they no longer arrive at former breeding grounds farther north. It seems likely, for example, that ancestors of the Broad-winged Hawks on various islands in the West Indies became trapped there and could not leave easily because of their dependency upon thermal soaring. Eventually they evolved into the various subspecies now found there (Heintzelman, 1975:293–300; Leck, 1980).

C H A P T E R 20

Hawk Counts as Indices to Raptor Population Trends

One of the most difficult tasks confronting wildlife biologists is the accurate determination of population levels of wild animals. Rarely is it possible to count all individuals of a species directly. The problem is particularly acute in respect to estimating population levels of birds of prey which, as breeding birds, are distributed over large geographic areas. Hence virtually no data are available providing accurate estimates of raptor population levels on a continental basis for long periods of time. However, some efforts have been made to estimate current regional population levels of certain endangered species or subspecies of birds of prey. Examples include California Condor field studies (Koford, 1953; Miller, McMillan, and McMillan, 1965; McMillan, 1968), various Peregrine Falcon populations (Hickey, et al., 1969), and various other diurnal birds of prey (Hickey, et al., 1969). The *entire* breeding raptor population of one area of modest size also was determined (Craighead and Craighead, 1956).

PURPOSE OF MIGRANT HAWK COUNTS

There are at least four basic reasons and purposes for making autumn and/or spring counts of migrating hawks: (1) conservation, (2) education, (3) recreation, and (4) research (Broun, 1949a; Heintzelman, 1983b; Senner, 1984b; Roberts, 1984). There is considerable overlap between each of these purposes, but in this book my discussion is restricted to research with possible conservation implications.

The basic research objective of making hawk counts, therefore, is to detect and identify short-term and long-term changes or trends in diurnal raptor populations (Heintzelman, 1975; Senner, 1984b; Roberts, 1984). A related secondary purpose is to develop and use conservation or management programs to halt or reverse declining raptor populations (Heintzelman, 1983b; Senner, 1984b).

METHODS AND ACCURACY OF MIGRANT HAWK COUNTS

The methods and accuracy of counts of migrating hawks, and their usefulness as indicators of trends in diurnal raptor populations, was reviewed by Fuller and Mosher (1981). They concluded that concentrations of migrants at key locations in North America and elsewhere allow counts to be made relatively rapidly, but that numerous unquantified variables inhibited interpretation of such counts.

Methods

As pointed out in Chapter One, and by Fuller and Mosher (1981), numerous and diverse methods are used to secure counts of migrant hawks and related data. Here I am concerned mostly with observation of migrations made at locations along established migration routes or known flight-lines. Particular emphasis is placed upon quantitative data and its analysis in this chapter.

More subjective methods also are used to detect short-term changes in raptor populations based upon counts of migrating hawks. Thus Hofslund (1976), Single and Copeland (1978), Robinson (1980), and others presented their general impressions of raptor population changes.

Accuracy

For counts of migrating hawks to reflect reasonably accurate changes and trends in raptor populations, at least two major conditions would have to remain constant every year: (1) hawks being counted must originate from the same geographic area, and (2) the same percentage of the population being sampled must be seen and detected at the site where hawk migration counts are made. Neither condition can be established with certainty. Indeed, very little is known regarding the precise geographic areas from which migrant hawks come. Moreover, it seems unlikely that the second condition prevails from year to year. Nevertheless, if hawk migration counts are used with caution, at least for certain selected species, perhaps with data from several locations pooled, changes or trends in some raptor populations might be detected (Fuller and Mosher, 1981).

There also are many additional variables that influence and determine the accuracy of counts of migrating hawks. They include: (1) observer ability, (2) observer skill and experience, (3) impact of multiple observers, (4) impact of various types of equipment, (5) number of lookouts or sites used simultaneously in a local area, (6) duplication of counts, (7) peer pressure among observers, and (8) general and local weather variables (Fuller and Mosher, 1981; Fuller, 1979).

While it is important that strict efforts be made at all hawk migration lookouts to make hawk counts as accurate as possible, Hawk Mountain in Pennsylvania plays an exceptional role in providing counts of migrant hawks to the scientific community because it is the best-known source of North American hawk migration data. Thus a critical review of the accuracy of its autumn hawk count data base is warranted.

Unfortunately substantial evidence indicates that autumn hawk migration data reported at Hawk Mountain contain serious flaws of several types. Broun (1949b: 7), for example, stated that hawk shooting along the Kittatinny Ridge did *not* significantly affect Sanctuary hawk counts, and that the numbers of hawks counted "held up impressively" until 1955 (Broun, 1965: 15). Later, however, he concluded that he failed to properly evaluate the impact of hawk shooting along the Kittatinny Ridge upon the hawk count data base at Hawk Mountain, and it was inappropriate to compare recent hawk counts from that site with data collected there during the shooting years (Broun, 1979b: 2–3). Broun (1965, 1966) also pointed out that the hawk counts from Hawk Mountain exhibited "great variation" because of the impact of general and local weather conditions and even pesticides for some species (Broun, 1965, 1966). Further flaws also were introduced into the data base at Hawk Mountain when the Sanctuary hawk counts included some observations of hawks at locations upridge from the Sanctuary (Broun, 1961: 9), and when new lookouts were added and used resulting in thousands of hawks being seen, counted, and added into the annual totals that previously would not have been reported at Hawk Mountain (Nagy, 1967, 1968; Harwood and Nagy, 1977; Wetzel, 1969). Additional factors that flaw the data base include more experienced hawk watchers helping to count on the lookouts, more hours spent on the lookouts, and competition among hawk watchers at the Sanctuary (Harwood and Nagy, 1977). Thus the statement by Senner (1984b: 43) that the Hawk Mountain data base is the "most consistently gathered set of raptor migration counts anywhere in the world" is incorrect and inconsistent with the documented flaws in those data. Indeed, warnings repeatedly were issued, over the years, against using hawk counts from Hawk Mountain as indices to changes in raptor populations (Broun, 1965, 1966, 1979b; Heintzelman, 1975; Harwood and Nagy, 1977).

In a similar manner, N. G. Smith (1985) concluded that the counts of migrating Broad-winged Hawks and Swainson's Hawks made by ground observers at concentration points in the neotropics, such as the Isthmus of Panama, are not accurate and may yield data that are not biologically meaningful because large numbers of birds pass aloft unseen because of the high altitudes at which they fly.

HAWK COUNT TRENDS

Changes in raptor populations are very difficult to identify and measure by direct means. Therefore some ornithologists used indirect methods to detect trends in diurnal raptor populations via analysis of counts of hawks seen migrating past autumn concentration points such as Hawk Mountain, Pennsylvania (Spofford, 1969; Nagy, 1977a). For reasons already discussed, however, it probably would be better to conclude that trends detected at Hawk Mountain (or other sites) merely represent changes in counts at that location rather then reflections of more widespread trends in diurnal raptor populations from larger geographic areas—at least until further research demonstrates otherwise. Fuller and Mosher (1981) concluded that a better method of comparison of hawk counts and other raptor census data is to pool information from many sites prior to making an analysis.

In the following discussions, analysis of autumn hawk migration counts reported at Bake Oven Knob, Pennsylvania (Heintzelman, 1963a–1968, 1969b, 1970d, 1982c, 1982i, 1983c, 1984e; Heintzelman and Armentano, 1964; Heintzelman and MacClay, 1972, 1973, 1975, 1976a, 1979; Heintzelman and Reed, 1982), are based upon the numbers of birds counted per hour during the period 1961–1984, whereas analysis of Hawk Mountain hawk count information is based upon the reports of Spofford (1969) and Nagy (1977a).

Vultures

Black Vultures were not seen at Bake Oven Knob prior to 1974 when a single bird appeared. The next vulture appeared in 1979, another in 1980, and since 1981 a yearly increase in numbers ranged between 0.4 and 0.9 birds per hour × 10. Because of the recent arrival of this species, however, it is too early to detect any possible long-term vulture count trends. Wilbur (1983) stated that more study is needed to determine Black Vulture trends throughout North America.

Turkey Vulture count trends at Bake Oven Knob, based on a sample of 4,856 birds, ranged between 3.2 and 9.4 birds per hour × 10 with a mean of 5.2 birds per hour × 10. There were two years (1968 and 1970) at the mean, 11 years of counts below the mean, and 11 years of counts above the mean. While very large yearly variations did not occur, it is possible that a weak trend of three-to-five year cycles exists in the Turkey Vulture counts. Wilbur (1983: 114) pointed out that Turkey Vulture trends are unclear in North America. Similar comparative data are not available for Hawk Mountain, Pennsylvania.

Ospreys and Harriers

Counts of Ospreys at Bake Oven Knob, based on a sample of 6,587 birds, ranged between 1.8 and 13.5 birds per hour × 10 with a mean of 7.3 birds per hour × 10. One year (1967) was at the mean, counts for 12 years were below the mean, and counts for 11 years were above the mean. Low counts in 1961, 1974, and 1982 suggested an approximate ten-year cycle. Curiously counts reported in the pre-1972 DDT era were not overly low, as would be expected (although mostly adults use the Kittatinny Ridge as an autumn migration route), perhaps because the birds came from unpolluted Canadian lakes (Peterson, 1966: 10), less competition on the wintering grounds occurred (Taylor, 1971: 3), some DDT-caused physiological effect on energy production in flight muscles during the pesticide era existed (Heintzelman, 1970c: 122), or simply the geographic differential migration pattern of this species did not reflect the serious reproductive failures of this species and low production of nestlings.

At Hawk Mountain, Pennsylvania, however, Spofford (1969) and Nagy (1977a) suggested that Osprey counts showed an increasing trend.

Northern Harriers passed Bake Oven Knob at a mean rate of 4.6 birds per hour × 10 with a range between 2.9 and 8.4 birds per hour × 10 based upon a sample of 4,422 birds. Two years (1973 and 1981) had counts at the mean, 14 years were below the mean, and 8 years had counts above the mean. A possible approximate ten-year cycle might exist with peaks in 1968 and 1977. Harrier counts at Hawk Mountain seemed to show an increasing trend according to Spofford (1969), whereas Nagy (1977a) concluded that a stable trend existed in counts there.

Eagles

A sample of 397 Bald Eagles was secured at Bake Oven Knob which produced a mean of 0.4 birds per hour × 10 with a range between 0.2 and 1.0 birds per hour × 10. One year's count (1983) was at the mean, counts for 13 years were below the mean, and counts for 10 years were above the mean. Bald Eagle counts generally were low prior to 1978, although several above-average counts appeared in the mid-1960s. There seemed to be a variable but generally increasing trend in eagle counts at the Knob after 1977—a trend that also appeared at Hawk Mountain, presumably because of less DDT contamination in the environment (Senner, 1984b: 44).

At Hawk Mountain, Pennsylvania, Spofford (1969) and Nagy (1977a) documented a marked increase in Bald Eagle counts between 1945 and 1950 followed by a sharp decline until about 1975 after which a modest increase tended to occur (Senner, 1984b: 44). Bald Eagles observed migrating southward past Bake Oven Knob and Hawk Mountain are mostly Florida birds returning to their breeding grounds after making northward post-breeding dispersals (Broley, 1947).

Perhaps more important than actual eagle counts seen at Bake Oven

Knob or Hawk Mountain were the documented changes in age ratios of migrating eagles as discussed earlier—changes that accurately reflected DDT pollution in Bald Eagle food chains and the environment.

A sample of 571 Golden Eagles counted passing Bake Oven Knob produced a mean of 0.5 birds per hour × 10 with a range between 0.3 and 0.8 birds per hour × 10. Seven years had counts at the mean, counts for six years were below the mean, and counts for 11 years were above the mean. The data suggest a relatively stable trend in Golden Eagle counts at the Knob. At Hawk Mountain, however, a slight declining trend in this species was reported (Spofford, 1969; Nagy, 1977a).

There were no major changes in age ratios of Golden Eagles seen at Bake Oven Knob since 1961, as occurred in Bald Eagles, thus indicating that these birds were not seriously affected by the DDT pollution in the environment that harmed Bald Eagles.

Accipiters

Sharp-shinned Hawks were counted at Bake Oven Knob at a mean rate of 8.0 hawks per hour with a range between 1.3 and 24.8 hawks per hour based upon a sample of 72,121 birds. Counts for 17 years were below the mean and counts for seven years were above the mean. Prior to 1975, all years were below the mean, whereas after 1975 only three years dropped below the mean. Indeed, the counts for 1977–1981 were much higher than the mean, although 1982–1984 counts were either below or close to the mean. Presumably the striking above-the-mean counts reported between 1977 and 1981 reflected a more pesticide-free environment after the 1972 ban on use of DDT in the United States (Heintzelman, 1982e), whereas weather variables may be responsible for lowered Sharp-shinned Hawk counts reported between 1982–1984 (Heintzelman, 1982i, 1984e).

At Hawk Mountain, a declining trend in Sharp-shinned Hawk counts was reported by Spofford (1969) and Nagy (1977a), although the latter also reported a slight rise after 1972.

Complicating the use of Sharp-shinned Hawk counts to detect count changes prior to 1957 was the impact of hawk shooting along the Kittatinny Ridge in Pennsylvania where this species was one of the primary targets of gunners. At first, Broun (1949b: 7, 1965: 15) stated that hawk shooting along the ridge did not have a significant effect on hawk counts made at Hawk Mountain. Later, however, he concluded that he did not properly evaluate the impact of shooting on the Hawk Mountain data base and that a comparison of hawk counts made during the shooting years (pre-1957) with post-shooting years was inappropriate (Broun, 1979b: 2–3).

The impact of DDT pollution upon accipiters also is complex and documented incompletely, but a careful analysis of DDE stress on Sharp-shinned Hawks strongly suggested that these hawks suffered serious population declines as a result of pesticide contamination during the DDT-use era (Snyder, et al., 1973). Eggshell thickness of Sharp-shinned Hawks decreased between 9 and 13 percent during the DDT-use era (Anderson and Hickey, 1972). Broun (1965: 15) also suggested a secondary pesticide contamination impact upon Sharp-shinned Hawk populations in the form of widespread poisoning and reduction of the food supply and implied that Sharp-shinned Hawk counts at Hawk Mountain reflected that environmental contamination.

Cooper's Hawk counts at Bake Oven Knob, based on a sample of 1,942 birds, ranged between 0.3 and 6.8 hawks per hour × 10 with a mean of 2.1 hawks per hour × 10. Sixteen years of counts were below the mean and counts for eight years were above the mean. All counts from years prior to 1975 were below the mean, and counts from three years after 1975 also dropped below the mean. Nevertheless, it appears that the generally higher Cooper's Hawk counts reported after 1975 reflected a more pesticide-free environment, less contaminated food chains and food webs, and perhaps a reversal of a pesticide-induced thin eggshell syndrome as documented by Anderson and Hickey (1972) and Snyder, et al. (1973). Cooper's Hawk counts at Hawk Mountain during the DDT-use era were depressed (Spofford, 1969; Nagy, 1977a). The Bake Oven Knob counts also may show a weak three-to-five year cycle (Heintzelman, 1982f).

Northern Goshawk counts at Bake Oven Knob, based on a sample of 1,693 birds, ranged between 0.3 and 5.4 birds per hour × 10 with a mean of 1.4 birds per hour × 10. Counts for 15 years fell below the mean, and counts for nine years were above the mean but no declining trend is obvious. However, the cyclical nature of Northern Goshawk populations in North America is well known (Broun, 1949a; Brown and Amadon, 1968; Hofslund, 1973; Keith, 1963; Lack, 1954; Mueller and Berger, 1967c; Mueller, Berger, and Allez, 1977; Speirs, 1939; Sutton, 1931). Forbush (1927) suggested a ten-year cycle, but inspection of the Bake Oven Knob counts suggest three-to-five year cycles—perhaps within ten-year cycles. Peaks occurred in 1962, 1965–1966, 1968, 1972–1973, 1975, 1978, 1981, and perhaps 1983, and lows in the counts occurred in 1964, 1967, 1971, 1974, 1976–1977, and 1979–1980 (Heintzelman, 1982h). A major Northern Goshawk invasion in 1972 (Heintzelman and MacClay, 1973) was followed in 1973 by a widespread echo flight (Heintzelman and MacClay, 1975; Hofslund, 1973).

Northern Goshawk counts at Hawk Mountain also showed no declining trend but cyclical eruptions appeared in 1935, 1953, 1957, 1961, 1965, 1968, and 1972 (Spofford, 1969; Nagy, 1977a).

Analysis of Northern Goshawk eggshell thickness and related ecological factors also did not demonstrate DDT stress for this species (Snyder, et al., 1973) although an eight percent reduction in the eggshell thickness index was reported (Anderson and Hickey, 1972).

Buteos

A sample of 4,217 Red-shouldered Hawks was secured at Bake Oven Knob with a range of counts between 1.7 and 9.7 hawks per hour × 10 and a mean of 4.4 hawks per hour × 10. Counts for 16 years were below the mean, whereas counts for eight years were above the mean. A possible ten-year cycle exists in the Red-shoulder counts with lows noted in 1962, 1972, and 1979 and peaks noted in 1968 and 1975. There also may be a slight declining trend in Red-shouldered Hawk counts.

At Hawk Mountain, a slight increase seemed to be present in the counts perhaps due to better identification skills (Spofford, 1969), but Nagy (1977a) suggested a declining trend in Red-shouldered Hawk counts.

Broad-winged Hawks are the most abundant raptor migrants seen at Bake Oven Knob. A sample of 175,454 birds produced counts that ranged between 6.5 and 41.2 hawks per hour with a mean of 19.9 hawks per hour. Counts for 13 years were below the mean and counts for 11 years were above the mean. Exceptional lows occurred in 1961, 1971, and 1983, whereas exceptional peaks in counts occurred in 1963, 1968, 1975, and 1978, thus suggesting possible three-to-five year cycles within ten-year cycles. This species is so dependent upon thermal soaring and cross-country flight, however, that many more decades of data are needed to confirm the presence of such cycles.

At Hawk Mountain, a slight increase in Broad-winged Hawk counts was reported (Spofford, 1969), whereas Nagy (1977a) concluded that counts of this species were stable.

Red-tailed Hawk counts at Bake Oven Knob, based on a sample of 48,597 birds, ranged between 2.8 and 8.9 hawks per hour with a mean of 5.1 hawks per hour. Counts for 12 years were below the mean and 12 years of counts were above the mean. There were very low counts in 1961, 1967, 1973, 1976, 1979, and 1984, thus suggesting a possible three-to-five year cycle. No obvious declining trend was noticed.

At Hawk Mountain, a stable hawk count trend existed for Red-tails after World War II, but at a lower level than prior to the war (Spofford, 1969), whereas Nagy (1977a) detected a slight increase in numbers of Red-tailed Hawks counted.

Changes in Rough-legged Hawk counts at Bake Oven Knob, based on a sample of 208 birds, ranged between 0 and 8.3 hawks per hour × 100 with a mean of 2.3 hawks per hour × 100. Counts for 12 years were below the mean and another 12 years of counts were above the mean. Low counts occurred in 1962, 1966–1968, 1972, 1979, 1982, and perhaps 1984 whereas in 1961, 1963, 1971, 1976, 1980, and 1983 peaks occurred. These counts seemed to confirm the cyclical nature of Rough-legged Hawks with approximate three-to-five year cycles appearing—perhaps within longer ten-year cycles.

At Hawk Mountain, cycles also appeared in the counts. Peaks occurred in 1939, 1948, 1952, 1956, 1961, 1964, and 1969 (Nagy, 1977a).

Falcons

Counts of American Kestrels at Bake Oven Knob, based on a sample of 4,381 birds, ranged between 1.7 and 9.6 birds per hour × 10 with a mean of 4.9 birds per hour × 10. There was one year (1969) at the mean, 11 years below the mean, and 12 years above the mean. There is no evidence of pesticide stress, but very low counts occurred in 1963, 1972, and 1984 suggesting an approximate ten-year cycle in counts.

American Kestrel counts at Hawk Mountain seemed to show an increasing trend (Spofford, 1969; Nagy, 1977a). It is not clear if approximate ten-year cycles are reflected in the Hawk Mountain kestrel counts.

Counts of Merlins at Bake Oven Knob, based on a sample of 224 birds, had a mean of 2.3 birds per hour × 100 with a range between 0 and 4.6 falcons per hour × 100. The count for 1982 was at the mean, counts for 10 years fell below the mean, and counts for 13 years were above the mean. Approximate three-year cycles may be reflected in the counts with peaks in 1962–1963, 1966, 1969, 1973, 1975, 1978, 1980, and 1983.

At Hawk Mountain, a slight decreasing trend appeared in Merlin counts (Nagy, 1977a). Not discussed, but also obvious in the Hawk Mountain counts, are approximate three-year cycles with peaks in 1936, 1939, 1941, 1950, 1952, 1954, 1957, 1960, 1963, 1966, and 1969.

Only 185 Peregrine Falcons form the Bake Oven Knob sample. The mean count was 2.0 falcons per hour × 100 with a range between 0 and 5.0 falcons per hour × 100. Counts for 13 years were below the mean count and counts for 11 years were above the mean count. Curiously, no clear pesticide stress appeared between 1962 and 1971, whereas counts for all years between 1972 and 1983 (except 1979) fell below the mean. A declining trend also seemed to appear in the counts.

At Hawk Mountain, a declining trend in Peregrine Falcon counts also appeared (Spofford, 1969; Nagy, 1977a).

APPENDIX

Scientific Names of Birds

The sixth edition (1983) of the American Ornithologists' Union's *Check-List of North American Birds* serves as the basis for the names and sequence of birds discussed in this book. Some of the names, and much of the sequence, differ considerably from those contained in past hawk migration literature. Nevertheless, it is necessary to adopt the sixth edition standards because they will appear in future scientific literature at least until the next edition of the Check-list appears.

Black Vulture	*Coragyps atratus*
Turkey Vulture	*Cathartes aura*
California Condor	*Gymnogyps californianus*
Osprey	*Pandion haliaetus*
Hook-billed Kite	*Chondrohierax uncinatus*
American Swallow-tailed Kite	*Elanoides forficatus*
Black-shouldered Kite	*Elanus caeruleus*
Snail Kite	*Rostrhamus sociabilis*
Mississippi Kite	*Ictinia mississippiensis*
Plumbeous Kite	*Ictinia plumbea*
Bald Eagle	*Haliaeetus leucocephalus*
White-tailed Eagle	*Haliaeetus albicilla*
Steller's Sea-Eagle	*Haliaeetus pelagicus*
Northern Harrier	*Circus cyaneus*
Sharp-shinned Hawk	*Accipiter striatus*
Cooper's Hawk	*Accipiter cooperii*
Northern Goshawk	*Accipiter gentilis*
Roadside Hawk	*Buteo magnirostris*
Red-shouldered Hawk	*Buteo lineatus*
Ridgway's Hawk	*Buteo ridgwayi*
Broad-winged Hawk	*Buteo platypterus*
Short-tailed Hawk	*Buteo brachyurus*
Swainson's Hawk	*Buteo swainsoni*

Zone-tailed Hawk	*Buteo albonotatus*
Hawaiian Hawk	*Buteo solitarius*
Red-tailed Hawk	*Buteo jamaicensis*
Common Buzzard	*Buteo buteo*
Ferruginous Hawk	*Buteo regalis*
Rough-legged Hawk	*Buteo lagopus*
Golden Eagle	*Aquila chrysaetos*
Eurasian Kestrel	*Falco tinnunculus*
American Kestrel	*Falco sparverius*
Merlin	*Falco columbarius*
Peregrine Falcon	*Falco peregrinus*
Gyrfalcon	*Falco rusticolus*
Prairie Falcon	*Falco mexicanus*
Great Horned Owl	*Bubo virginianus*
Horned Lark	*Eremophila alpestris*

LITERATURE CITED

Able, K. P.
  1965   Hawk and Common Loon Flight Over Bernheim Forest. *Kentucky War-bler*, 41 (4): 64.
Adolphson, D. G.
  1974   Birds of Prey Banding in South Dakota 1965 to 1973. *South Dakota Bird Notes*, 26 (3): 50–52.
Albrecht, C. J.
  1922   Migration of Hawks. *Murrelet*, 3 (3): 5.
Allen, J. A.
  1871   The Migration of Hawks. *American Naturalist*, 5 (1): 173.
Allen, R. P.
  1932   Hawk Flights and Law Enforcement at Cape May September 15–October 29, 1932. Unpublished report.
  1933   Cape May Hawk Flights/The Situation: 1933. Unpublished report.
Allen, R. P., and R. T. Peterson
  1936   The Hawk Migrations at Cape May Point, New Jersey. *Auk*. 53: 393–404.
Amadon, D.
  1967   Birds of Prey in the Collection of the American Museum of Natural History. *Raptor Research News*, 1 (3): 53–54.
American Ornithologists' Union
  1957   *Check-list of North American Birds*. Fifth Edition. Port City Press, Inc., Baltimore, Md.
  1983   *Check-List of North American Birds*. Sixth Edition. Allen Press, Inc., Lawrence, Kan.
Anderson D. W., and J. J. Hickey
  1972   Eggshell Changes in Certain North American Birds. *Proc. XVth International Ornithological Congress*, Leiden, Netherlands. Pp. 514–540.
Anderson, K. S., and K. Powers
  1979   Offshore Sightings of Birds of Prey. *HMANA Newsletter*, 4 (2): 16.
Andrews, M.
  1942   Hawks and Herons. *Feathers*, 4 (5): 37.
Anonymous
  1875   Great Flight of Hawks. *Rod and Gun*, 7: 3.
  1953   Hawk Count—1952. *New Hampshire Bird News*, 6 (1): 19.
  1954   Autumn Hawk Migration Report 1953. *News Letter Thunder Bay Field Naturalists Club*, 8 (1): 4.
  1965   April 1965 Hawk Migrations. *Prothonotary*, 31 (5): 52–54.
  1966   New Hampton Hawk Watch. *New Hampshire Audubon Quarterly*, 19 (3): 98.
  1973   Meet Dave Evans—Hawk Bander. *Hawk Ridge Nature Reserve Newsletter*, 1: 5.
  1974   STRI Biologist Studies Spectacular Migrations. *Smithsonian Institution Research Reports*, 7: 1–2.
  1975a   The Report of the Smirnoff Hawk Patrol. *J. Hawk Migration Assn. North America*, 1 (1): 13.
  1975b   Summary of Hawk Migration. *News Letter 47*. Hawk Mountain Sanctuary Assn., Kempton, Pa.

287

1976 Summary of Hawk Migration. *News Letter 48*. Hawk Mountain Sanctuary Assn., Kempton, Pa.

1977a Summary of Hawk Migration. *News Letter 49*. Hawk Mountain Sanctuary Assn., Kempton, Pa.

1977b Hawk Report Whitehorse Mountain, Cornwall-on-Hudson, Orange County, New York. Mimeographed report.

1978 The Migration. *Hawk Mountain News*, 50: 21–24.

1979 The Migration Hawk Mountain 1978. *Hawk Mountain News*, 50 [*sic;* should be 51]: 30–34.

1981 Summary of Hawk Migration. *Hawk Mountain News*, 55: 21–28.

1983 The First Successful Satellite Tracking Radio Tag for Whale or Dolphin. *J. Ceta-Research*, 4: 1.

1984 Summary of the 1984 Hawk Migration. *Derby Hill Newsletter*, 7 (1): 1–4.

Appell, G. N.
1970 Hawk Migrations Over the Harpswells. *Maine Nature*, 2: 26.

1975 Casco Bay, Maine. *Proc. North American Hawk Migration Conference 1974*. Pp. 8–11.

Apps, K.
1975 High Mountain Hawk Counts—Autumn 1974. *New Jersey Nature News*, 30 (1): 18–19.

Armistead, H. T.
1981 Middle Atlantic Coast Region. *American Birds*, 35 (2): 166–169.

1983a Middle Atlantic Coast Region. *American Birds*, 37 (2): 164–167.

1983b Middle Atlantic Coast Region. *American Birds*, 37 (5): 855–858.

Arvin, J. C.
1982 South Texas Region. *American Birds*, 36 (5): 871–873.

Atherton, L. S., and B. H. Atherton
1980 Florida Region. *American Birds*, 34 (2): 152–155.

1981 Florida Region. *American Birds*, 35 (2): 172–176.

1982 Florida Region. *American Birds*, 36 (2): 168–171.

1983 Florida Region. *American Birds*, 37 (2): 170–173.

Austin, O. L., Jr.
1932 The Birds of Newfoundland Labrador. *Memoirs No. 7*. Nuttall Ornithological Club, Cambridge, Mass.

Axtell, H. A.
1970 An Unusual Hawk Migration at Fort Erie, Ontario. *Prothonotary*, 36 (10): 128–129.

Bagg, A. C., and S. A. Eliot, Jr.
1937 *Birds of the Connecticut Valley in Massachusetts*. Hampshire Bookshop, Northampton, Mass.

Bagg, A. M.
1947 Watch New England Hawk Flights! *Bulletin Massachusetts Audubon Society*, 31: 81–82.

1949 Flight Over the Valley. Connecticut Valley Hawk Flights—September, 1948. *Bulletin Massachusetts Audubon Society*, 33: 135–137.

1950 Minimum Temperatures and Maximum Flights. Connecticut Valley Hawk Flights—September, 1949. *Bulletin Massachusetts Audubon Society*, 34: 75–80.

1970 A Summary of the 1969 Fall Migration Season, with Special Attention to Eruptions of Various Boreal and Montane Species and an Analysis of Correlations between Wind Flows and Migration. *Audubon Field Notes*, 24 (1): 4–13.

Bagg, A. M., and R. P. Emery
  1962   Northeastern Maritime Region. *Audubon Field Notes,* 16 (1): 7–12.
  1966   Northeastern Maritime Region. *Audubon Field Notes,* 20 (1): 14.
Bailey, A. M., and R. J. Niedrach
  1965   *Birds of Colorado.* Volume one. Denver Museum of Natural History, Denver, Colorado.
Bailey, B. H.
  1912   A Remarkable Flight of Broad-winged Hawks. *Proc. Iowa Academy of Science,* 19: 195–196.
Bailey, S. F.
  1967   Fall Hawk Watch at Mt. Peter. *Kingbird,* 17 (3): 129–142.
  1969   1968 Hawk Watch at Mt. Peter. *Kingbird,* 19 (4): 200–203.
Baker, C. A.
  1983   Summary of the 1983 Hawk Migration. *Derby Hill Newsletter,* 6 (1): 1–4.
Balph, D. F., and M. H. Balph
  1983   On the Psychology of Watching Birds: the Problem of Observer-Expectancy Bias. *Auk,* 100 (3): 755–757.
Barbour, R.
  1908   A Large Migration of Hawks. *Auk,* 25 (1): 82–84.
Barrows, W. B.
  1912   *Michigan Bird Life.* Michigan Agricultural College, East Lansing, Mich.
Bartelt, P. E., and C. J. Orde
  1976   A Raptor Survey and Trapping Techniques in Clay County, South Dakota. *South Dakota Bird Notes,* 28 (2): 24–30, 36.
Bates, M. R.
  1975   Mount Tom, Massachusetts. *Proc. North American Hawk Migration Conference 1974.* Pp. 24.
Baughman, J. L.
  1947   A Very Early Notice of Hawk Migrations. *Auk,* 64: 304.
Baumgartner, F. M.
  1955   Southern Great Plains. *Audubon Field Notes,* 9 (1): 36–38.
Baumgartner, J.
  1979   First Annual Report. Whitefish Point Bird Observatory 1978–1979.
  1980   Second Annual Report. Whitefish Point Bird Observatory 1979–1980.
  1981   Whitefish Point Bird Observatory Third Annual Report 1980–1981
  1982   Whitefish Point Bird Observatory Fourth Annual Report 1981–1982
  1983   Whitefish Point Bird Observatory Fifth Annual Report 1982–1983
Baumgartner, J. E., and A. H. Kelley
  1981   Summary of the Whitefish Point Bird Observatory Annual Report for 1979. *Jack-Pine Warbler,* 59 (2): 53–59.
Baxter, J. R.
  1969   [Hawk Flight Summary]. *Sandpiper,* 11 (4): 54–55.
  1970a   Hawk Flights, Spring 1970. *Sandpiper,* 12 (4): 50–51.
  1970b   Eagle Observation at Girard, Pennsylvania. *Sandpiper,* 13 (3): 43.
  1971   Hawk Flight Summary. *Sandpiper,* 13 (4): 62–63.
  1972   Hawk Flight. *Sandpiper,* 14 (4): 61–62.
Beamish, P., and S. Carroll
  1983   Acoustics of Theo Before and After Attachment of a Satellite Tracking Tag. *J. Ceta-Research,* 4: 2.
Beaton, R. J.
  1951   Hawk Migration at South Mountain. *Atlantic Naturalist,* 6: 166–168.

Beck, R. A.
  1972   Milepost 92: Where Birders Meet. *Virginia Society of Ornithology Newsletter*, 18 (5): 2–3.
Bednarz, J.
  1978   Red-tailed Hawk Migration in Western Iowa. *Iowa Bird Life*, 48 (4): 141–142.
Beebe, F. L.
  1974   Field Studies of the Falconiformes of British Columbia. *Occasional Paper Series No. 17*. British Columbia Provincial Museum, Victoria, B. C.
Beebe, R.
  1933   Influence of the Great Lakes on the Migration of Birds. *Wilson Bulletin*, 45 (3): 118–121.
Beer, J. R.
  1966   The Pigeon Hawk in Minnesota. *Loon*, 38 (4): 129–131.
Behrend, F. W.
  1950a   Broad-winged Hawks over Hump Mountain. *Migrant*, 21: 10–11.
  1950b   Hawk Migration in Upper East Tennessee. *Migrant*, 21: 70–72.
  1951   Fall Migrations of Hawks in 1951. *Migrant*, 22: 53–57.
  1952   Fall Migrations of Hawks in 1952. *Migrant*, 23: 62–65.
  1953   Hawk Migration-Fall 1953. *Migrant*, 24: 69–73.
  1954a   Plans for Hawk Watching—1954. *Migrant*, 25: 24–25.
  1954b   1954 Fall Migration Count of Hawks. *Migrant*, 25: 69–72.
Belknap, J. B.
  1962   Raptors at Point Peninsula. *Kingbird*, 12 (1): 22.
Bennett, G. M., M. Mitchell, and W. W. H. Gunn
  1958   Ontario-Western New York. *Audubon Field Notes*, 12 (1): 26–30.
Benz, S.
  1980a   Migration 1979. *Hawk Mountain News*, 53: 42–53.
  1980b   Northern Appalachians. *HMANA Newsletter*, 5 (2): 20–22.
  1981a   Northern Appalachians. *HMANA Newsletter*, 6 (1): 15–16.
  1981b   Northern Appalachians. *HMANA Newsletter*, 6 (2): 20–22.
  1982a   The Fall Season Migration: A New Emphasis. *Hawk Mountain News*, 57: 20–27.
  1982b   Northern Appalachians. *HMANA Newsletter*, 7 (2): 23–24.
  1982c   Northern Appalachians. *HMANA Newsletter*, 7 (1): 17.
  1983a   The Fall Season: Migration Report. *Hawk Mountain News*, 59: 14–21.
  1983b   Northern Appalachians. *HMANA Newsletter*, 8 (2): 21–24.
  1983c   Northern Appalachians. *HMANA Newsletter*, 8 (1): 14.
  1984a   The Fall Season Migration Report. *Hawk Mountain News*, 61: 27–35.
  1984b   Northern Appalachians. *HMANA Newsletter*, 9 (2): 23–27.
Benz, S., and D. Dereamus
  1984   Northern Appalachians. *HMANA Newsletter*, 9 (1): 17–19.
Berger, A. J.
  1981   *Hawaiian Birdlife*. Second Edition. University of Hawaii Press, Honolulu, Hawaii.
Berger, D. D.
  1954   Hawk Banding at Cedar Grove, Wisconsin. *Flicker*, 26 (3): 90–93.
Bergey, A.
  1975   Hawk Migration Behavior at the Susquehanna River. *Cassinia*, 55: 37–38.
Bernard, R. F.
  1966a   Western Great Lakes Region. *Audubon Field Notes*, 20 (1): 45–46, 50–53.
  1966b   Western Great Lakes Region. *Audubon Field Notes*, 20 (4): 511–513.

1967   Western Great Lakes Region. *Audubon Field Notes,* 21 (4): 510–513.
1968   Western Great Lakes Region. *Audubon Field Notes,* 22 (4): 529–531.
Berry, R. B.
1971   Peregrine Falcon Population Survey, Assateague Island, Maryland, Fall, 1969. *Raptor Research News,* 5: 31–43.
Beske, A. E.
1982   Local and Migratory Movements of Radio-Tagged Juvenile Harriers. *Raptor Research,* 16 (2): 39–53.
Beveridge, T.
1954   Hawk Migration. *Blue Jay,* 12 (3): 9.
Bierly, M. L.
1973   A Flock of Mississippi Kites in Western Kentucky. *Kentucky Warbler,* 49 (4): 72.
Bigg, M., I. MacAskie, and G. Ellis
1983   Photo-identification of Individual Killer Whales. *Whalewatcher,* 17 (1): 3–5.
Bihun, A., Jr.
1967   The Montclair Hawk Lookout Sanctuary 1967 Hawk Watch. *New Jersey Nature News,* 22: 146–147.
1968   Montclair Hawk Lookout Sanctuary 1968 Hawk Watch. *New Jersey Nature News,* 23: 134–135.
1969   Montclair Hawk Lookout Sanctuary 1969 Hawk Watch. *New Jersey Nature News,* 24: 151–153.
1970   Montclair Hawk Lookout Sanctuary 1970 Hawk Watch. *New Jersey Nature News,* 25: 146–148.
1972   Montclair Bird Club Hawk Watch. *New Jersey Nature News,* 27 (3): 115–118.
1973   Montclair Bird Club Hawk Watch Daily Record. *New Jersey Nature News,* 28 (2): 85–88.
1974   Montclair Bird Club Hawk Watch Daily Record. *New Jersey Nature News,* 29 (1): 31–35.
1975a   Montclair Bird Club Hawk Watch Daily Record. *New Jersey Nature News,* 30 (1): 6–9.
1975b   Montclair Bird Club Hawk Watch Daily Record. Xeroxed Report for 1975.
1976   Montclair Bird Club Hawk Watch Daily Record. Xeroxed Report for 1976.
1977   Montclair Bird Club Hawk Watch Daily Record. Xeroxed Report for 1977.
1978   Montclair Bird Club Hawk Watch Daily Record. Xeroxed Report for 1978.
1979   Montclair Bird Club Hawk Watch Daily Record. Xeroxed Report for 1979.
1980a   1980 Spring Hawk Watch Montclair Hawk Lookout Sanctuary, Montclair, New Jersey. Mimeographed report.
1980b   Montclair Bird Club Hawk Watch Daily Record. Xeroxed Report for 1980.
1981   Montclair Bird Club Hawk Watch Daily Record. Xeroxed Report for 1981.
1982   Montclair Bird Club Hawk Watch Daily Record. Xeroxed Report for 1982.
1983   Montclair Bird Club Hawk Watch Daily Record. Xeroxed Report for 1983.

1984 Montclair Bird Club Spring Hawk Watch. Mimeographed Report for 1984.

Bildstein, K. L., et al.
1984 Sex and Age Differences in Fall Migration of Northern Harriers. *J. Field Ornithology*, 55 (2): 143–150.

Binford, L. C.
1965 Unusual Bird Records from the Upper Peninsula of Michigan. *Jack-Pine Warbler*, 43 (3): 144–145.
1979 Fall Migration on Diurnal Raptors at Point Diablo, California. *Western Birds*, 10: 1–16.

Birkeland, H.
1934 Migration of Hawks in Story County. *Iowa Bird Life*, 4 (4): 52.

Bittner, J. D.
1975 Bald Eagle Ridge, Pennsylvania. *Proc. North American Hawk Migration Conference*, 1974. Pp. 5–6.

Black, G. B.
1969a Turkey Vulture Migration at Red Rock. *Iowa Bird Life*, 39 (2): 27–29.
1969b 1969 Turkey Vulture Observations at Red Rock Refuge. *Iowa Bird Life*, 39 (4): 78–80.

Bohall, P. G., and M. W. Collopy
1984 Seasonal Abundance, Habitat Use, and Perch Sites of Four Raptor Species in North-Central Florida. *J. Field Ornithology*, 55 (2): 181–189.

Bond, J.
1956 *Check-List of Birds of the West Indies*. Academy of Natural Sciences of Philadelphia, Philadelphia, Pa.
1970 *Native and Winter Resident Birds of Tobago*. Academy of Natural Sciences of Philadelphia, Philadelphia, Pa.
1980 *Birds of the West Indies*. Fourth Edition. Houghton Mifflin Company, Boston, Mass.

Boon, E. C.
1961 Migratory Congregations of Hawks. *Blue Jay*, 19 (1): 23.

Borneman, J. C.
1976 California Condors Soaring into Opaque Clouds. *Auk*, 93 (3): 636.

Boyajian, N. R.
1968a Hudson-St. Lawrence Region. *Audubon Field Notes*, 22 (1): 13–18.
1968b Hudson-St. Lawrence Region. *Audubon Field Notes*, 22 (4): 507–510.
1971 Hudson-St. Lawrence Region. *American Birds*, 25 (1): 31–35.

Boyer, M. J.
1972 Bounty Payments in Lehigh County, 1819–1839. *Proc. Lehigh County Historical Society*, 29: 77–81.

Boyle, W. J., Jr., R. O. Paxton, and D. A. Cutler
1982 Hudson-Delaware Region. *American Birds*, 36 (5): 833–836.
1983 Hudson-Delaware Region. *American Birds*, 37 (5): 850–855.

Brady, A.
1965 Early Hawk Flight at Cape May. *Cassinia*, 48: 36–37.

Bray, T.
1984 Raptor Migration. *Nebraska Bird Review*, 52(1): 23.

Breck, G. W.
1959 Operation Hawk-Watch 1959—Final Report on 3 Across-The-State Hawk-Watches. Mimeographed Report.
1960a Operation Hawk-Watch 1960—Results of "3-Week" Watch in Upper Montclair, N. J. Mimeographed Report.

1960b   The Montclair Hawk Lookout. *Linnaean News-Letter,* 14: No. 5.

1960c   Operation Hawk-Watch 1959. *Urner Field Observer,* 8 (2): 15–20.

Breck, G. W., and R. A. Breck

1964   The Montclair Hawk Lookout Sanctuary. *New Jersey Nature News,* 19: 149–150.

1965   The Montclair Hawk Lookout Sanctuary 1965 Hawk Watch. *New Jersey Nature News,* 21: 164–165.

1966   The Montclair Hawk Lookout Sanctuary 1966 Hawk Watch. *New Jersey Nature News,* 21: 164–165.

Breck, R. A.

1962   Montclair Hawk Lookout Sanctuary. *New Jersey Nature News,* 17: 138–139.

1963   Montclair Hawk Lookout Sanctuary. *New Jersey Nature News,* 18: 149–150.

Brendel, E., and D. D. Roby

1984   Mississippi Kite in Philadelphia, Pennsylvania. *Cassinia,* 60: 62.

Breninger, G. F.

1897   Nocturnal Flights of the Turkey Vulture. *Osprey,* 2: 54–55.

Brett, J.

1981   The School in the Clouds, Continued. *Hawk Mountain News,* 56: 4–11.

1984a   American Conservation's "Glorious Joan of Arc". *Hawk Mountain News,* 62: 4–14.

1984b   The First Years on the Mountain: Maurice Broun's Journals. *Hawk Mountain News,* 62: 20–28.

Brett, J. J., and A. C. Nagy

1973   *Feathers in the Wind.* Hawk Mountain Sanctuary Assn., Kempton, Pa.

Brock, K. J.

1984   An Impressive Peregrine Flight at Miller Beach. *Indiana Audubon Quarterly,* 62 (1): 20–22.

Broley, C. L.

1947   Migration and Nesting of Florida Bald Eagles. *Wilson Bulletin,* 59: 3–20.

Bronoel, J. K.

1948   Duluth Bird Club Project for the Conservation of Hawks, Owls, and Song Birds. *Flicker,* 20 (1): 28–29.

1954   Duluth Club Participates in Hawk Conservation. *Flicker,* 26 (3): 94–95.

Brookfield, C. M.

1953   Florida Region. *Audubon Field Notes,* 7 (1): 12–13.

Brooks, M.

1944   A Check-list of West Virginia Birds. *Bulletin 316.* West Virginia Agricultural Experiment Station, Morgantown, W. Va.

1956   Appalachian Region. *Audubon Field Notes,* 10 (1): 22–25.

1958   Appalachian Region. *Audubon Field Notes,* 12 (1): 30–32.

1959   Appalachian Region. *Audobon Field Notes,* 13 (1): 28–30.

1971   Hawk Migration Along Monterey Mountain. *Raven,* 42 (4): 62.

Brophy, T.

1982   Eagle Valley. *HMANA Newsletter,* 7 (1): 1, 3.

Broun, M.

1935   The Hawk Migration During the Fall of 1934, Along the Kittatinny Ridge in Pennsylvania. *Auk,* 52: 233–248.

1936   Three Seasons at Hawk Mountain Sanctuary. *Publication 61.* Emergency Conservation Committee.

1939   Fall Migration of Hawks at Hawk Mountain, Pennsylvania, 1934–1938. *Auk*, 56: 429–441.

1940   No title. *News Letter to Members No. 3*. Hawk Mountain Sanctuary Assn., Kempton, Pa.

1946   Report of the Curator. *News Letter to Members No. 15*. Hawk Mountain Sanctuary Assn., Kempton, Pa.

1947   Hawk Mountain Sanctuary Autumn News from the Curator October, November and December, 1946. *News Letter to Members No. 16*. Hawk Mountain Sanctuary Assn., Kempton, Pa.

1948   Hawk Mountain Sanctuary/The Curator's Bird Report for 1947. *News Letter to Members No. 17*. Hawk Mountain Sanctuary Assn., Kempton, Pa.

1949a   *Hawks Aloft: The Story of Hawk Mountain*. Dodd, Mead Co., New York, N. Y.

1949b   Curator's Report 1948. *News Letter to Members No. 18*. Hawk Mountain Sanctuary Assn., Kempton, Pa.

1950   The Curator's Report. *News Letter to Members No. 19*. Hawk Mountain Sanctuary Assn., Kempton, Pa.

1951a   The Curator's Report. *News Letter to Members No. 20*. Hawk Mountain Sanctuary Assn., Kempton, Pa.

1951b   Hawks and the Weather. *Atlantic Naturalist*, 6: 105–112.

1952   The Curator's Report. *News Letter to Members No. 21*. Hawk Mountain Sanctuary Assn., Kempton, Pa.

1953   The Curator's Report. *News Letter to Members No. 22*. Hawk Mountain Sanctuary Assn., Kempton, Pa.

1954   The Curator's Report. *News Letter to Members No. 23*. Hawk Mountain Sanctuary Assn., Kempton, Pa.

1955   The Curator's Report, *News Letter to Members No. 24*. Hawk Mountain Sanctuary Assn., Kempton, Pa.

1956a   The Curator's Report. *News Letter to Members No. 25*. Hawk Mountain Sanctuary Assn., Kempton, Pa.

1956b   Pennsylvania's Bloody Ridges. *Nature Magazine*, 49: 288–292.

1957   The Curator's Report. *News Letter to Members No. 26*. Hawk Mountain Sanctuary Assn., Kempton, Pa.

1958a   The Curator's Report. *News Letter to Members No. 27*. Hawk Mountain Sanctuary Assn., Kempton, Pa.

1958b   Hawk Mountain Sanctuary Deer-Hunting Regulations. *News Letter to Members No. 28*. Hawk Mountain Sanctuary Assn., Kempton, Pa.

1959   The Curator's Report. *News Letter to Members No. 29*. Hawk Mountain Sanctuary Assn., Kempton, Pa.

1960   Curator's Report—1959. *News Letter to Members No. 30*. Hawk Mountain Sanctuary Assn., Kempton, Pa.

1961   Curator's Report—1960. *News Letter to Members No. 31*. Hawk Mountain Sanctuary Assn., Kempton, Pa.

1962   Curator's Report—1961. *News Letter to Members No. 32*. Hawk Mountain Sanctuary Assn., Kempton, Pa.

1963a   *Hawk Migrations and the Weather*. Hawk Mountain Sanctuary Assn., Kempton, Pa.

1963b   Curator's Report—Highlights of 1962. *News Letter to Members No. 34*. Hawk Mountain Sanctuary Assn., Kempton, Pa.

1964   Curator's Report—1963. *News Letter to Members No. 35*. Hawk Mountain Sanctuary Assn., Kempton, Pa.

1965 Curator's Report—1964. *News Letter to Members No. 36.* Hawk Mountain Sanctuary Assn., Kempton, Pa.

1966 Curator's Report—1965. *News Letter to Members No. 37.* Hawk Mountain Sanctuary Assn., Kempton, Pa.

1979 The Importance of Being Sober/Are We Suffering from Diplopia? *Osprey,* 2 (1): 2–3.

Broun, M., and B. V. Goodwin
1943 Flight-Speeds of Hawks and Crows. *Auk.,* 60: 487–492.

Brown, B. P., Jr.
1966 Chicago Lake Front Migration—Fall 1966. *Audubon Bulletin,* 140: 11–15.

Brown, J., and D. Brown
1972 A Note About Red-tailed Hawks. *Iowa Bird Life,* 42 (2): 51.

Brown, L.
1977 *Birds of Prey/Their Biology and Ecology.* A&W Publishers, Inc., New York, N.Y.

Brown, L. and D. Amadon
1968 *Eagles, Hawks and Falcons of the World.* Two Volumes. McGraw-Hill Book Co., New York, N.Y.

Brown, W. H.
1978 Migrating Broad-winged Hawks in Lexington. *Kentucky Warbler,* 54 (4): 73.

Brudenell-Bruce, P. G. C.
1975 *The Birds of the Bahamas.* Taplinger Publishing Co., New York, N.Y.

Bryant, P. J., and S. K. Lafferty
1983 Photo-identification of Gray Whales. *Whalewatcher,* 17 (1): 6–9.

Buckley, P. A., R. O. Paxton, and D. A. Cutler
1978 Hudson-Delaware Region. *American Birds,* 32 (2): 182–189.

Beuchner, H. K., F. C. Craighead, Jr., and J. J. Craighead
1971 Satellites for Research on Free-Roaming Animals. *BioScience,* 21 (24): 1201–1205.

Bull, J.
1964 *Birds of the New York Area.* Harper & Row, New York, N.Y.
1974 *Birds of New York State.* Doubleday/Natural History Press, Garden City, N.Y.

Burleigh, T.D.
1958 *Georgia Birds.* University of Oklahoma Press, Norman, Okla.
1972 *Birds of Idaho.* Caxton Printers, Ltd., Caldwell, Ida.

Burns, F. L.
1911 A Monograph of the Broad-winged Hawk *(Buteo platypterus). Wilson Bulletin,* 23 (3 & 4): 143–320.

Burrows, R.
1981 *A Birdwatcher's Guide to Atlantic Canada.* Volume one. Private publication, Canada.

Burton, D. E., and J. Woodford
1960 Ontario-Western New York Region. *Audubon Field Notes,* 14 (4): 383–386.

Bussjaeger, L. J., et al.
1967 Turkey Vulture Migration in Veracruz, Mexico. *Condor,* 69: 425–426.

Callaway, S., and A. Callaway
1940 A Flock of Migrating Swainson Hawks Visits Fairbury, Jefferson County. *Nebraska Bird Review,* 8 (1): 22.

Cameron, E.
1964   Hawk Cliff. *Blue Bill*, 11 (3): 31–33.
Cant, G.
1946   Observations of a Black Gyrfalcon at Raccoon Ridge, Warren County. *Urner Field Observer*, 1 (6): 5.
Carleton, G.
1962   Hudson-St. Lawrence Region. *Audubon Field Notes*, 16 (1): 12–15.
1963   Hudson-St. Lawrence Region. *Audubon Field Notes*, 17 (1): 15–18.
Carpenter, F. S.
1934   Flight of Broad-winged Hawks. *Kentucky Warbler*, 10 (4): 19–20.
Carpenter, M.
1949   A Hawk Flight Over Reddish Knob. *Raven*, 20 (9 & 10): 58–59.
1960   Observations on Hawk Migrations on the Shenandoah Mountain, Virginia, 1949–1959. *Raven*, 31 (9 & 10): 88–90.
Carrier, P.
n.d.   *A Guide for Hawks Seen in the North East.* New England Hawk Watch Study, Portland, Conn.
Carroll, J. R.
1980   1979 Fall Hawk Migration at the Near Trapps, Shawangunk Mountains, N.Y. *Kingbird*, 30 (2): 76–84.
Carson, R. D.
1960   A Migratory Congregation of Swainson's Hawks. *Blue Jay*, 18 (4): 158.
Carter, J. H. III
1974   Osprey Migration at Wrightsville Beach, N.C. *Chat*, 38 (2): 39.
Carter, J. L.
1978   Migrating Hawks in Bedford County, Virginia. *Raven*, 49 (2): 33–34.
Cary, M.
1899   A Phenomenal Flight of Hawks. *Auk*, 16: 352–353.
Chamberlain, B. R.
1953   Southern Atlantic Coast Region. *Audubon Field Notes*, 7 (4): 267–268.
1956   Southern Atlantic Coast Region. *Audubon Field Notes*, 10 (1): 15–18.
1958   Southern Atlantic Coast Region. *Audubon Field Notes*, 12 (1): 19–21.
1960b   Southern Atlantic Coast Region. *Audubon Field Notes*, 14 (4): 377–379.
1961   Southern Atlantic Coast Region. *Audubon Field Notes*, 15 (4): 399–402.
1962   Southern Atlantic Coast Region. *Audubon Field Notes*, 16 (1): 18–21.
Chandik, T., and A. Baldridge
1968   Middle Pacific Coast Region. *Audubon Field Notes*, 22 (1): 83–88.
1969   Middle Pacific Coast Region. *Audubon Field Notes*, 23 (1): 99–106.
Chapman, F. M.
1933   The Migration of Turkey Buzzards as Observed on Barro Colorado Island, Canal Zone. *Auk*, 50: 30–34.
Chapman, H. F.
1953   Broad-winged Hawks in Eastern So. Dak. *South Dakota Bird Notes*, 5 (3): 46–47.
Chase, T., Jr., and R. O. Paxton
1966   Middle Pacific Coast Region. *Audubon Field Notes*, 20 (1): 87–90.
Choate, E. A.
1965   Hawk Flights Cape May Point 1965. Unpublished manuscript.
1972   Spectacular Hawk Flight at Cape May Point, New Jersey, on 16 October 1970. *Wilson Bulletin*, 84: 340–341.
Choate, E. A., and F. Tilly
1973   The 1970 Autumn Hawk Count at Cape May Point, New Jersey. *Science*

Notes No. 11. New Jersey State Museum, Trenton, N.J.

Christie, D. S.
1980   Hawk Migration Studies. *New Brunswick Naturalist*, 10 (2): 48–49.
1981a   Hawk Migration in Atlantic Canada. *Atlantic Canada News*, 1: 2–10.
1981b   Atlantic Canada. *HMANA Newsletter*, 6 (2): 11.
1982a   Atlantic Canada. *HMANA Newsletter*, 7 (1): 11.
1982b   Atlantic Canada. *HMANA Newsletter*, 7 (2): 13–14.
1983a   Atlantic Canada. *HMANA Newsletter*, 8 (1): 8–9.
1983b   Atlantic Canada. *HMANA Newsletter*, 8 (2): 16.

Cink, C. L.
1972   Prairie Falcon in Lancaster County. *Nebraska Bird Review*, 40 (2): 44–45.

Cinquina, J.
1975   Highlands Audubon Society Mt. Peter, N.Y.—Hawk Watch 1974. *Highlands Audubon Society Newsletter*, 5 (2): 5–7.
1976   Highlands Audubon Society Mt. Peter, N.Y.—Hawk Watch 1975. *Highlands Audubon Society Newsletter*, 6 (1): 4–6.
1977   Highlands Audubon Society Mt. Peter, N.Y.—Hawk Watch 1976. *Highlands Audubon Society Newsletter*, 7 (2): 7–9.
1984   Spring at Mount Peter. *Highlands Audubon Society Newsletter*, 14 (6).

Cinquina, J., and A. Martin
1979   Mount Peter Hawk Watch—1978. *Highlands Audubon Society Newsletter*, 9 (1): 6–8.
1980   Mount Peter. *Highlands Audubon Society Newsletter*, 10 (1): 7–9.
1981   Mount Peter. *Highlands Audubon Society Newsletter*, 11 (1): 4–6.
1982   Mount Peter. *Highlands Audubon Society Newsletter*, 12 (1): 4–6.
1983   Mount Peter—1982. *Highlands Audubon Society Newsletter*, 13 (1).
1984   Mount Peter. *Highlands Audubon Society Newsletter*, 14 (1).

Clark, W. S.
1968   Migration Trapping of Hawks at Cape May, N.J. *EBBA News*, 31: 112–114.
1969   Migration Trapping of Hawks at Cape May, N.J.—Second Year. *EBBA News*, 32: 69–77.
1970   Migration Trapping of Hawks (and Owls) at Cape May, N.J.—Third Year. *EBBA News*, 33: 181–189.
1971   Migration Trapping of Hawks (and Owls) at Cape May, N.J.—Fourth Year. *EBBA News*, 34: 160–169.
1972   Migration Trapping of Hawks (and Owls) at Cape May, N.J.—Fifth Year. *EBBA News*, 35: 121–131.
1973   Cape May Point Raptor Banding Station—1972 Results. *EBBA News*, 36: 150–165.
1975a   Analysis of Data. *Proc. North American Hawk Migration Conference 1974*. Pp. 87–97.
1975b   Coastal Plain. *J. Hawk Migration Assn. North America*, 1 (1): 25–29.
1976   Cape May Point Raptor Banding Station—1974 Results. *North American Bird Bander*, 1 (1): 5–13.
1978a   Spring Hawk Movement at Sandy Hook, N.J.—1977. *New Jersey Audubon*, 4 (2): 43–47.
1978b   1977 Raptor Banding Project Summary. *Peregrine Observer*, 2 (2): 5.
1984   Raptor Banding Project, 1983. *Peregrine Observer*, 7 (1): 16.

Clark, W. S., and M. E. Pramstaller
1980   *Field I.D. Pamphlet for North American Raptors*. Raptor Information Center, National Wildlife Federation, Washington, D.C.

Clement, R. C.
1958   Broadwings in Rhode Island. *Narragansett Naturalist,* 1 (4): 118–119.
Clendinning, A. E.
1954   Notes on Fall Hawk Migration—1954. *Cardinal,* 15: 17–24.
Cleveland, N. J., et al.
1980   Birder's Guide to Southeastern Manitoba. *Eco Series No. 1.* Manitoba Naturalists Society, Winnipeg, Manitoba.
Cochran, W. W.
1972   A Few Days of the Fall Migration of a Sharp-shinned Hawk. *Hawk Chalk,* 11 (1): 39–44.
1975   Following A Migrating Peregrine From Wisconsin to Mexico. *Hawk Chalk,* 14 (2): 28–37.
Coffey, B. B., Jr.
1981   Prairie Falcon at Memphis. *Migrant,* 52 (1): 18.
Cogswell, H. L.
1956   Middle Pacific Coast Region. *Audubon Field Notes,* 10 (1): 50–54.
Cogswell, H. L., and R. H. Pray
1955   Middle Pacific Coast Region. *Audubon Field Notes,* 9 (1): 50–54.
Collins, H. H., Jr.
1933   Hawk Slaughter at Drehersville. *Bulletin No. 3. Annual Report of the Hawk and Owl Society.*
1935   A Hawk Sanctuary. *Nature Magazine,* 25: 84–86. February issue.
Cone, C. D., Jr.
1961   The Theory of Soaring Flight in Vortex Shells—Part I. *Soaring,* 25 (No. 4).
1962a   Thermal Soaring of Birds. *American Scientist,* 50: 180–209.
1962b   The Soaring Flight of Birds. *Scientific American,* 260 (April): 130–134, 136, 138, 140.
Cook, A. J.
1893   Birds of Michigan. *Bulletin 94.* Agricultural Experiment Station, State Agricultural College, Lansing, Michigan.
Cooper, D.
1983   HMANA Report Forms—Common Mistakes and Corrections. *HMANA Newsletter,* 8 (2): 10–11.
Cope, J. B.
1953   Fall Migration Studies of the Broad-winged Hawk. *Indiana Audubon Quarterly,* 31 (1): 14–15.
Copeland, D. W.
1977a   Eastern Great Lakes. *HMANA Newsletter,* 2 (1): 8–9.
1977b   Eastern Great Lakes. *HMANA Newsletter,* 2 (2): 18–21.
1977c   The Spring 1976 Hawk Watch at the Beamer Conservation Area. *Prothonotary,* 43 (2): 32–33.
1978a   Eastern Great Lakes. *HMANA Newsletter,* 3 (1): 10–13.
1978b   Eastern Great Lakes. *HMANA Newletter,* 3 (2): 26–28.
Cory, C. B.
1909   The Birds of Illinois and Wisconsin. *Pub. 131.* Field Museum of Natural History, Chicago, Ill.
Cote, R.
1982   Northern New England. *HMANA Newsletter,* 7 (2): 14.
1983   Northern New England. *HMANA Newsletter,* 8 (2): 17–18.

Craighead, F. C.
   1933   Notes on the Hawk Migration at Cape May, September 22 and 23, 1933. Unpublished manuscript.
Craighead, F. C., Jr., et al.
   1972   Satellite and Ground Radiotracking of Elk. In *Animal Orientation and Navigation*. National Aeronautics and Space Administration, Washington, D. C. Pp. 99–111.
Craighead, J. J.
   1971   Satellite Monitoring of Black Bear. *BioScience*, 21 (24): 1206–1212.
Craighead, J. J., and F. C. Craighead, Jr.
   1956   *Hawks, Owls and Wildlife*. Stackpole Co., Harrisburg, Pa.
Craighead, F. C., Jr., and T. C. Dunstan
   1976   Progress Toward Tracking Migrating Raptors by Satelllite. *Raptor Research*, 10 (4): 112–119.
Cross, A. A.
   1927   Observations on Hawk-Banding with Records and Recoveries. *Bulletin Northeastern Bird-Banding Assn.*, 3: 29–33.
Crossan, D. F., and R. A. Stevenson, Jr.
   1956   Hawk Migrations Along the Middle Eastern Seaboard, Delaware to North Carolina. *Chat*, 20 (1): 2.
Crowell, J. B., Jr., and H. B. Nehls
   1968   Northern Pacific Coast Region. *Audubon Field Notes*, 22 (1): 78–83.
   1969   Northern Pacific Coast Region. *Audubon Field Notes*, 23 (1): 94–99.
   1970   Northern Pacific Coast Region. *Audubon Field Notes*, 24 (1): 82–88.
   1971   Northern Pacific Coast Region. *American Birds*, 25 (4): 787, 791–793.
   1972   Northern Pacific Coast Region. *American Birds*, 26 (1): 107–111.
   1973   Northern Pacific Coast Region. *American Birds*, 27 (1): 105–110.
   1975   Northern Pacific Coast Region. *American Birds*, 29 (1): 105–112.
   1977   Northern Pacific Coast Region. *American Birds*, 31 (2): 212–216.
Cruickshank, A. D.
   1937   A Swainson's Hawk Migration. *Auk*, 54: 385.
Cruickshank, A. D., and H. G. Cruickshank
   1980   *The Birds of Brevard County, Florida*. Florida Press, Inc., Orlando, Fla.
Cruickshank, H. G.
   1941   *Bird Islands Down East*. Macmillan Company, New York, N. Y.
Crumb, D. W., and B. Peebles
   1977   To Find A Fall Hawk Lookout, *Kingbird*, 27 (2): 82–85.
Cryder, R. F.
   1928   Broad-winged Hawk Migration. *Audubon Bulletin*, 19: 19.
Cunningham, R. L.
   1964   Florida Region. *Audubon Field Notes*, 18 (1): 24–28.
   1965   Florida Region. *Audubon Field Notes*, 19 (1): 28–33.
   1966   Florida Region. *Audubon Field Notes*, 20 (4): 496–501.
Currie, N. W.
   1978   *Hawk Migration 1977 Report*. Privately published.
   1979   *Hawk Migration 1978 Report*. Privately published.
   1980   Southern New England/L. I. *HMANA Newsletter*, 5 (2): 9–11.
   1982   Southern New England/L. I. *HMANA Newsletter*, 7 (2): 15–16.
Currie, N. W., and R. Cote
   1981   *1981 Hawk Migration New England Hawk Watch*. Connecticut Audubon Council, Seymour, Conn.

1982   *1982 Hawk Migration New England Hawk Watch.* Audubon Council of Connecticut, Sharon, Conn.

1983   *1983 Hawk Migration New England Hawk Watch.* Audubon Council of Connecticut, Sharon, Conn.

Currie, N. W., and D. MacRae

1979   *1979 Hawk Migration New England Hawk Watch.* Connecticut Audubon Council, Seymour, Conn.

1980   *1980 Hawk Migration New England Hawk Watch.* Connecticut Audubon Council, Seymour, Conn.

Cutler, B. D., and E. A. Pugh

1960   Middle Pacific Coast Region. *Audubon Field Notes,* 14 (1): 67–70.

Czajkowski, E. H.

n. d.   *A School Yard Guide to Hawk-Watching.* Concord Union School District, Concord, New Hampshire.

Dales, J. H.

1959   The Lazy Broad-wings. *South Peel Naturalist,* 1 (6): 1–7.

Dapper, J. R.

1953   Hawks and Butterflies. *Chat,* 17 (1): 24.

Darrow, H. N.

1963   Direct Autumn Flight-Line from Fire Island, Long Island, to the Coast of Southern New Jersey. *Kingbird,* 13 (1): 4–12.

1982   Turkey Vultures in the Florida Keys. *HMANA Newsletter,* 7 (2): 6–7.

1983   Late Fall Movements of Turkey Vultures and Hawks in the Florida Keys. *Florida Field Naturalist,* 11 (2): 35–39.

David, N., and M. Gosselin

1976   Quebec. *American Birds,* 30 (1): 37.

1982a   Quebec Region. *American Birds,* 36 (2): 155–157.

1982b   Quebec Region. *American Birds,* 36 (5): 831–832.

Davis, R.

1976   Broad-winged Hawks at Pea Island N. W. R. *Chat,* 40 (2): 44–45.

Deane, R.

1905   An Unusual Flight of Hawks in 1858. *Wilson Bulletin,* 17 (1): 13–14.

DeGarmo, W. R.

1953   A Five-Year Study of Hawk Migration. *Redstart,* 20 (3): 39–54.

1967   Notes on Hawk Watching. *Delmarva Ornithologist,* 4 (1): 9–11.

Dekker, D.

1970   Migrations of Diurnal Birds of Prey in the Rocky Mountain Foothills West of Cochrane, Alberta. *Blue Jay,* 28 (1): 20–24.

DeSante, D., and V. Remsen

1973   Middle Pacific Coast Region. *American Birds,* 27 (1): 110–119.

Dickey, D. R., and A. J. Van Rossem

1938   The Birds of El Salvador. *Field Museum Zoological Series,* 23: 1–609.

Dinsmore, J. J., et al.

1984   *Iowa Birds.* Iowa State University Press, Ames, Iowa.

DiSilvestro, R.

1976   Merlin. *Nebraska Bird Review,* 44 (2): 34.

Dodge, J. R.

1981   The Beginning of the Spring Raptor Migration at Braddock Bay, February 1981. *Goshawk,* 37 (5): 65–66.

Donohue, G. S.

1978a   South-Central. *HMANA Newsletter,* 3 (1): 7–8.

1978b   South-Central. *HMANA Newsletter*, 3 (2): 16–18.
1979a   South-Central. *HMANA Newsletter*, 4 (1): 10–12.
1979b   South-Central. *HMANA Newsletter*, 4 (2): 23–28.
1980   South-Central. *HMANA Newsletter*, 5 (1): 7–10.
1981a   South-Central. *HMANA Newsletter*, 6 (1): 13–15.
1981b   South-Central. *HMANA Newsletter*, 6 (2): 18–20.
1982   South-Central. *HMANA Newsletter*, 7 (1): 14–17.
Doolittle, E. A.
1919   A Flight of Broad-winged Hawks and Roughlegs in Lake Co., Ohio. *Auk*, 36: 568.
Dorow, H., and C. Kurtz
1975   East Central Counties. *Iowa Bird Life*, 45 (1): 15–20.
Drennan, S. R.
1981   *Where to Find Birds in New York State.* Syracuse University Press, Syracuse, N.Y.
Dumont, P. A.
1934a   Fall Migration Notes of Red-tailed Hawks in Iowa. *Iowa Bird Life*, 4 (2): 18–19.
1934b   Additional Observations of Harlan's, Krider's, and the Western Red-tailed Hawks in Iowa. *Iowa Bird Life*, 4 (4): 51–52.
1935   An Autumnal Flight of Broad-winged Hawks in Eastern Iowa. *Iowa Bird Life*, 5 (1): 5.
Dunbar, R. J.
1950   Hawk Flights in Knox County. *Migrant*, 21: 74.
Duncan, B. W.
1981   Cooper's Hawks Banded at Hawk Cliff, Ontario, 1971–1980. *Ontario Bird Banding*, 14 (2): 21–32.
1982a   Vinemount Raptor Banding Station 1981 Report. *Ontario Bird Banding*, 15 (1): 16–23.
1982b   Sharp-shinned Hawks Banded at Hawk Cliff, Ontario, 1971–1981: Analysis of Data. *Ontario Bird Banding*, 15 (2): 24–38.
1983   Red-tailed Hawks Banded at Hawk Cliff, Ontario: 1971–1982. *Ontario Bird Banding*, 16 (2): 20–29.
Duncan, R. A.
1980a   Southeast. *HMANA Newsletter*, 5 (1): 7.
1980b   Southeast. *HMANA Newsletter*, 5 (2): 16–19.
1981   Southeast. *HMANA Newsletter*, 6 (2): 16–18.
Duncan, R., and L. Duncan
1975   A Substantial Hawk Migration in Northwest Florida. *Florida Field Naturalist*, 3 (1): 2–4.
1978a   Southeast. *HMANA Newsletter*, 3 (1): 6–7.
1978b   Southeast. *HMANA Newsletter*, 3 (2): 13–16.
1979a   Southeast. *HMANA Newsletter*, 4 (1): 9–10.
1979b   Southeast. *HMANA Newsletter*, 4 (2): 21–23.
Dunn, C. G.
1895   Migration of Hawks in Rhode Island. *Observer*, 6: 192.
Dunne, P. J.
1977   Spring Hawk Movement Along Raccoon Ridge, N.J.—1976. *New Jersey Audubon*, 3 (2): 19–29.
1978a   Spring Hawk Movement Along Raccoon Ridge, N.J.—1977. *New Jersey Audubon*, 4 (1): 39–48.

1978b   Season's Passage The Autumn Hawk Watch. *Peregrine Observer*, 2 (2): 3–4.

1978c   Coastal Plain. *HMANA Newsletter*, 3 (2): 9–13.

1979a   Coastal Plain. *HMANA Newsletter*, 4 (1): 9.

1979b   Coastal Plain. *HMANA Newsletter*, 4 (2): 16–21.

1980a   Coastal Plain. *HMANA Newsletter*, 5 (2): 12–16.

1980b   Expanded Hawk Watch, 1979 Projects: Clouds (and) Circle Game. *Peregrine Observer*, 3 (1): 7.

1981   Coastal Plain. *HMANA Newsletter*, 6 (2): 15–16.

1982   Coastal Plain. *HMANA Newsletter*, 7 (2): 16–18.

1983   Coastal Plain. *HMANA Newsletter*, 8 (2): 20–21.

1984   A Gift of Vision. *Peregrine Observer*, 7 (1): 6–10.

Dunne, P., et al.

1981   Cape May Point Hawk Watch Totals and Peak Flights, Autumn 1980. *Peregrine Observer*, 4 (1): 11.

Dunne, P. J., and W. S. Clark

1977   Fall Hawk Movement at Cape May Point, N.J.—1976. *New Jersey Audubon*, 3 (7–8): 114–124.

Dunne, P., D. Keller, and R. Kochenberger

1984   *Hawk Watch/A Guide for Beginners.* Cape May Bird Observatory, Cape May Point, N.J.

Dunne, P., and H. LeGrand

1984   Coastal Plain. *HMANA Newsletter*, 9 (1): 16–17.

Dunstan, T. C.

1969   First Recovery of Bald Eagle Banded in Minnesota. *Loon*, 41 (3): 92.

1972   Radio-Tagging Falconiform and Strigiform Birds. *Raptor Research*, 6: 93–102.

1975   The Use of Telemetry in Studying Hawk Migration. *Proc. North American Hawk Migration Conference 1974*. Pp. 41–46.

Eastwood, E.

1967   *Radar Ornithology.* Methuen & Co., Ltd., London.

Eaton, E. H.

1910   *Birds of New York. Part 1.* Memoir 12. New York State Museum, Albany, N.Y.

1914   *Birds of New York. Part 2.* Memoir 12. New York State Museum, Albany, N.Y.

Eckert, K. R.

1980   Western Great Lakes Region. *American Birds*, 34 (5): 778–781.

1981   Western Great Lakes Region. *American Birds*, 35 (5): 825–828.

Eddy, G.

1948   Turkey Vulture Concentration. *Murrelet*, 29 (1): 11.

Edge, R.

1939a   No title. *News Letter to Members No. 1.* Hawk Mountain Sanctuary Assn., Kempton, Pa.

1939b   No title. *News Letter to Members No. 2.* Hawk Mountain Sanctuary Assn., Kempton, Pa.

1940a   No title. *News Letter to Members No. 4.* Hawk Mountain Sanctuary Assn., Kempton, Pa.

1940b   No Title. *News Letter to Members No. 5.* Hawk Mountain Sanctuary Assn., Kempton, Pa.

1942   No title. *News Letter to Members No. 8.* Hawk Mountain Sanctuary Assn., Kempton, Pa.

1943   No title. *News Letter to Members No. 11*. Hawk Mountain Sanctuary Assn., Kempton, Pa.

1945   No title. *News Letter to Members No. 13*. Hawk Mountain Sanctuary Assn., Kempton, Pa.

1956   No title. *News Letter to Members No. 25*. Hawk Mountain Sanctuary Assn., Kempton, Pa.

Edscorn, J. B.
1974   Florida Region. *American Birds*, 28 (1): 40–44.
1975   Florida Region. *American Birds*, 29 (1): 44–48.
1976   Florida Region. *American Birds*, 30 (1): 54–58.
1977   Florida Region. *American Birds*, 31 (2): 166–169.
1978   Florida Region. *American Birds*, 32 (2): 193–197.
1979   Florida Region. *American Birds*, 33 (2): 169–171.

Edwards, E. P.
1972   *A Field Guide to the Birds of Mexico*. Published by the Author, Sweet Briar, Va.

Edwards, J. L.
1939   General Observations of Hawk Migration in New Jersey. *Bulletin* 1 (May): 8–11. Urner Ornithological Club, Newark, N. J.

Einspahr, H. M., and E. J. Meeham
1975   The First Record of the Occurrence of Prairie Falcon in the State of Alabama. *Alabama Birdlife*, 23 (1–2): 14–15.

Eisenmann, E.
1963   Is the Black Vulture Migratory? *Wilson Bulletin*, 75: 244–249.

Elkins, F. T.
1956   Hawk Flights at Port Stanley. *Bulletin Massachusetts Audubon Society*, 40 (7): 377–378, 382.
1962   Spring Broad-winged Hawk Flight at Mexico, N. Y. *Kingbird*, 12 (2): 79–80.
1974   Notes on Hawks. *Bird Observer of Eastern Massachusetts*, 2 (4): 103–112.

Elkins, F., and H. Elkins
1941   Waiting for the Hawks to Fly. *Bulletin Massachusetts Audubon Society*, 25 (6): 123–126.

Elliott, J. J.
1960   Falcon Flights on Long Island. *Kingbird*, 10 (4): 155–157.

Elliott, L.
1967   Birds of Prey, 1960. *South Dakota Bird Notes*, 19 (1): 4–6.

Emanuel, V. L.
1982   South Texas Region. *American Birds*, 36 (2): 194–197.

Enderson, J. H.
1964   A Study of the Prairie Falcon in the Central Rocky Mountain Region. *Auk*, 81 (3): 332–352.
1965   A Breeding and Migration Survey of the Peregrine Falcon. *Wilson Bulletin*, 77: 327–339.

Erwin, C.
1974   An Account of Six Golden Eagle Sightings in Dougherty County, Ga. *Oriole*, 39 (1): 8–10.

Essar, M.
1961   Migratory Congregations of Hawks. *Blue Jay*, 19 (1): 23.

Eubanks, T. R.
1971   Unusual Flight of Mississippi Kites in Payne County, Oklahoma. *Bulletin Oklahoma Ornithological Society*, 4 (4): 33.

Evans, D. L.
 1974   1973 Hawk Ridge Banding Counts. *Hawk Ridge Nature Reserve Newsletter*,
         3: 3.
 1975   Banding Station Report. *Hawk Ridge Nature Reserve Newsletter*, 5: 8.
 1976   1975 Hawk Ridge Banding Report. *Hawk Ridge Nature Reserve Annual
         Report, April, 1976.* Pp. 13–14.
 1978   Hawk Ridge Research Station Report—1977. *Hawk Ridge Nature Reserve
         Newsletter,* 11: 8–9.
 1979   Hawk Ridge Research Station Report—1978. *Hawk Ridge Nature Reserve
         Annual Report April, 1979.* Pp. 7–8.
 1980   Banding Station Report. *Hawk Ridge Nature Reserve Annual Report March
         1980.* Pp. 6.
Evans, D. L., and C. R. Sindelar
 1974   First Record of the Goshawk for Louisiana—A Collected, Banded Bird.
         Bird-Banding, 45 (3): 270.
Eynon, A. E.
 1941   Hawk Migration Routes in the New York City Region. *Proc. Linnaean
         Society New York,* 52–53: 113–116.
 1947   Hawk Flights at Montclair Quarry March–April 1947. *Urner Field Ob-
         server,* 2 (2): 7.
Fables, D., Jr.
 1955   *Annotated List of New Jersey Birds.* Urner Ornithological Club, Newark, N.
         J.
Farrand, J., Jr.
 1983   *The Audubon Society Master Guide to Birding.* Volume one. Alfred A.
         Knopf, New York, N. Y.
Ferguson, A. L., and H. L. Ferguson
 1922   The Fall Migration of Hawks as Observed at Fishers Island, N. Y. *Auk,* 39:
         488-496.
Ferry, J. F.
 1896   Hawk Flights Noticed at Lake Forest, Ill. *Oologist,* 13 (4): 36.
ffrench, R.
 1973   *A Guide to the Birds of Trinidad and Tobago.* Livingston Publishing Com-
         pany, Wynnewood, Pa.
Fichter, E.
 1939   A Flight of Swainson's Hawks at Lincoln, Lancaster County. *Nebraska Bird
         Review,* 7 (1): 11.
Field, M.
 1970   Hawk-Banding on the Northern Shore of Lake Erie. *Ontario Bird Ban-
         ding,* 6 (4): 52–69.
 1971   Hawk Cliff Raptor Banding Station First Annual Report, 1971. *Ontario
         Bird Banding,* 7 (3): 56–75.
Field, M., and D. Field
 1979   Hawk Cliff Raptor Banding Station Seventh Annual Report: 1977. *On-
         tario Bird Banding,* 12 (1): 2–28.
 1980a   Hawk Cliff Raptor Banding Station Eighth Annual Report: 1978. *Ontario
         Bird Banding,* 13 (1): 2–29.
 1980b   Hawk Cliff Raptor Banding Station Ninth Annual Report: 1979. *Ontario
         Bird Banding,* 13 (3): 2–27.
Field, M., and W. Rayner
 1974   Hawk Cliff Raptor Banding Station Third Annual Report, 1973. *Ontario
         Bird Banding,* 9 (2): 26–58.

1976   Hawk Cliff Raptor Banding Station Fourth Annual Report 1974. *Ontario Bird Banding,* 10 (2): 21–49.
1977   Hawk Cliff Raptor Banding Station Fifth Annual Report 1975. *Ontario Bird Banding,* 11 (1): 1–28.
1978   Hawk Cliff Raptor Banding Station Sixth Annual Report 1976. *Ontario Bird Banding,* 12 (1): 1–27.
Finch, D. W.
1969   Northeastern Maritime Region. *Audubon Field Notes,* 23 (1): 13–22.
1970   Northeastern Maritime Region. *Audubon Field Notes,* 24 (4): 578–582.
1972   Northeastern Maritime Region. *American Birds,* 26 (1): 31–37.
1973   Northeastern Maritime Region. *American Birds,* 27 (1): 24–30.
1974   Northeastern Maritime Region. *American Birds,* 28 (1): 111–121.
1976   Northeastern Maritime Region. *American Brids,* 30 (1): 29–36.
Fingerhood, E. D.
1984a   The Lancaster County Gyrfalcon Incursions of 1981–1983. *Cassinia,* 60: 34–40.
1984b   Gyrfalcon Records in Pennsylvania: Part Two. *Cassinia,* 60: 41–46.
Fingerhood, E. D., and S. Lipschutz
1982   Gyrfalcon *(Falco rusticolus)* Records in Pennsylvania. *Cassinia,* 59: 68–76.
Finucane, T. W.
1956   1955 Fall Migration of Hawks. *Migrant,* 27: 10–12.
1957   Annual Autumn Hawk Count 1956. *Migrant,* 28: 1–3.
1958   Annual Autumn Hawk Count 1957. *Migrant,* 29: 1–5.
1959   Annual Autumn Hawk Count 1958. *Migrant,* 30: 1–5.
1960   Annual Autumn Hawk Count 1959. *Migrant,* 31: 1–10.
1961a   Annual Autumn Hawk Count 1961 [*sic;* should be 1960]. *Migrant,* 32: 22–28.
1961b   Annual Aut. an Hawk Count 1961. *Migrant,* 32: 57–64.
1963   Annual Autumn Hawk Count 1962. *Migrant,* 34: 1–7.
1964   Annual Autumn Hawk Count, 1963. *Migrant,* 35: 7–13.
1965   Annual Autumn Hawk Count 1964. *Migrant,* 36: 1–7.
1966   Annual Autumn Hawk Count 1965. *Migrant,* 37: 1–3.
1967   Annual Autumn Hawk Count 1966. *Migrant,* 38: 6–8.
1968   Annual Autumn Hawk Count, 1967. *Migrant,* 39: 27–29.
1969   Annual Autumn Hawk Count, 1968. *Migrant,* 40: 28–31.
1970   Annual Autumn Hawk Count. *Migrant,* 41: 14–18.
1971   Annual Autumn Hawk Count. *Migrant,* 42: 1–4.
1972   Annual Autumn Hawk Count. *Migrant,* 43: 35–37, 41.
1973   Annual Autumn Hawk Count. *Migrant,* 44: 34–36.
1974   Annual Autumn Hawk Count. *Migrant,* 45: 32–35.
1975a   Annual Autumn Hawk Count. *Migrant,* 46: 52–54.
1975b   Southern Appalachians. *J. Hawk Migration Assn. North America,* 1 (1): 43–46.
1976a   Annual Autumn Hawk Count. *Migrant,* 47: 25–28.
1976b   Southern Appalachians. *HMANA Newsletter,* 1 (2): 14–16.
1977a   Annual Autumn Hawk Count. *Migrant,* 48: 86–88.
1977b   Southern Appalachians. *HMANA Newsletter,* 2 (2): 16–18.
1978a   Southern Appalachians. *HMANA Newsletter,* 3 (1): 10.
1978b   Southern Appalachians. *HMANA Newsletter,* 3 (2): 25–26.
1978c   Annual Autumn Hawk Count. *Migrant,* 49 (3): 49–52.
1979a   Annual Autumn Hawk Count, 1978. *Migrant,* 50 (2): 25–26.
1979b   Southern Appalachians. *HMANA Newsletter,* 4 (2): 33–35.

1980a   Southern Appalachians. *HMANA Newsletter,* 5 (1): 12.
1980b   Southern Appalachians. *HMANA Newsletter,* 5 (2): 23–24.
1980c   Annual Autumn Hawk Count, 1979. *Migrant,* 51 (4): 81–83.
1981a   Southern Appalachians. *HMANA Newsletter,* 6 (1): 16–17.
1981b   Southern Appalachians. *HMANA Newsletter,* 6 (2): 22–23.
Fischer, D. L.
1979   Mount Tom State Reservation. *Bird Observer of Eastern Massachusetts,* 7 (4): 129–132.
Fitch, F. W., Jr.
1965   Swallow-tailed Kite at Rock Eagle. *Oriole,* 30 (3): 94.
Fitch, H. S., H. A. Stephens, and R. O. Bare
1973   Road Counts of Hawks in Kansas. *Kansas Ornithological Society Bulletin,* 24 (4): 33–35.
Flahaut, M. R., and Z. M. Schultz
1955   North Pacific Coast Region. *Audubon Field Notes,* 9 (4): 350–353.
1956   Northern Pacific Coast Region. *Audubon Field Notes,* 10 (1): 47–50.
Floyd, H. C.
1983   New Frontiers in Hawkwatching: Hawk Migration Conference IV. *Bird Observer of Eastern Massachusetts,* 11 (3): 143–147.
Forbes, H. S., and H. B. Forbes
1927   An Autumn Hawk Flight. *Auk,* 44: 101–102.
Forbush, E. H.
1927   *Birds of Massachusetts and Other New England States.* Volume two. Massachusetts Dept. of Agriculture, Boston, Mass.
Forster, R. A.
1982   Sharp-shinned Hawks on Outer Cape Cod. *HMANA Newsletter,* 7 (2): 6.
Foster, G. H.
1955   Thermal Air Currents and Their Use in Bird-Flight. *British Birds,* 48 (6): 241–253.
Fowler, D., and S. Fowler
1981   Hawk Cliff Raptor Banding Station Tenth Annual Report: 1980. *Ontario Bird Banding,* 14 (2): 3–14.
1982   Hawk Cliff Raptor Banding Station Eleventh Annual Report: 1981. *Ontario Bird Banding,* 15 (2): 3–15.
Fowler, L. J.
1982   Autumn Hawk Flights 1981. *Migrant,* 53 (3): 58–62.
Fox, R. P.
1956   Large Swainson's Hawk Flight in South Texas. *Auk,* 73: 281–282.
Frey, E. S.
1940   Hawk Notes from Sterrett's Gap, Pennsylvania. *Auk,* 57: 247–250.
1943   *Centennial Check-List of the Birds of Cumberland County, Pennsylvania and Her Borders.* Published privately, Lemoyne, Pa.
Friedmann, H.
1950   *The Birds of North and Middle America. Bull. 50.* United States National Museum, Washington, D. C.
Freidmann, H., L. Griscom, and R. T. Moore
1950   Distributional Check-List of the Birds of Mexico. Part I. *Pacific Coast Avifauna* No. 29. Cooper Ornithological Club.
Fuller, M. R.
1979   Hawk Count Research. *HMANA Newsletter,* 4 (2): 1–2.

Fuller, M. R., and J. A. Mosher
  1981   Methods of Detecting and Counting Raptors: A Review. *Studies in Avian Biology*, 6: 235–246.
Fuller, M. R., and C. S. Robbins
  1979   Report Forms. *HMANA Newsletter*, 4 (2): 1–4.
Gabrielson, I. N., and S. G. Jewett
  1970   *Birds of the Pacific Northwest*. Dover Publications, Inc., New York, N. Y.
Gabrielson, I. N., and F. C. Lincoln
  1959   *The Birds of Alaska*. Stackpole Company, Harrisburg, Pa.
Gagnon, T. H.
  1974   1974 Hawk Migration at Mt. Tom State Reservation. *Bird News of Western Massachusetts*, 14 (6): 84.
Gammel, A. M., and H. S. Huenecke
  1956   Northern Great Plains Region. *Audubon Field Notes*, 10 (1): 32–35.
Ganier, A. F.
  1951   Migrating Turkey Vultures. *Migrant*, 22: 70.
Gates, D.
  1976   Turkey Vultures. *Nebraska Bird Review*, 44 (2): 34.
George J. R., and L. E. Schreffler
  1984   The Cornwall Fire Tower Hawk Watch. In *A Guide to the Birds of Lancaster County, Pennsylvania*. Lancaster County Bird Club, Lancaster, Pa. Pp. 25–29.
Gerow, J.
  1943   Hawk Takes Horned Lark in Beam of Auto Headlights. *Murrelet*, 24 (1): 11.
Gerrard, J. M., et al.
  1978   Migratory Movements and Plumage of Subadult Saskatchewan Bald Eagles. *Canadian Field-Naturalist*, 92 (4): 375–382.
Gerrard, J. M., and P. N. Gerrard
  1982   Spring Migration of Bald Eagles near Saskatoon. *Blue Jay*, 40 (2): 97–104.
Gerrard, J. M., and R. M. Hatch
  1983   Bald Eagle Migration Through Southern Saskatchewan and Manitoba and North Dakota. *Blue Jay*, 41 (3): 146–154.
Gerrard, P., et al.
  1974   Post-Fledging Movements of Juvenile Bald Eagles. *Blue Jay*, 32 (4): 218–226.
Gibson, D. D.
  1969   Alaska Region. *Audubon Field Notes*, 23 (1): 91–94
  1970   Alaska Region. *Audubon Field Notes*, 24 (1): 79–82.
  1971   Alaska Region. *American Birds*, 25 (1): 91–94.
  1972   Alaska Region. *American Birds*, 26 (1): 104–106.
  1979.   Alaska Region. *American Birds*, 33 (2): 204–206.
  1981   Alaska Region. *American Birds*, 35 (5): 852–854.
  1983   Alaska Region. *American Birds*, 36 (5): 884–885.
  1984   Alaska Region. *American Birds*, 38 (2): 234–236.
Gibson, D. D., and G. V. Byrd
  1975   Alaska Region. *American Birds*, 29 (4): 894–896.
Gillespie, J. A.
  1944   Birds of Rittenhouse Square. *Cassinia*, 33: 24–26.
Gillespie, M.
  1960   Long Distance Flyer—the Ospreys. *EBBA News*, 23 (3): 55–62.

Giraud, J. P.
1844   *The Birds of Long Island.* Wiley & Putnam, New York, N.Y.
Glandon, E. W.
1935   Large Flights of Swainson's Hawks in the Nebraska Sandhills. *Nebraska Bird Review*, 3 (3): 84.
Glockner-Ferrari, D. A.
1982   Photoidentification of Humpback Whales. *Whalewatcher,* 16 (4): 9–11.
Godfrey, W. E.
1979   *The Birds of Canada.* National Museum of Natural Sciences, Ottawa, Canada.
Goldman, H.
1970   Wings Over Brattleboro. *Yankee,* 34 (10): 100–105.
1974   Autumn Hawk Watch. *Yankee,* 38 (10): 80–85, 141–142, 145.
Goldman, L. C., and F. G. Watson
1953a   South Texas Region. *Audubon Field Notes,* 7 (1): 39–40.
1953b   South Texas Region. *Audubon Field Notes.* 7 (4): 282–285.
Gollop, J. B.
1967   Visible Migration—Saskatoon, 1966. *Blue Jay,* 25 (1): 20–22.
Goodwin, C. E.
1965   Ontario-Western New York Region. *Audubon Field Notes,* 19 (4): 466–468.
1968   Ontario-Western New York Region. *Audubon Field Notes,* 22 (1): 31–35.
1969   Ontario-Western New York Region. *Audubon Field Notes,* 23 (1): 41–46.
1970   Ontario-Western New York Region. *Audubon Field Notes,* 24 (4): 595–599.
1971a   Ontario-Western New York Region. *American Birds,* 25 (1): 49–54.
1971b   Ontario-Western New York Region. *American Birds,* 25 (4): 735–739.
1972a   Ontario-Western New York Region. *American Birds,* 26 (1): 54–59.
1972b   Ontario-Western New York Region. *American Birds,* 26 (4): 754–758.
1975   Ontario Region. *American Birds,* 29 (1): 48–53.
1976   Ontario Region. *American Birds,* 30 (1): 59–64.
1977a   Ontario Region. *American Birds,* 31 (2): 169–173.
1977b   Ontario Region. *American Birds,* 31 (5): 993–996.
1978   Ontario Region. *American Birds,* 32 (5): 997–1001.
1979a   Ontario Region. *American Birds,* 33 (2): 171–174.
1979b   Ontario Region. *American Birds,* 33 (5): 765–768.
1980   Ontario Region. *American Birds,* 34 (2): 155–158.
1982a   Ontario Region. *American Birds,* 36 (2): 171–174.
1982b   *A Bird-Finding Guide to Ontario.* University of Toronto Press, Toronto, Ontario, Canada.
Gooley, M. P.
1978   The 1978 Fall Hawk Migration at Quabbin Reservoir. *Chickadee,* 48: 28–30.
Gould, C. N.
1921   A Flock of Hawks. *Proc. Oklahoma Academy Science,* 1: 33.
Graber, R. R.
1962   Middlewestern Prairie Region. *Audubon Field Notes,* 16 (1): 36.
Graham, E. W.
1972   Swainson's Hawk at Bake Oven Knob, Pennsylvania. *Cassinia,* 53:43.
Graham, F., Jr.
1984   Battle of Hawk Mountain. *Audubon,* 86 (5): 28–31.
Grant, G. S.
1967   Osprey Migration over Topsail Island, N.C. *Chat,* 31 (4): 96.

Grantham, J.
1973 Heavy Spring Migration of Hawks Over Longwood Gardens, *Cassinia*, 54: 29.

Gray, L.
1961 Banding Sharpshins at Point Pelee. *EBBA News*, 24 (2): 25–26.

Green, J. C.
1962a 1961 Fall Hawk Migration, Duluth. *Flicker*, 34 (1): 30–31.
1962b 1962 Fall Hawk Migration, Duluth. *Flicker*, 34: 121, 124–125.
1963 Western Great Lakes Region. *Audubon Field Notes*, 17 (4): 404–407.
1964a Western Great Lakes Region. *Audubon Field Notes*, 18 (1): 33–34, 39–42.
1964b 1963 Fall Hawk Migration, Duluth. *Loon*, 36 (1): 30–31.
1965a 1964 Fall Hawk Migration, Duluth. *Loon*, 37 (2): 88–89.
1965b 1965 Fall Hawk Migration, Northwestern Minn. *Loon*, 37 (4): 126–130.
1975 Western Great Lakes, *J. Hawk Migration Assn. North America*, 1 (1): 47–51.

Green, J. C., and R. B. Janssen
1975 *Minnesota Birds: Where, When and How Many.* University of Minnesota Press, Minneapolis, Minn.

Greenberg, R., and R. Stallcup
1974 Middle Pacific Coast Region. *American Birds*, 28 (4): 845–850.

Griffin, C. R., J. M. Southern, and L. D. Frenzel
1980 Origins and Migratory Movements of Bald Eagles Wintering in Missouri. *J. Field Ornithology*, 51 (2): 161–167.

Griffin, D. R.
1969 The Physiology and Geophysics of Bird Navigation. *Quart. Review Biology*, 44: 255–276.
1973 Oriented Bird Migration in or Between Opaque Cloud Layers. *Proc. American Philosophical Society*, 117 (2): 117–141.

Griffin, W. W.
1941 Four Unpublished Golden Eagle Records from Georgia. *Oriole*, 6 (1): 12.

Grzybowski, J. A.
1966 April 1966 Hawk Migrations. *Prothonotary*, 32 (5): 63–64.
1983 Gyrfalcon in Oklahoma City: Southernmost Record for North America. *Bulletin Oklahoma Ornithological Society*, 16 (4): 27–29.

Guiguet, C. J.
1952 An Unusual Occurrence of Turkey Vultures on Vancouver Island. *Murrelet*, 33 (1): 11.

Gunderson, H. L.
1951 A Report of the Fall Hawk Migration. *Flicker*, 23 (4): 87.

Gunn, W. W. H.
1954 Hints for Hawk Watchers. *Bulletin Federation Ontario Naturalists*, 65: 16–19.
1958 Ontario-Western New York. *Audubon Field Notes*, 12 (4): 348–352.

Haas, F. C.
1984 The Birds of Ridley Creek State Park. *Cassinia*, 60: 23–33.

Hackman, C. D.
1954 A Summary of Hawk Flights over White Marsh, Baltimore County, Maryland. *Maryland Birdlife*, 10 (2–3): 19–26.

Hackman, C. D., and C. J. Henny
1971 Hawk Migration over White Marsh, Maryland. *Chesapeake Science*, 12 (3): 137–141.

Hagar, J. A.
  1937a   Hawks at Mount Tom. *Bulletin Massachusetts Audubon Society,* 21 (April): 5–8.
  1937b   More Hawks at Mount Tom. *Bulletin Massachusetts Audubon Society,* 21 (October): 5–8.
Hall, G. A.
  1963   Appalachian Region. *Audubon Field Notes,* 17 (4): 401–404.
  1964a   Fall Migration on the Allegheny Front. *Redstart,* 31 (2): 30–53.
  1964b   Appalachian Region. *Audubon Field Notes,* 18 (1): 30–33.
  1965   Appalachian Region. *Audubon Field Notes,* 19 (4): 470–473.
  1970   Appalachian Region. *Audubon Field Notes,* 24 (1): 47–51.
  1972   Appalachian Region. *American Birds,* 26 (1): 62–66.
  1975   Appalachian Region. *American Birds,* 29 (4): 850–854.
  1977   Appalachian Region *American Birds,* 31 (2): 176–179.
  1979   Appalachian Region. *American Birds,* 33 (2): 176–178.
  1983a   West Virginia Birds/Distribution and Ecology. *Special Publication Carnegie Museum of Natural History No. 7.* Carnegie Museum of Natural History, Pittsburgh, Pa.
  1983b   Appalachian Region. *American Birds,* 37 (2): 179–182.
  1984   Appalachian Region. *Amerian Birds,* 38 (2): 200–204.
Hamer, F., et al.
  1982   Cape May Point Hawk Watch Totals and Peak Flights, Autumn 1981. *Peregrine Observer,* 5 (1): 10.
Hancock, D.
  1970   *Adventure with Eagles.* Wildlife Conservation Centre, Saanichton, B. C.
Hanley, W.
  1968   Last Call for Hawks. *Narragansett Naturalist,* 10 (1): 2–12.
Harden, W. D.
  1972   Predation by Hawks on Bats at Vickery Bat Cave. *Bulletin Oklahoma Ornithological Society,* 5 (1): 4–5.
Harris, B.
  1973   Notes on the 1972 Hawk Migration in Northeastern South Dakota with Observations on Gyrfalcon, Prairie Falcon and Goshawks. *South Dakota Bird Notes,* 25 (2): 24–26.
Harris, W. C.
  1980   Prairie Provinces Region. *American Birds,* 34 (2): 172–174.
  1981   Prairie Provinces Region. *American Birds,* 35 (2): 195–196.
  1983   Prairie Provinces Region. *American Birds,* 37 (2): 192–194.
Harrison, G. H.
  1976   *Roger Tory Peterson's Dozen Birding Hot Spots.* Simon and Schuster, New York, N. Y.
Harwood, M.
  1973   *The View from Hawk Mountain.* Charles Scribner's Sons, New York, N. Y.
  1980   New Directions for Hawkwatching. *J. Hawk Migration Assn. North America,* 2 (1): 1–4.
Harwood, M., and A. C. Nagy
  1977   Assessing the Hawk Counts at Hawk Mountain. *Transactions North American Osprey Research Conference.* Pp. 69–75.
Hatch, D. R. M.
  1966   Northern Great Plains Region. *Audubon Field Notes,* 20 (1): 61–64.
  1967   Northern Great Plains Region. *Audubon Field Notes,* 21 (1): 47–51.
  1968   Northern Great Plains Region. *Audubon Field Notes,* 22 (1): 55.

Hatch, P. L.
1892   *Notes on the Birds of Minnesota*. Geological and Natural History Survey, Minneapolis, Minn.

Haugh, J. R.
1966   Some Observations on the Hawk Migration at Derby Hill. *Kingbird*, 16 (1): 5–16.
1972   A Study of Hawk Migration in Eastern North America. *Search*, 2 (16): 1–60.

Haugh, J. R., and T. J. Cade
1966   The Spring Hawk Migration Around the Southeastern Shore of Lake Ontario. *Wilson Bulletin*, 78: 88–110.

Haverschmidt, F.
1968   *Birds of Surinam*. Oliver & Boyd, Edinburgh, Scotland.

Hawkes, R. S.
1958   Hawk Migration in Nova Scotia. *Maine Field Observer*, 3 (11): 123.

Hebard, F. V.
1960   *The Land Birds of Penobscot Bay*. Portland Society of Natural History, Portland, Maine.

Heimerdinger, H. O.
1974   1971 Hawk Watch on Allegheny Front Mountain at Bear Rocks. *Redstart*, 41 (4): 119–120.

Heintzelman, D S.
1963a   Bake Oven Hawk Flights. *Atlantic Naturalist*, 18: 154–158.
1963b   Bake Oven Knob Migration Observations. *Cassinia*, 47: 39–40.
1966   Bake Oven Knob Autumn Hawk Migration Observations (1964 and 1965). *Cassinia*, 49: 33.
1968   Bake Oven Knob Autumn Hawk Migration Observations, 1966 and 1967. *Cassinia*, 50: 26–27.
1969a   The Black Vulture in Pennsylvania. *Pennsylvania Game News*, 40 (5): 17–19.
1969b   Autumn Birds of Bake Oven Knob. *Cassinia*, 51: 11–32.
1970a   Wings Over Hawk Mountain. *National Wildlife*, 8 (5): 22–27.
1970b   Autumn Hawk Watch. *Frontiers*, 35 (1): 16–21.
1970c   Speculation on DDT and Altered Osprey Migrations. *Raptor Research News*, 4: 120–124.
1970d   Bake Oven Knob Autumn Hawk Migration Observations (1957 and 1969). *Cassinia*, 52: 37.
1970e   The Hawks of New Jersey. *Bulletin 13*. New Jersey State Museum, Trenton, N.J.
1972a   The 1971 Autumn Hawkwatch at Catfish Fire Tower, New Jersey. *New Jersey Nature News*, 27 (1): 19–21.
1972b   Some Autumn Bird Records from the Catfish Fire Tower, New Jersey. *Science Notes No. 6*. New Jersey State Museum, Trenton, N.J.
1972c   The Importance of Record Keeping When Watching Hawk Migrations. *California Condor*, 7 (4): 6–7.
1972d   *A Guide to Northeastern Hawk Watching*. Privately published, Lambertville, N.J.
1972e   Hawks Across New Jersey. *New Jersey Nature News*, 27: 100–104.
1972f   Speculation on the Possible Origin of some Sharp-shinned and Red-tailed Hawk Flights along the Kittatinny Ridge—Autumn 1971. *Science Notes No. 10*. New Jersey State Museum, Trenton, N.J.

1973   The 1972 Autumn Hawk Count at Tott's Gap, Pennsylvania. *Science Notes No. 12*. New Jersey State Museum, Trenton, N.J.

1975   *Autumn Hawk Flights: The Migrations in Eastern North America*. Rutgers University Press, New Brunswick, N.J.

1976a   *A Guide to Eastern Hawk Watching*. Pennsylvania State University Press, University Park, Pa.

1976b   Bird Survey on Aves Island. *Explorers Journal*, 54 (2): 65.

1978   *North American Ducks, Geese, & Swans*. Winchester Press, New York, N.Y.

1979a   *A Guide to Hawk Watching in North America*. Pennsylvania State University Press, University Park, Pa.

1979b   *Hawks and Owls of North America*. Universe Books, New York, N.Y.

1979c   Where Birders Enjoy the Raptor Parade. *Defenders*, 54 (5): 261–264.

1980a   Some Recent Raptor Conservation Efforts in Pennsylvania. *Sierra Club Newsletter* (Pennsylvania Chapter), April Issue, pages 1, 3, and 6.

1980b   *The Illustrated Bird Watcher's Dictionary*. Winchester Press, Tulsa, Okla.

1981   Diurnal Raptors in Amazonia. *Explorers Journal*, 59 (3): 122–123.

1982a   How About Hawk Watching. *Ranger Rick's Nature Magazine*, 16 (10): 6–10.

1982b   Daily Rhythms of Migrating Golden Eagles and Bald Eagles in Autumn at Bake Oven Knob, Pennsylvania. *Cassinia*, 59: 65–67.

1982c   Hours of Observation at the Bake Oven Knob, Pennsylvania Hawk Lookout (1957–1981). *Cassinia*, 59: 78.

1982d   More Autumn Bird Records from Bake Oven Knob, Pennsylvania (1976–1981). *Cassinia*, 59: 85.

1982e   Variations in Numbers of Sharp-shinned Hawks Migrating Past Bake Oven Knob, Pennsylvania, in Autumn (1961 to 1981). *American Hawkwatcher*, 1: 1–4.

1982f   Variations in Numbers of Cooper's Hawks Migrating Past Bake Oven Knob, Pennsylvania, in Autumn (1961 to 1981). *American Hawkwatcher*, 2: 1–4.

1982g   Variations in Utilization of the Kittatinny Ridge in Eastern Pennsylvania in Autumn by Migrating Golden Eagles and Bald Eagles (1968–1981). *American Hawkwatcher*, 3: 1–4.

1982h   Variations in Numbers of Northern Goshawks Migrating Past Bake Oven Knob, Pennsylvania in Autumn (1961–1981). *American Hawkwatcher*, 4: 1–4.

1982i   The 1982 Autumn Hawk Count at Bake Oven Knob, Pennsylvania. *American Hawkwatcher*, 5: 1–6.

1983a   Variations in Numbers of, and Influence of Intersecting Diversion-Lines Upon, Ospreys Migrating Along the Kittatinny Ridge in Eastern Pennsylvania in Autumn. *American Hawkwatcher*, 6: 1–4.

1983b   An Interdisciplinary Comparison of Recreational Hawk Watching in Pennsylvania with Recreational Whale Watching in California. *American Hawkwatcher*, 7: 1–4.

1983c   The 1983 Autumn Hawk Count at Bake Oven Knob, Pennsylvania. *American Hawkwatcher*, 8: 1–6.

1983d   *The Birdwatcher's Activity Book*. Stackpole Books, Harrisburg, Pa.

1984a   Bake Oven Knob Hawk Counts Produce Significant Information. *Pennsylvania Wildlife*, 4 (6): 50.

1984b   National Birds of Prey Conservation Week: America's Potential New Raptor Conservation Tool. *Eyas*, 7 (1): 8.

1984c   *Guide to Owl Watching in North America.* Winchester Press, Piscataway, N.J.
1984d   Bear Rocks: Comments on a Name. *Keystone Trails Assn. Newsletter,* spring issue. Pg. 8.
1984e   The 1984 Autumn Hawk Count at Bake Oven Knob, Pennsylvania. *American Hawkwatcher,* 9: 1–5.
1985   Remarks for the First National Birds of Prey Conservation Week 7 October 1984. *Bake Oven Knob Newsletter,* 4: 1–2.

Heintzelman, D. S., and T. V. Armentano
1964   Autumn Bird Migration at Bake Oven Knob, Pa. *Cassinia,* 48: 2–18.

Heintzelman, D. S., and R. MacClay
1971   An Extraordinary Autumn Migration of White-breasted Nuthatches. *Wilson Bulletin,* 83: 129–131.
1972   The 1970 and 1971 Autumn Hawk Counts at Bake Oven Knob, Pennsylvania. *Cassinia,* 53: 3–23.
1973   The 1972 Autumn Hawk Count at Bake Oven Knob, Pennsylvania. *Cassinia,* 54: 3–9.
1974   Turkey Vultures Thermal Soaring into Opaque Clouds. *Auk,* 91 (4): 849.
1975   The 1973 and 1974 Autumn Hawk Counts at Bake Oven Knob, Pennsylvania. *Cassinia,* 55: 17–28.
1976a   The 1975 Autumn Hawk Count at Bake Oven Knob, Pennsylvania. *Cassinia,* 56: 15–21.
1976b   Autumn Bird Records from Bake Oven Knob, Pennsylvania (1969–1975). *Cassinia,* 56: 23–24.
1976c   Another Exceptional Autumn Migration of White-breasted Nuthatches. *Cassinia,* 56: 31.
1979   The 1976, 1977, and 1978 Autumn Hawk Counts at Bake Oven Knob, Pennsylvania. *Cassinia,* 57: 19–20.

Heintzelman, D. S., and B. Reed
1982   The 1979, 1980, and 1981 Autumn Hawk Counts at Bake Oven Knob, Pennsylvania. *Cassinia,* 59: 62–64.

Henderson, D. H.
1941   The Mississippi Kite *(Ictinia misisippiensis)* in Richmond County. *Oriole,* 6 (2): 24.

Henderson, N.
1962   Spring Hawk Migration Through the Cleveland, Ohio, Region. *Cleveland Bird Calendar,* 58 (3): 28–33.
1963   Spring 1963 Hawk Migration Over Cleveland. *Cleveland Bird Calendar,* 59 (3): 28–31.

Henderson, N., and C. Margolis
1959   Hawk-Watching From the Terminal Tower. *Cleveland Bird Calendar,* 55 (2): 15.

Henny, C. J., and W. S. Clark
1982   Measurements of Fall Migrant Peregrine Falcons from Texas and New Jersey. *J. Field Ornithology,* 53 (4): 326–332.

Henny, C. J., and W. T. Van Velzen
1972   Migration Patterns and Wintering Localities of American Ospreys. *J. Wildlife Management,* 36 (4): 1133–1141.

Henshaw, H. W.
1901   Birds of Prey as Ocean Waifs. *Auk,* 18: 162–165.

Herndon, L. R.
1949   Golden Eagle at Hump Mountain, Tenn.–N.C. *Migrant,* 20: 17.

Heuser, E. P.
1940   Bald Eagles at Guttenberg. *Iowa Bird Life,* 10 (2): 29.
Hickey, J. J. (Ed.)
1969   *Peregrine Falcon Populations/Their Biology and Decline.* University of Wisconsin Press, Madison, Wis.
Hicks, D. L., D. T. Rogers, Jr., and G. I. Child
1966   Autumnal Hawk Migration Through Panama. *Bird-Banding,* 37: 121–123.
Hicks, L. E.
1935   An Indiana Hawk Migration. *Indiana Audubon Year Book.* Pp. 85–86.
Hill, R. K.
1957   Hawk Watching. *New Hampshire Bird News,* 10 (3): 64–67.
HMANA
n.d.   *Instructions for Daily Report Forms.* Hawk Migration Assn. of North America, Washington, Conn.
Hochbaum, H. A.
1955   *Travels and Traditions of Waterfowl.* Charles T. Branford Co., Newton, Mass.
Hoffman, J.
1979   Turkey Vulture Flight Over Downtown Cleveland. *Cleveland Bird Calendar,* 75 (2): 18–19.
Hoffman, R.
1982   1982 Spring Hawk Count in Southern Sauk County. *Passenger Pigeon,* 44 (3): 123–124.
Hoffman, S.
1981a   Western Hawkwatching. *HMANA Newsletter,* 6 (1): 1–4.
1981b   Southwest. *HMANA Newsletter,* 6 (2): 27–29.
1981c   Southwest. *HMANA Newsletter,* 6 (1): 22–23.
1982a   Southwest. *HMANA Newsletter,* 7 (1): 22–24.
1982b   Western Hawkwatching. *HMANA Newsletter,* 7 (2): 10–12.
1982c   Southwest. *HMANA Newsletter,* 7 (2): 30–32.
1983a   Southwest. *HMANA Newsletter,* 8 (1): 21.
1983b   Southwest. *HMANA Newsletter,* 8 (2): 35–37.
1984a   Southwest. *HMANA Newsletter,* 9 (1): 26–27.
1984b   Southwest. *HMANA Newsletter,* 9 (2): 35–36.
Hofslund, P. B.
1952   The 1952 Hawk Migration Count at Duluth. *Flicker,* 24 (4): 162.
1954a   The Hawk Pass at Duluth, Minnesota. *Wilson Bulletin,* 66: 224.
1954b   The Hawk Pass at Duluth. *Flicker,* 26 (3): 96–99.
1955   Hawk Flyway. *Flicker,* 27 (4): 158–159.
1958   Bird Migration in the Duluth Area. *Flicker,* 30 (2): 53–57.
1962   The Duluth Hawk Flyway—1951–1961. *Flicker,* 34: 88–92.
1966   Hawk Migration Over the Western Tip of Lake Superior. *Wilson Bulletin,* 78: 79–87.
1973   An Invasion of Goshawks. *Raptor Research,* 7 (3–4): 107–108.
1976   The Continental Summary: Autumn, 1975. *HMANA Newsletter,* 1 (2): 2–3.
Holland, G.
1981   Manitoba Hawk Watch. *Manitoba Naturalists Society Bulletin,* 5 (4): 12–13.
1982   1982—Fourth Annual Fall Hawk Watch—September 11. *Manitoba Naturalists Society Bulletin,* 6:8–9.

Holt, J. B., Jr., and R. Frock, Jr.
  1980   Twenty Years of Raptor Banding on the Kittatinny Ridge. *Hawk Mountain News*, 54: 8–32.
Holthuijzen, A. M. A., and L. Oosterhuis
  1982   Telemetry Studies on Migrating Female Sharp-shinned Hawks *(Accipiter striatus)* at Cape May Point, New Jersey. *Peregrine Observer*, 5 (1): 6–8.
Honaman, A.
  1955   An Exciting Hawk Flight. *Audubon Magazine*, 57: 198–199.
Hopkins, D. A., and G. S. Mersereau
  1971   *Fall Hawk Migration 1971.* Privately published.
  1972   *Hawk Migration 1972.* Privately published.
  1973a   *Hawk Migration Spring 1973.* Privately published.
  1973b   *Hawk Migration 1973 Report.* Privately published.
  1974a   *Hawk Migration Spring 1974.* Privately published.
  1974b   *Hawk Migration 1974 Report.* Privately published.
  1975a   *Hawk Migration Spring 1975.* Privately published.
  1975b   *Hawk Migration 1975 Report.* Privately published.
  1975c   Comments on the Motor-Glider Use in Hawk Migration Study. *J. Hawk Migration Assn. North America*, 1 (1): 23–25.
  1976   *Hawk Migration 1976 Report.* Privately published.
Hopkins, D. A., G. S. Mersereau, and W. A. Welch
  1975   *The Report of the Smirnoff Hawk Patrol.* Connecticut Audubon Council, Windsor, Conn.
Hopkins, D. A., J. Mitchell, and W. A. Welch
  1978   *The Hawk Patrol 1977.* Connecticut Audubon Council, Windsor, Conn.
Hopkins, D. A., et al.
  1978   *The Hawk Patrol 1978/The Aerial Observation of the Broad-winged Hawk Migration.* Privately published.
  1979   *Motor-Glider and Cine-Theodolite Study of the 1979 Fall Broad-winged Hawk Migration in Southern New England.* Connecticut Audubon Council, Tariffville, Conn.
Hopkins, M., Jr.
  1966   Bird Notes from a River Cruise in South Georgia. *Oriole*, 31 (2): 29–30.
  1968   Golden Eagle in Irwin County, Georgia. *Oriole*, 33 (4): 50.
Horsfall, B.
  1908   Migration of Hawks. *Auk*, 25: 474–475.
Houston, C. S.
  1967   Recoveries of Red-tailed Hawks Banded in Saskatchewan. *Blue Jay*, 25 (3): 109–111.
  1968a   Recoveries of Marsh Hawks Banded in Saskatchewan. *Blue Jay*, 26 (1): 12–13.
  1968b   Recoveries of Swainson's Hawks Banded in Saskatchewan. *Blue Jay*, 26 (2): 86–87.
  1971a   Northern Great Plains Region. *American Birds*, 25 (1): 71–74.
  1971b   Northern Great Plains Region. *American Birds*, 25 (4): 758–761, 764.
  1972a   Northern Great Plains. *American Birds*, 26 (1): 78–80.
  1972b   Northern Great Plains. *American Birds*, 26 (4): 774–777.
  1973   Northern Great Plains. *American Birds*, 27 (1): 75–78.
  1974   South American Recoveries of Franklin's Gulls and Swainson's Hawks Banded in Saskatchewan. *Blue Jay*, 32 (3): 156–157.
Houston, C. S., and S. J. Shadick
  1974   Northern Great Plains. *American Birds*, 28 (4): 814–817.

Howell, A. H.
1932 *Florida Bird Life*. Coward-McCann, Inc., New York, N.Y.

Howland, K. V. S.
1893 Hawk Migration. *Forest and Stream*, 40: 513.

Hubbard, J. P.
1977 Raptor Watching on Sierra Grande. *New Mexico Ornithological Society Bulletin*, 5 (1): 7.

Huber, R. L.
1964 Ferruginous Hawk Sight Record for Traverse County. *Loon*, 36 (3): 108.

Hundley, M. H.
1967 Migration of Swainson's Hawks and Turkey Vultures Through Guatemala. *Florida Naturalist*, 40 (1): 29.

Hunn, E. S., and P. W. Mattocks, Jr.
1979 Northern Pacific Coast Region. *American Birds*, 33 (2): 206–209.
1981 Northern Pacific Coast Region. *American Birds*, 35 (2): 216–219.
1982 Northern Pacific Coast Region. *American Birds*, 36 (2): 209.
1983 Northern Pacific Coast Region. *American Birds*, 37 (2): 214–218.

Hunt, W. G., R. R. Rogers, and D. J. Slowe
1975 Migratory and Foraging Behavior of Peregrine Falcons on the Texas Coast. *Canadian Field-Naturalist*, 89 (2): 111–123.

Hurley, G.
1970 Fall Hawk Migration Along Peters Mountain in Monroe County, West Virginia. *Redstart*, 37 (3): 82–86.
1975 Fall Hawk Migration Along Peters Mountain, Part 2. *Redstart*, 42 (4): 114–117.

Imhof, T. A.
1962 Central Southern Region. *Audubon Field Notes*, 16 (1): 43–47.
1966 Central Southern Region. *Audubon Field Notes*, 20 (4): 515–519.
1967 Central Southern Region. *Audubon Field Notes*, 21 (4): 514–517.
1972 Central Southern Region. *American Birds*, 26 (4): 769–774.
1976 *Alabama Birds*. Second Edition. University of Alabama Press, University, Ala.
1980 Central Southern Region. *American Birds*, 34 (5): 785–787.
1981 Central Southern Region. *American Birds*, 35 (5): 832–834.
1983 Central Southern Region. *American Birds*, 37 (5): 878–882.

Ingram, T. N.
1965 *A Field Guide for Locating Bald Eagles at Cassville, Wisconsin*. Southwestern Wisconsin Audubon Club.
1966 The Bald Eagle in Illinois. *Audubon Bulletin*, 138: 26–28.

Ingram, T., T. Brophy, and D. Sherman
1982 Raptor Migrations Over Eagle Valley Nature Preserve. *Proc. '82 Bald Eagle Days*. Eagle Valley Environmentalists Inc., Apple River, Ill. Pp. 106–112.

Irwin, O. F.
1962 Annual Autumn Hawk Count 1961 at Memphis, Tenn. *Inland Bird Banding News*, 34 (4): 49–50.

Isaacs, F. B., and L. A. Hennigar
1980 Spring Hawk Migration at Brockway Mountain, Keweenaw County, Michigan. *J. Hawk Migration Assn. North America*, 2: 24–33.

Jackson, J. A.
1983 A Record of the Black Vulture from Bimini, Bahamas. *Florida Field Naturalist*, 11 (2): 17.

James, D.
    1966   Central Southern Region. *Audubon Field Notes*, 20 (1): 55–61.
Janssen, R. B.
    1975   Western Great Lakes. *American Birds*, 29 (4): 854–858.
Jeheber, P.
    1976   Hawks Over the Hudson Highlands. *Kingbird*, 26 (3): 136–140.
Johnsgard, P. A.
    1980   An Analysis of Migration Schedules of Non-Passerine Birds in Nebraska. *Nebraska Bird Review*, 48 (2): 26–36.
Johnson, D.
    1977   First Organized Hawk Watch in Northeastern Illinois. *Illinois Audubon Bulletin*, 181: 34–35.
Johnson, D. A.
    1984   National Birds of Prey Week Scheduled Oct. 7–13. *Leader*, 5 (8): 11–12.
Johnson, W. M.
    1950   Migrating Hawks in the Blue Ridge. *Migrant*, 21: 72–74.
Johnston, D. W.
    1957   Kites in the Macon Area. *Oriole*, 22 (2): 16–18.
Johnston, I. H.
    1923   *Birds of West Virginia*. State Department of Agriculture, Charleston, W. Va.
Jones, A. H.
    1939   Another Migrating Flock of Swainson Hawks. *Nebraska Bird Review*, 7 (1): 11.
Jung, C. S.
    1935   Migration of Hawks in Wisconsin. *Wilson Bulletin*, 47: 75–76.
    1964   Weather Conditions Affecting Hawk Migrations. *Lore*, 14 (4): 134–142.
Kale, H. W. III
    1970   Florida Region. *Audubon Field Notes*, 24 (4): 591–595.
    1972   Florida Region. *American Birds*, 26 (4): 751–754.
    1973   Florida Region. *American Birds*, 27 (4): 761–765.
    1976   Florida Region. *American Birds*, 30 (4): 828–832.
    1977   Florida Region. *American Birds*, 31 (5): 988–992.
    1978   Florida Region. *American Birds*, 32 (5): 993–997.
    1979   Florida Region. *American Birds*, 33 (5): 762–765.
    1980   Florida Region. *American Birds*, 34 (5): 767–770.
    1981   Florida Region. *American Birds*, 35 (5): 814–817.
    1983   Florida Region. *American Birds*, 37 (5): 860–863.
Kale, H. W., and J. D. Almand
    1963   Golden Eagle in Franklin County, Georgia. *Oriole*, 28 (1): 17–18.
Kane, R., and P. A. Buckley
    1975   Hudson-St. Lawrence Region. *American Birds*, 29 (1): 29–34.
Katona, S. K., et al.
    1980   *Humpback Whale: A Catalogue of Individuals Identified in the Western North Atlantic Ocean by Means of Fluke Photographs*. Second Edition. College of the Atlantic, Bar Harbor, Maine.
Katona, S. K., J. A. Becard, and K. C. Balcomb
    1982   The Atlantic Humpback Fluke Catalogue. *Whalewatcher*, 16 (4): 3–8.
Keith, L. B.
    1963   *Wildlife's Ten-Year Cycle*. University of Wisconsin Press, Madison, Wis.
Keller, C. E., S. A. Keller, and T. C. Keller

1979 *Indiana Birds and Their Haunts/A Checklist and Finding Guide.* Indiana University Press, Bloomington, Ind.

Keller, D.

1983a High School Hawk Watch Project. *Peregrine Observer,* 6 (1): 11.

1983b The High School Hawkwatch Project. *HMANA Newsletter,* 8 (2): 6.

1983c High School Hawk Watch Project. *Peregrine Observer,* 6 (2): 15.

1984a High School Hawkwatch. *HMANA Newsletter,* 9 (1): 1–3.

1984b High School Hawk Watch Wrap-Up. *Peregrine Observer,* 7 (1): 12–13.

Kelley, A. H.

1965a Michigan Bird Survey, Spring 1965. *Jack-Pine Warbler,* 43 (4): 156–161.

1965b Michigan Bird Survey, Summer 1965. *Jack-Pine Warbler,* 43 (4): 164–168.

1967 Michigan Bird Survey, Fall 1966. *Jack-Pine Warbler,* 45 (1): 18–24.

1972a Michigan Bird Survey, Fall 1971. *Jack-Pine Warbler,* 50 (1): 2–10.

1972b Spring Migration at Whitefish Point, 1966–1971. *Jack-Pine Warbler,* 50 (3): 69–75.

1974 Michigan Bird Survey, Fall 1973. *Jack-Pine Warbler,* 52 (1): 33–43.

1975 Michigan Bird Survey, Fall 1974. *Jack-Pine Warbler,* 53 (1): 24–31.

1976 Michigan Bird Survey, Fall 1975. *Jack-Pine Warbler,* 54 (1): 7–15.

Kellogg, S.

1983 Continental Summary. *HMANA Newsletter,* 8 (2): 12–16.

Kellogg, S., and D. MacRae

1980 Northern New England. *HMANA Newsletter,* 5 (2): 7–9.

Kelly, D.

1983 Photo-Identification of Bottlenose Dolphins in Southern California. *Whalewatcher,* 17 (2): 6–8.

Kennedy, R. S.

1973 Notes on the Migration of Juvenile Ospreys from Maryland and Virginia. *Bird-Banding,* 44 (3): 180–186.

1975 The South. *J. Hawk Migration Assn. North America,* 1 (1): 30–33.

1976a The South. *HMANA Newsletter,* 1 (1): 3–4.

1976b The South. *HMANA Newsletter,* 1 (2): 8–9.

1977a The South. *HMANA Newsletter,* 2 (1): 5–6.

1977b The South. *HMANA Newsletter,* 2 (2): 10–12.

Kerlinger, P.

1980 A Tracking Radar Study of Migrating Hawks. *J. Hawk Migration Assn. North America,* 2 (1): 34–42.

1981 Delaware Bay: An Obstacle to Hawk Migration? *Peregrine Observer,* 4 (1): 7–8.

1983 HMANA Research. *HMANA Newsletter,* 8 (1): 7.

1985 Water-crossing Behavior of Raptors During Migration. *Wilson Bulletin,* 97 (1): 109–113.

Kerlinger, P., and M. Bennett

1977 Hawk Migration at Oneonta. *Kingbird,* 27: 74–79.

Kerlinger, P., J. D. Cherry, and K. D. Powers

1983 Records of Migrant Hawks from the North Atlantic Ocean. *Auk,* 100 (2): 488–490.

Kerlinger, P., and S. A. Gauthreaux, Jr.

1983 Avian Migration Mobile Research Laboratory Comes to Cape May. *Peregrine Observer,* 6 (1): 14.

Kerlinger, P., and P. H. Lehrer

1982 Anti-Predator Responses of Sharp-shinned Hawks. *Raptor Research,* 16 (2): 33–36.

Kessler, E.
1975   Soaring Vultures Use a Dust Devil to Gain Altitude. *Wilson Bulletin*, 87
        (1): 113–114.
Klabunde, W.
1976   Spring Hawk Watch at Beamer Conservation Area—March 13 thru April
        30. *Prothonotary*, 42 (12): 166–167.
1979a   Eastern Great Lakes. *HMANA Newsletter*, 4 (1): 13–14.
1979b   1979 Spring Hawk Migration on South Shore of Lake Ontario. *Prothono-
        tary*, 45 (7): 105–111.
1980a   Eastern Great Lakes. *HMANA Newsletter*, 5 (1): 12–14.
1981a   Eastern Great Lakes. *HMANA Newsletter*, 6 (1): 17–19.
1981b   Spring Hawk Migration—South Shore of Lake Ontario 1981. *Prothono-
        tary*, 47 (8): 100–105.
1982a   Eastern Great Lakes. *HMANA Newsletter*, 7 (1): 18–20.
1982b   1982 Hawk Migration at Grimsby, Ontario. *Prothonotary*, 48 (12): 153–
        157.
1983a   Eastern Great Lakes. *HMANA Newsletter*, 8 (1): 16–17.
1983b   Spring Hawk Migration Season at Grimsby, Ontario Beamer Con-
        servation Area. *Prothonotary*, 49 (8): 126–131.
1984a   Eastern Great Lakes. *HMANA Newsletter*, 9 (1): 20–21.
1984b   A Quest for Hawkwatch Lookouts in the Hamburg Area. *Prothonotary*, 50
        (1): 12–13.
Klabunde, W., and W. Burch
1980   Hawk Watchers at Grimsby. *Prothonotary*, 46 (5): 76.
Kleen, V. M.
1969   Banding North American Migrants in Panama. *EBBA News*, 32: 170–
        173.
1974   Middlewestern Prairie Region. *American Birds*, 28 (1): 58–63.
1975a   Middlewestern Prairie Region. *American Birds*, 29 (1): 64–68.
1975b   Middlewestern Prairie Region. *American Birds*, 29 (4): 858–862.
1977   Middlewestern Prairie Region. *American Birds*, 31 (2): 182–186.
1978a   Middlewestern Prairie Region. *American Birds*, 32 (2): 210-215.
1978b   Middlewestern Prairie Region. *American Birds*, 32 (5): 1012–1017.
1979a   Middlewestern Prairie Region. *American Birds*, 33 (2): 181–185.
1979b   Middlewestern Prairie Region. *American Birds*, 33 (5): 776.
1980a   Middlewestern Prairie Region. *American Birds*, 34 (2): 166–169.
1980b   Middlewestern Prairie Region. *American Birds*, 34 (5): 781–785.
Kleen, V. M., and L. Bush
1973a   Middlewestern Prairie Region. *American Birds*, 27 (1): 66–70.
1973b   Middlewestern Prairie Region. *American Birds*, 27 (4): 777–781.
Kleiman, J. P.
1966   Migration of Rough-legged Hawks over Lake Erie. *Wilson Bulletin*, 78:
        122.
1979   Eastern Great Lakes. *HMANA Newsletter*, 4 (2): 36–37.
1980   Eastern Great Lakes. *HMANA Newsletter*, 5 (2): 24–26.
1981   Eastern Great Lakes. *HMANA Newsletter*, 6 (2): 23–25.
1982   Eastern Great Lakes, *HMANA Newsletter*, 7 (2): 27–29.
1983   Eastern Great Lakes. *HMANA Newsletter*, 8 (2): 28–30.
Knight, R. L.
1979   Observations on Wintering Bald and Golden Eagles on the Columbia
        River, Washington. *Murrelet*, 60: 99–105.

Knighton, G. E.
1970   Another Swallow-tailed Kite at Augusta. *Oriole,* 35 (1): 14–15.
Kochenberger, R.
1984   1983 Expanded Hawk Watch. *Peregrine Observer,* 7 (1): 10–11.
Koebel, T. D.
1970   Bearfort Tower Lookout. New Jersey Highland Hawk Observation Notes—1968–1969. *Urner Field Observer,* 12 (1): 9–21.
1983   The Origin and History of Raccoon Ridge. *Urner Field Observer,* 18 (1): 49–52.
Koford, C. B.
1953   *The California Condor.* Research Report No. 4. National Audubon Society, New York, N. Y.
Kohlbry, M.
1980a   1979 Fall Migration Summary. *Hawk Ridge Nature Reserve Annual Report March 1980:* 7–13.
1980b   Western Great Lakes. *HMANA Newsletter,* 5 (2): 27.
Korkoerle, G. V.
1960a   Hawk Flight at Marathon, Ontario. *News Letter Thunder Bay Field Naturalists Club,* 14 (1): 12.
1960b   Gyrfalcon. *News Letter Thunder Bay Field Naturalists Club,* 14 (1): 13.
Kotz, M.
1973   Final Report 1972 Waggoners Gap Hawk Watch. *Appalachian Audubon News,* January issue. Pg. 4.
Kranick, K.
1982   Spring Hawk Flights at Blue Mountain. *Special Publication Lehigh Valley Audubon Society—June 1982:* 1–6.
1984   Spring Non-Passerine Migration at Baer Rocks [*sic;* should be Bear Rocks], Pennsylvania, 1983. *Cassinia,* 60: 19–22.
Krause, H.
1961   Northern Great Plains Region. *Audubon Field Notes,* 15 (1): 51–54.
Kuerzi, R. G.
1937   Witmer Stone Wildlife Sanctuary Report for 1937. Unpublished report.
Kunkle, D. E.
1976   Some Bird Observations On, Over, and Around Delaware Bay. *Urner Field Observer,* 16 (1): 13–25.
Kuser, D.
1962   *Annotated List of Birds of High Point State Park and Stokes State Forest.* New Jersey Dept. Conservation and Economic Development, Trenton, N. J.
Kuyava, G. C.
1958   Sight Record of Swallow-tailed Kite. *Flicker,* 30 (1): 35.
Kuyt, E.
1961   Migratory Congregations of Hawks. *Blue Jay,* 19 (1): 23.
1967   Two Banding Returns for Golden Eagle and Peregrine Falcon. *Bird-Banding,* 38: 78–79.
Lack, D.
1954   *The Natural Regulation of Animal Numbers.* Oxford University Press, London.
Lakela, O.
1946   Destruction of Hawks Within Duluth City Limits. *Flicker,* 18 (4): 108–109.
1948   Bird Notes from Duluth. *Flicker,* 20 (1): 29.

La Mar, K. E.
1937   Migrating Hawks and Other Birds. *Iowa Bird Life*, 7 (1): 7.
Land, H. C.
1970   *Birds of Guatemala.* Livington Publishing Co., Wynnewood, Pa.
Lang, E. B.
1943   Hawk Watching in the Watchungs. *Audubon Magazine*, 45: 346–351.
Larrison, E. J.
1977   A sighting of the Broad-winged Hawk in Washington. *Murrelet*, 58 (1): 18.
Lasley, G. W.
1984   South Texas Region. *American Birds*, 38 (2): 221–223.
Laurie, P., J. W. McCord, and N. C. Jenkins
1981   Autumn Hawk Migrations at Fort Johnson, Charleston, S. C. *Chat*, 45 (4): 85–90.
Lawrence, S.
1984   *The Audubon Society Field Guide to the Natural Places of the Mid-Atlantic States: Coastal.* Pantheon Books, New York, N. Y.
Lawrence, S., and B. Gross
1984   *The Audubon Society Field Guide to the Natural Places of the Mid-Atlantic States: Inland.* Pantheon Books, New York, N. Y.
Laymon, S. A., and W. D. Shuford
1980   Middle Pacific Coast Region. *American Birds*, 34 (2): 195–199.
Layne, J. N.
1982   Analysis of Florida-Related Banding Data for the American Kestrel. *North American Bird Bander*, 7 (3): 94–99.
Leck, C. F.
1975   *Birds of New Jersey/Their Habits and Habitats.* Rutgers University Press, New Brunswick, N. J.
1980   Establishment of New Population Centers with Changes in Migration Patterns. *J. Field Ornithology*, 51 (2): 168–173.
1984   *The Status and Distribution of New Jersey's Birds.* Rutgers University Press, New Brunswick, N. J.
Lee, B. B., and D. S. Lee
1978   Observations of the Autumn Hawk Migrations Along North Carolina's Outer Banks. *ASB Bulletin*, 25 (2): 53–54.
Lee, D. S.
1977   Field Studies from a Rooftop: Monitoring Autumn Hawk Migration. *American Biology Teacher*, 39 (1): 17–20.
Lee, D. S., and Z. B. Sykes
1975   Autumn Hawk and Jay Migration Studies at Towson, 1973. *Maryland Birdlife*, 31 (1): 5–12.
LeGrand, H. E., Jr.
1979   Southern Atlantic Coast Region. *American Birds*, 33 (2): 165–168.
1980a   Southern Atlantic Coast Region. *American Birds*, 34 (2): 149–152.
1980b   Two Records of the Prairie Falcon for Northwestern South Carolina. *Chat*, 44 (4): 104–105.
1981a   Southern Atlantic Coast Region. *American Birds*, 35 (2): 170–172.
1981b   Southern Atlantic Coast Region. *American Birds*, 35 (5): 812–814.
1982   Southern Atlantic Coast Region. *American Birds*, 36 (2): 165–168.
1983a   1983 Spring Hawk Watch at Sandy Hook, New Jersey. *Peregrine Observer*, 6 (2): 16–17.

1983b   Southern Atlantic Coast Region. *American Birds*, 37 (2): 167–170.
1984   The Fall 1983 Hawk Flight at Cape May: Diary of a Distraught Hawk-watcher. *Peregrine Observer*, 7 (1): 3–6.

LeGrand, H. E., Jr., and P. Dunne
1983   The Record-Breaking Fall Peregrine Falcon Flight at Cape May Point. *Peregrine Observer*, 6 (1): 2–3.

LeGrand, H. E., Jr., et al.
1983   Cape May Point Hawk Watch Totals and Peak Flights for Autumn, 1982. *Peregrine Observer*, 6 (1): 11.

Leopold, A.
1942   A Raptor Tally in the Northwest. *Condor*, 44: 37–38.

LeValley, R., and D. Roberson
1983   Middle Pacific Coast Region. *American Birds*, 37 (2): 218–222.

Lincoln, F. C.
1936   Recoveries of Banded Birds of Prey. *Bird-Banding*, 7: 38–45.

Lister, R.
1964   Northern Great Plains Region. *Audubon Field Notes*, 18 (4): 461.
1965   Northern Great Plains Region. *Audubon Field Notes*, 19 (1): 48–53.

Listman, W., D. Wolf, and D. Bieber
1949   Spring Hawk Migration, 1949. *Goshawk*, 2 (3): 3–4.

Littlefield, A. D., S. P. Thompson, and B. D. Ehlers
1984   History and Present Status of Swainson's Hawks in Southeast Oregon. *Raptor Research*, 18 (1): 1–5.

Loetscher, F. W., Jr.
1955   North American Migrants in the State of Veracruz, Mexico: A Summary. *Auk*, 72: 14–54.

Loftin, H.
1967   Hawks Delayed by Weather on Spring Migration Through Panama. *Florida Naturalist*, 40 (1): 29.

Looney, M. W.
1972   Predation on Bats by Hawks and Owls. *Bulletin Oklahoma Ornithological Society*, 5 (1): 1–4.

Lowery, G. H., Jr.
1974   *Louisiana Birds.* Third Edition. Louisiana State University Press, Baton Rouge, La.

Lowery, G. H., Jr., and R. J. Newman
1953   Central Southern Region. *Audubon Field Notes*, 7 (1): 20–23.
1954   *The Birds of the Gulf of Mexico.* Fishery Bulletin 89. United States Fish and Wildlife Service, Washington, D. C. Pp. 519–540.

Luttringer, L. A., Jr.
1938   Outlaws of the Air—Or Are They? *Pennsylvania Game News*, 9 (4): 10–11.

MacRae, D.
1978   Northern New England. *HMANA Newsletter*, 3 (2): 6–7.
1979a   Northern New England. *HMANA Newsletter*, 4 (2): 10–13.
1979b   Fall 1978 Hawk Migration Through New Hampshire. *New Hampshire Audubon Annual 1979:* 30–31.
1979c   Northern New England. *HMANA Newsletter*, 4 (1): 7.
1981   Northern New England. *HMANA Newsletter*, 6 (2): 11–13.
1983   Pacific. *HMANA Newsletter*, 8 (2): 37–38.
1984   Pacific. *HMANA Newsletter*, 9 (1): 27–28.

Magee, M. J.
1922   Hawk Migration Route at Whitefish Point, Upper Peninsula of Michigan. *Auk*, 39: 257–258.

Maley, A.
1972   Western Great Lakes Region. *American Birds,* 26 (4): 763–765.
1974a   Western Great Lakes Region. *American Birds,* 28 (1): 56–58.
1974b   Western Great Lakes. *American Birds,* 28 (4): 804–806.
1975   Western Great Lakes. *American Birds,* 29 (1): 62–64.
Manning, R.
1979   No title. *Nebraska Bird Review,* 47 (4): 66.
Mans, M.
1963   Middle Pacific Coast Region. *Audubon Field Notes,* 17 (1): 61–66.
Martin, D. W.
1959   Hawk Migration at Curlew, Union County. *Kentucky Warbler,* 35 (1): 12–13.
Martin, P.
1978   Highlands Audubon Society Mt. Peter, N. Y.—Hawk Watch 1977. *Highlands Audubon Society Newsletter,* 8 (1): 5–7.
Marx, W. B.
1971   The 1970 Spring Migration of Hawks in the Central Highlands of New Jersey. *Urner Field Observer,* 13 (1): 11–17.
Mate, B. R.
1983   Movements and Dive Characteristics of a Satellite Monitored Humpback Whale. *J. Ceta-Research,* 4: 2.
Mathisen, J. and A. Mathisen
1959   Population Dynamics of Diurnal Birds of Prey in the Panhandle of Nebraska. *Nebraska Bird Review,* 27 (1): 2–5, 12–16.
Mattocks, P. W., Jr., and E. S. Hunn
1980   Northern Pacific Coast Region. *American Birds,* 34 (2): 191–194.
May, J. B.
1950   A-Hawking We Shall Go. *Bulletin Massachusetts Audubon Society,* 34 (6): 239–243.
Mayfield, G. R.
1980   Vast Broadwing Hawk Migration. *Kentucky Warbler,* 56 (1): 23.
Mays, L. P.
1969   Swainson's Hawk Migration. *Scissortail,* 19 (4): 73.
Mazzeo, R.
1955   Notes on a Fall Migration at Matinicus Rock, Maine. *Auk,* 72: 348–354.
McCaskie, G.
1980a   Southern Pacific Coast Region. *American Birds,* 34 (2): 199–204.
1980b   Southern Pacific Coast Region. *American Birds,* 34 (5): 814–817.
1983   Southern Pacific Coast Region. *American Birds,* 37 (2): 223–226.
McIlhenny, E. A.
1939   An Unusual Migration of Broad-winged Hawks. *Auk,* 56: 182–183.
McIntosh, J.
1957   Hawk Flight in Minneapolis. *Flicker,* 29 (1): 39.
McIntyre, C.
1982a   1981 Higbee Beach Raptor Count. *Peregrine Observer,* 5 (1): 9–10.
1982b   Sandy Hook Hawk Watch Totals and Peak Flight Dates, Spring 1982. *Peregrine Observer,* 5 (2): 10.
McKinney, R. G.
1984   A Lingering Sharp-shinned Hawk. *Goshawk,* 40 (12): 97.
McLaughlin, F. W.
1973   Swallow-tailed Kite in Cape May County. *Cassinia,* 54: 26.
McMillan, I.
1968   *Man and the California Condor.* E. P. Dutton & Co., New York, N. Y.

Mellinger, E. O.
1973   Hawk Flight in Rabun County. *Oriole*, 38 (1): 12.
Melquist, W. E., D. R. Johnson, and W. D. Carrier
1978   Migration Patterns of Northern Idaho and Eastern Washington Ospreys. *Bird-Banding*, 49 (3): 234–236.
Mengel, R. M.
1965   *The Birds of Kentucky*. Ornithological Monographs No. 3. American Ornithologists' Union.
Menzel, K.
1974   Hawk Concentration at Valentine. *Nebraska Bird Review*, 42 (1): 19.
Merriam, C. H.
1877   A Review of the Birds of Connecticut with Remarks on their Habits. *Trans. Connecticut Academy*, 4: 1–150.
Merriam, R. D.
1954a   Hawk Count, Fall 1953. *Bird Survey of the Detroit Region*, 1953.
1954b   Detroit Region Hawk Observations, Fall 1954.
1956   Hawk Migration, Fall 1954. *Bird Survey of the Detroit Region*, 1954.
Mersereau, G. S.
1975   New England Maritimes. *J. Hawk Migration Assn. North America*, 1 (1): 6–13.
1976   New England/Maritimes. *HMANA Newsletter*, 1 (2): 3–7.
1977   New England. *HMANA Newsletter*, 2 (1): 4–5.
Miller, A. D.
1952   Hawk Count, 1951. *Bird Survey of the Detroit Region*. 1951.
Miller, A. H., I. I. McMillan, and E. McMillan
1965   *The Current Status and Welfare of the California Condor*. Research Report No. 6. National Audubon Society, New York, N. Y.
Miller, B. L.
1941   *Lehigh County Pennsylvania Geology and Geography*. Bulletin C39. Fourth Series. Pennsylvania Geological Survey, Harrisburg, Pa.
Miller, J. C.
1979   Reverse Movements of Sharp-shinned Hawks. *Cassinia*. 57: 43.
Mills, E. and L. Mills
1971   Hook Mountain, N. Y. Hawk Watch. *California Condor*, 6 (4): 12.
Mills, S., S. Alden, and J. Hubbard
1975   Southwest Region. *American Birds*, 29 (1): 98–103.
Millsap, B. A., and J. R. Zook
1983   Effects of Weather on Accipiter Migration in Southern Nevada. *Raptor Research*, 17 (2): 43–56.
Mindell, D. P., and M. H. Mindell
1984   Raptor Migration in Northwestern Canada and Eastern Alaska, Spring 1982. *Raptor Research*, 18 (1): 10–15.
Mohr, C. E.
1969   Our Handsome Birds of Prey: Hawks and Eagles. *Delaware Conservationist*, 13 (4): 3–13.
Mollhoff, W. J.
1979   Hawk Concentration near Albion. *Nebraska Bird Review*, 47 (4): 66.
Monroe, B. L., Jr.
1968   *A Distributional Survey of the Birds of Honduras*. Ornithological Monographs No. 7. American Ornithologists' Union.
Monson, G.
1956   Southwest Region. *Audubon Field Notes*, 10 (1): 44–47.

1958   Southwest Region. *Audubon Field Notes,* 12 (1): 49–51.
1972   Southwest Region. *American Birds,* 26 (1): 100–104.
1973   Southwest Region. *American Birds,* 27 (4): 803–806.
Monson, G., and A. R. Phillips
1981   *Annotated Checklist of the Birds of Arizona.* Second Edition/Revised and Expanded. University of Arizona Press, Tucson, Ariz.
Moon, L. W.
1977a   Raptor Migration Braddock's Bay Hawk Lookout Spring 1977. *Goshawk,* 33 (5): 34.
1977b   April 1977 Braddock Bay Hawk Lookout. *Goshawk,* 33 (6): 42–43.
1977c   May 1977 Braddock Bay Hawk Lookout. *Goshawk,* 33 (7): 51–52.
1977d   Complete Summary of Raptor Migration, Braddock Bay Hawk Lookout, Monroe County, Rochester, N. Y., Spring 1977. *Goshawk,* 33 (10): 83–84.
1978a   Beginning of the Raptor Migration: February–March 1978. *Goshawk,* 34 (5): 47–50.
1978b   Braddock Bay Raptor Migration Report April 1978. *Goshawk,* 34 (6): 61–64.
1978c   Braddock Bay Raptor Migration Report May 1978. *Goshawk,* 34 (7): 79–82.
1978d   Final Report of the Spring Hawk Migration at Braddock Bay June 1978. *Goshawk,* 34 (8): 91–96.
1978e   Complete Summary of Raptor Migration, Braddock Bay Hawk Lookout, Monroe County, Rochester, N. Y., Spring 1978. *Goshawk,* 34 (9): 100.
1984   Hawk Observation Platform Dedicated. *Goshawk,* 40 (6): 54.
Moon, N.
1984   Observation Platform Being Built at Hawk Lookout. *Goshawk,* 40 (2): 10.
Moon, L. W., and N. S. Moon
1979a   The Beginning of the Spring Raptor Migration at Braddock Bay, February 27–March 31, 1979. *Goshawk,* 35 (5): 45–48.
1979b   Braddock Bay Raptor Migration Report April 1979. *Goshawk,* 35 (6): 59–63.
1979c   Braddock Bay Raptor Migration Report May 1979. *Goshawk,* 35 (7): 75–78.
1979d   Summary of the Spring Raptor Migration at Braddock Bay, February 27–June 30, 1979. *Goshawk,* 35 (8): 83–89.
1980a   The Beginning of the Spring Raptor Migration at Braddock Bay, February 27–March 31, 1980. *Goshawk,* 36 (5): 37–40.
1980b   Braddock Bay Raptor Migration Report, April 1980. *Goshawk,* 36 (6): 51–54.
1980c   Braddock Bay Raptor Migration Report, May 1980. *Goshawk,* 36 (7): 63–66.
1980d   Braddock Bay Raptor Migration Season's Summary (and June Report) February 25–June 30, 1980. *Goshawk,* 36 (8): 73–78.
1981a   Braddock Bay Raptor Migration Report, March 1981. *Goshawk,* 37 (5): 61–64.
1981b   Braddock Bay Raptor Migration Report, April 1981. *Goshawk,* 37 (6): 75–78.
1981c   Braddock Bay Raptor Migration Report, May 1981. *Goshawk,* 37 (7): 87–90.
1981d   Complete Summary of Raptor Migration Braddock Bay Hawk Lookout, Monroe County (near Rochester), N. Y., Spring 1981. *Goshawk,* 37 (8): 95–100.

1982a The Beginning of the Spring Raptor Migration at Braddock Bay, February–March 1982. *Goshawk*, 38 (5): 35–38.

1982b Braddock Bay Raptor Migration Report, April 1982. *Goshawk*, 38 (6): 51–54.

1982c Braddock Bay Raptor Migration Report, May 1982. *Goshawk*, 38 (7): 61–64.

1982d Braddock Bay Raptor Migration Report Season's Summary (and June Report) 26 February –30 June, 1982. *Goshawk*, 38 (8): 71–76.

1982e Braddock Bay Hawk Lookout. *HMANA Newsletter*, 7 (2): 1–3.

1983a The Beginning of the Spring Raptor Migration at Braddock Bay, February–March 1983. *Goshawk*, 39 (5): 37–40.

1983b Braddock Bay Raptor Migration Report, April 1983. *Goshawk*, 39 (6 & 7): 51–54.

1983c Braddock Bay Raptor Migration Report Season's Summary 20 February–30 June 1983. *Goshawk*, 39 (8): 59–66.

1984a The Beginning of the Spring Raptor Migration at Braddock Bay February–March 1984. *Goshawk*, 40 (5): 41–44.

1984b Braddock Bay Raptor Migration Report April 1984. *Goshawk*, 40 (6): 51–53.

1984c Braddock Bay Raptor Migration Report Season's Summary 19 February–30 June 1984. *Goshawk*, 40 (7 & 8): 65–72.

Moore, G. E.
1951 Hawk Migration Count. *Bluebird*, 18 (9): 1–2.

Moore, M.
1982 *A Beginner's Guide to Hawkwatching*. Hawk Migration Assn. of North America, Medford, Mass.

Morris, L.
1978 Prairie Falcons. *Nebraska Bird Review*, 46 (3): 63.

Morrissey, T. J.
1968 Notes of Birds in the Davenport Area. *Iowa Bird Life*, 38 (3): 67–85.

Morton, R.
1982 Spring Migration at Hawk Mountain. *Hawk Mountain News*, 58: 7–10.
1983 Spring Migration at Hawk Mountain. *Hawk Mountain News*, 60: 10–12.

Mosher, J. A., et al.
1978 Raptors of the Uinta National Forest, Utah. *Great Basin Naturalist*, 38 (4): 438–446.

Mueller, H. C., and D. D. Berger
1961 Weather and Fall Migration of Hawks at Cedar Grove, Wisconsin. *Wilson Bulletin*, 73: 171–192.

1965 A Summer Movement of Broad-winged Hawks. *Wilson Bulletin*, 77: 83–84.

1966 Analyses of Weight and Fat Variations in Transient Swainson's Thrushes. *Bird-Banding*, 37: 83–112.

1967a Wind Drift, Leading Lines, and Diurnal Migrations. *Wilson Bulletin*, 79: 50–63.

1967b Fall Migration of Sharp-shinned Hawks. *Wilson Bulletin*, 79: 397–415.

1967c Some Observations and Comments on the Periodic Invasions of Goshawks. *Auk*, 84 (2): 183–191.

1968 Sex Ratios and Measurements of Migrant Goshawks. *Auk*, 85: 431–436.

1973 The Daily Rhythm of Hawk Migrations at Cedar Grove, Wisconsin. *Auk*, 90: 591–596.

Mueller, H. C., D. D. Berger, and G. Allez
1977 The Periodic Invasions of Goshawks. *Auk*, 94 (4): 652–663.
1979 The Identification of North American Accipiters, *American Birds*, 33 (3): 236–240.
1981 Age and Sex Differences in Wing Loading and Other Aerodynamic Characteristics of Sharp-shinned Hawks. *Wilson Bulletin*, 93 (4): 491–499.
Muir, D. G.
1978 Summary of the 1978 Hawk Migration. *Derby Hill Newsletter*, 1 (1): 1–2.
1979 Summary of the 1979 Hawk Migration. *Derby Hill Newsletter*, 2 (1): 1–4.
1980 Summary of the 1980 Hawk Migration. *Derby Hill Newsletter*, 3 (1): 1–4.
1981 Summary of the 1981 Hawk Migration. *Derby Hill Newsletter*, 4 (1): 1–4.
1982 Summary of the 1982 Hawk Migration. *Derby Hill Newsletter*, 5 (1): 1–4.
Mumford, R. E.
1959a Middlewestern Prairie Region. *Audubon Field Notes*, 13 (1): 33–37.
1959b Midwestern Prairie Region. *Audubon Field Notes*, 13 (4): 373–376.
1960 Middlewestern Prairie Region. *Audubon Field Notes*, 14 (1): 38–41.
1961a Middlewestern Prairie Region. *Audubon Field Notes*, 15 (1): 44–46.
1961b Middlewestern Prairie Region. *Audubon Field Notes*, 15 (4): 413–416.
Murphy, R. C.
1933 A Great Hawk Flight in Westchester County, New York. *Bird Lore*, 35: 320–321.
Murray, B. G., Jr.
1964 A Review of Sharp-shinned Hawk Migration Along the Northeastern Coast of the United States. *Wilson Bulletin*, 76: 257–264.
1969 Sharp-shinned Hawk Migration in the Northeastern United States. *Wilson Bulletin*, 81: 119–120.
Murray, J. J.
1946 Some North Carolina Bird Notes. *Chat*, 10: 32–34.
Nagy, A. C.
1967 Curator's Report—1966. *News Letter to Members No. 39.* Hawk Mountain Sanctuary Assn., Kempton, Pa.
1968 Curator's Report. *News Letter to Members No. 40.* Hawk Mountain Sanctuary Assn., Kempton, Pa.
1970 Curator's Report. *News Letter to Members No. 42.* Hawk Mountain Sanctuary Assn., Kempton, Pa.
1972 1971 Curator's Report. *News Letter to Members No. 44.* Hawk Mountain Sanctuary Assn., Kempton, Pa.
1973 1972 Curator's Report. *News Letter to Members No. 45.* Hawk Mountain Sanctuary Assn., Kempton, Pa.
1976 Northern Appalachians. *HMANA Newsletter*, 1 (2): 11–14.
1977a Population Trend Indices Based on 40 Years of Autumn Counts at Hawk Mountain Sanctuary in North-eastern Pennsylvania. *Proc. World Conference on Birds of Prey, 1975:* 243–253.
1977b Northern Appalachians. *HMANA Newsletter*, 2 (2): 12–16.
1977c Northern Appalachians. *HMANA Newsletter*, 2 (1): 7–8.
1978a Northern Appalachians. *HMANA Newsletter*, 3 (1): 9–10.
1978b Northern Appalachians. *HMANA Newsletter*, 3 (2): 22–25.
1979a Northern Appalachians. *HMANA Newsletter*, 4 (1): 12–13.
1979b Northern Appalachians. *HMANA Newsletter*, 4 (2): 29–33.
1979c Miracle Day. *Hawk Mountain News*, 50 [*sic;* should be 51]: 25–29.
1980 Northern Appalachians. *HMANA Newsletter*, 5 (1): 11.
1984 Labor Day Bald Eagles. *Hawk Mountain News*, 62: 37–41.

Nero, R. W.
  1961   Northern Great Plains Region. *Audubon Field Notes*, 15 (4): 419–422.
  1962a Northern Great Plains Region. *Audubon Field Notes*, 16 (1): 47–50.
  1962b Northern Great Plains Region. *Audubon Field Notes*, 16 (4): 423–426.
  1963   Northern Great Plains Region. *Audubon Field Notes*, 17 (1): 40–44.
Newman, D.
  1977   Migrating Hawks. *Cleveland Bird Calendar*, 73 (2): 22.
Newman, R. J.
  1956a Central Southern Region. *Audubon Field Notes*, 10 (1): 29–32.
  1956b Central Southern Region. *Audubon Field Notes*, 10 (4): 338–341.
  1958a Central Southern Region. *Audubon Field Notes*, 12 (1): 36–39.
  1958b Central Southern Region. *Audubon Field Notes*, 12 (4): 358–362.
  1959   Central Southern Region. *Audubon Field Notes*, 13 (1): 37–41.
  1960   Central Southern Region. *Audubon Field Notes*, 14 (1): 41–47.
  1961   Central Southern Region. *Audubon Field Notes*, 15 (1): 46–51.
Newman, R. J., and S. L. Warter
  1959   Central Southern Region. *Audubon Field Notes*, 13 (4): 376–380.
Nichols, C. K.
  1953a Hudson-St. Lawrence Region. *Audubon Field Notes*, 7 (1): 6–9.
  1953b Hudson-St. Lawrence Region. *Audubon Field Notes*, 7 (4): 262–265.
  1955   Hudson-St. Lawrence Region. *Audubon Field Notes*, 9 (1): 10–14.
  1956   Hudson-St. Lawrence Region. *Audubon Field Notes*, 10 (1): 8–12.
  1958   Hudson-St. Lawrence Region. *Audubon Field Notes*, 12 (4): 336–340.
  1959   Hudson-St. Lawrence Region. *Audubon Field Notes*, 13 (1): 13–17.
  1960   Hudson-St. Lawrence Region. *Audubon Field Notes*, 14 (1): 17–20.
Nikula, B.
  1982   Northeastern Maritime Region. *American Birds*, 36 (5): 827–831.
  1983   Northeastern Maritime Region. *American Birds*, 37 (5): 844–849.
Nolan, V., Jr.
  1953   Middlewestern Prairie Region. *Audubon Field Notes*, 7 (1): 18–20.
  1955   Middlewestern Prairie Region. *Audubon Field Notes*, 9 (1): 28–31.
  1956   Middlewestern Prairie Region. *Audubon Field Notes*, 10 (1): 27–29.
  1958a Middlewestern Prairie Region. *Audubon Field Notes*, 12 (1): 33–36.
  1958b Middlewestern Prairie Region. *Audubon Field Notes*, 12 (4): 356–358.
Norse, W. J.
  1979a *Vermont Hawk Watch Spring 1979*. Vermont Institute of Natural Science, Woodstock, Vt.
  1979b *Vermont Hawk Watch Fall 1979*. Vermont Institute of Natural Science, Woodstock, Vt.
  1980a *Vermont Hawk Watch Spring 1980*. Vermont Institute of Natural Science, Woodstock, Vt.
  1980b *Vermont Hawk Watch Fall 1980*. Vermont Institute of Natural Science, Woodstock, Vt.
  1981a *Vermont Hawk Watch Spring 1981*. Vermont Institute of Natural Science, Woodstock, Vt.
  1981b *Vermont Hawk Watch Fall 1981*. Vermont Institute of Natural Science, Woodstock, Vt.
  1982a *Vermont Hawk Watch Spring 1982*. Vermont Institute of Natural Science, Woodstock, Vt.
  1982b *Vermont Hawk Watch Fall 1982*. Vermont Institute of Natural Science, Woodstock, Vt.

1983a   *Vermont Hawk Watch Spring 1983.* Vermont Institute of Natural Science, Woodstock, Vt.

1983b   *Vermont Hawk Watch Fall 1983.* Vermont Institute of Natural Science, Woodstock, Vt.

Oberholser, H. C.
1938   The Bird Life of Louisiana. *Bulletin 28.* Louisiana Department of Conservation, New Orleans, La.

1974   *The Bird Life of Texas.* Volume one. University of Texas Press, Austin, Texas.

O'Connor, M. F.
1977   Peregrine Falcons at Indian Cave State Park. *Nebraska Bird Review,* 45 (4): 57–58.

Odom, T.
1966   Summary of Broad-winged Hawk Flights Across Tennessee from 1951 through 1964. *J. Tennessee Academy Science,* 41: 95–96.

Ogden, J. C.
1974   The Short-tailed Hawk in Florida. I. Migration, Habitat, Hunting Techniques, and Food Habits. *Auk,* 91 (1): 95–110.

O'Hara, R. D.
1964   Gyrfalcon Sight Record. *Loon,* 36 (1): 25–26.

Ohlander, B. G.
1976   Gyrfalcon Taken in Nebraska. *Nebraska Bird Review,* 44 (1): 3.

1980   Gyrfalcon, Prairie Falcons. *Nebraska Bird Review,* 48 (2): 44.

Olson, R.
1952   The Possibilities of Establishing A Hawk and Owl Banding and Photographic Station at Duluth, Minnesota. *Flicker,* 24 (3): 111–115.

Otto, B. and B. Otto
1984   No title. *Nebraska Bird Review,* 52 (1): 23.

Packard, C. M.
1962   Migrating Broad-wings. *Maine Field Naturalist,* 18 (11): 163–164.

Paff, W. A.
1927   A Hawk Flight. *Auk,* 44: 420.

Palmer, R. S.
1949   Maine Birds. *Bulletin Museum Comparative Zoology,* 102.

Panzer, A.
1976   Summary of the 1975 Banding Season at the Kittatinny Ridge. *Urner Field Observer,* 16 (1): 4–5.

Parker, T.
1974   Southwest Region. *American Birds,* 28 (1): 87–90.

Parks, T.
1957   Hawk Migration. *Chat,* 21 (4): 88–89.

Parmalee, P. W.
1954   The Vultures: Their Movements, Economic Status, and Control in Texas. *Auk,* 71: 443–453.

Parmalee, P. W. and B. G. Parmalee
1967   Results of Banding Studies of the Black Vulture in Eastern North America. *Condor,* 69: 146–155.

Parnell, J. F.
1965   Southern Atlantic Coast Region. *Audubon Field Notes* 19 (1): 25–28.

1967   Southern Atlantic Coast Region. *Audubon Field Notes.* 21 (1): 19–22.

1968   Southern Atlantic Coast Region. *Audubon Field Notes,* 22 (1): 22–25.

Parrish, J. R., D. T. Rogers, Jr., and F. P. Ward
    1983   Identification of Natal Locales of Peregrine Falcons *(Falco peregrinus)* by Trace-Element Analysis of Feathers. *Auk,* 100 (3): 560–567.
Parrish, J. R., and E. W. Wischusen
    1980   New Banding Record for Peregrine Falcons in Alabama. *Alabama Birdlife,* 28 (1,2): 3–6.
Patton, F. J.
    1940   Two Unusual Hawk Records for Blue Springs, Gage County. *Nebraska Bird Review,* 8 (2): 97.
    1951   Golden Eagles Seen in Gage County. *Nebraska Bird Review,* 19 (2): 31.
Paulson, D. R., and H. M. Stevenson
    1962   Florida Region. *Audubon Field Notes,* 16 (4): 398–404.
Paxton, R. O., W. J. Boyle, Jr., and D. A. Cutler
    1981   Hudson-Delaware Region. *American Birds,* 35 (5): 804–808.
    1983   Hudson-Delaware Region. *American Birds,* 37 (2): 160–164.
    1984   Hudson-Delaware Region. *American Birds,* 38 (2): 181–185.
Paxton, R. O., P. A. Buckley, and D. A. Cutler
    1976   Hudson-Delaware Region. *American Birds,* 30 (1): 39–46.
Paxton, R. O., and T. A. Chandik
    1967   Middle Pacific Coast Region. *Audubon Field Notes,* 21 (1): 72–76.
Paxton, R. O., K. C. Richards, and D. A. Cutler
    1980   Hudson-Delaware Region. *American Birds,* 34 (2): 143–147.
Paxton, R. O., P. W. Smith, and D. A. Cutler
    1979   Hudson-Delaware Region. *American Birds,* 33 (2): 159–163.
Payne, R. B.
    1983   A Distributional Checklist of the Birds of Michigan. *University Michigan Museum Zoology Misc. Publications,* 164: 1–71.
Peacock, J., and M. Myers
    1981   Western Great Lakes. *HMANA Newsletter,* 6 (1): 19–21.
    1982a   Western Great Lakes. *HMANA Newletter,* 7 (1): 21.
    1982b   Western Great Lakes. *HMANA Newsletter,* 7 (2): 29–30.
    1983a   Western Great Lakes. *HMANA Newsletter* 8 (1): 18–20.
    1983b   Western Great Lakes. *HMANA Newsletter,* 8 (2): 30–32.
Peakall, D. B.
    1962   No title. *Kingbird,* 12 (2): 80.
Pearson, L., and B. H. Warren
    1897   *Diseases and Enemies of Poultry.* Clarence M. Busch, State Printer of Pennsylvania.
Pearson, T. G., C. S. Brimley, and H. H. Brimley
    1942   *Birds of North Carolina.* North Carolina Department of Agriculture, Raleigh, N. C.
Pease, R. and C. Goodrich
    1978   Birding in Provincetown. *Bird Observer of Eastern Massachusetts,* 6 (2): 40–47.
Peet, M. M.
    1905   The Fall Migration of Birds at Washington Harbor, Isle Royale, in 1905. In *An Ecological Survey of Isle Royale, Lake Superior.* Wynkoop Hallenbeck, Crawford Co., Lansing, Mich.
Pennycuick, C. J.
    1975   The Use of Motor-Gliders for Studying Raptor Migration. *Proc. North American Hawk Migration Conference 1974.* Pp. 65–71.

Perkins, J. P.
   1964   17 Flyways Over the Great Lakes. Part 1. *Audubon Magazine,* 66 (5): 294–299.
   1975   Discussion of Telemetry and Radar. *Proc. North American Hawk Migration Conference 1974.* Pg. 59.

Peterjohn, B. G.
   1982a   Middlewestern Prairie Region. *American Birds,* 36 (2): 182–186.
   1982b   Middlewestern Prairie Region. *American Birds,* 36 (5): 857–861.
   1983a   Middlewestern Prairie Region. *American Birds,* 37 (2): 185–189.
   1983b   Middlewestern Prairie Region. *American Birds,* 37 (5): 874–878.

Peters, H. S., and T. D. Burleigh
   1951   *The Birds of Newfoundland.* Houghton Mifflin Co., Boston Mass.

Petersen, P. C., Jr.
   1965   Middlewestern Prairie Region. *Audubon Field Notes,* 19 (1): 44–46.
   1966a   Middlewestern Prairie Region. *Audubon Field Notes,* 20 (1): 53–55.
   1966b   Middlewestern Prairie Region. *Audubon Field Notes,* 20 (4): 513–515.
   1968a   Middlewestern Prairie Region. *Audubon Field Notes,* 22 (1): 48–50.
   1968b   Middlewestern Prairie Region. *Audubon Field Notes,* 22 (4): 531–533.
   1969   Middlewestern Prairie Region. *Audubon Field Notes,* 23 (1): 64–65.
   1970a   Middlewestern Prairie Region. *Audubon Field Notes,* 24 (1): 54–55.
   1970b   Middlewestern Prairie Region. *Audubon Field Notes,* 24 (4): 608, 613–615.
   1971   Middlewestern Prairie Region. *American Birds,* 25 (1): 64–66.

Peterson, R. T.
   1966   Tribute to Maurice and Irma Broun. *News Letter to Members No. 38.* Hawk Mountain Sanctuary Assn., Kempton, Pa.
   1985   Hawk Mountain Celebrates its Fiftieth Anniversary. *Bird Watcher's Digest,* 7 (3): 83–92.

Peterson, R. T., and E. L. Chalif
   1973   *A Field Guide to Mexican Birds and Adjacent Central America.* Houghton Mifflin Co., Boston, Mass.

Pettingill, O. S., Jr.
   1962   Hawk Migrations Around the Great Lakes. *Audubon Magazine,* 64 (1): 44–45, 49.
   1977   *A Guide to Bird Finding East of the Mississippi.* Second Edition. Oxford University Press, New York, N. Y.
   1981   *A Guide to Bird Finding West of the Mississippi.* Second Edition. Oxford University Press, New York, N. Y.

Phillips, A., J. Marshall, and G. Monson
   1964   *The Birds of Arizona.* University of Arizona Press, Tucson, Ariz.

Phillips, J. H. H., and W. L. Putnam
   1961   Hawk Flight in Lake Superior Park. *News Letter Thunder Bay Field Naturalists Club,* 15 (4): 64.

Phinney, E. W.
   1977   Northern New England. *HMANA Newsletter,* 2 (2): 6–7.

Pierson, E. C., and J. E. Pierson
   1981   *A Birder's Guide to the Coast of Maine.* Down East Books, Camden, Maine.

Pinchon, P. R., and C. Vaurie
   1961   The Kestrel *(Falco tinnunculus)* in the New World. *Auk,* 78: 92–93.

Pistorius, A.
   1977   *Vermont Hawk Migration Fall 1977.* Vermont Institute of Natural Science, Woodstock, Vt.

1978 *Vermont Hawk Migration Fall 1978.* Vermont Institute of Natural Science, Woodstock, Vt.

Platt, S.
1974 Mid-October Hawk Migration in Payne County, Oklahoma. *Bulletin Oklahoma Ornithological Society,* 7 (3): 55.

Poole, E. L.
1934 The Hawk Migration Along the Kittatinny Ridge in Pennsylvania. *Auk,* 51: 17–20.
1964 *Pennsylvania Birds/An Annotated List.* Delaware Valley Ornithological Club, Philadelphia, Pa.

Postupalsky, S.
1976a Bald Eagle Migration Along the South Shore of Lake Superior. *Jack-Pine Warbler,* 54 (3): 98–104.
1976b Banded Northern Bald Eagles in Florida and Other Southern States. *Auk,* 93 (4): 835–836.

Potter, E. F. and P. W. Sykes, Jr.
1980 A Probable Winter Record of Swainson's Hawk from Tyrrell County, N. C., with Comments on a Fall 1965 Sighting from the Outer Banks. *Chat,* 44 (3): 76–78.

Potter, J. K.
1949 Hawks Along the Delaware. *Cassinia,* 37: 13–16.

Potter, J. K., and J. J. Murray
1953 Middle Atlantic Coast Region. *Audubon Field Notes,* 7 (1): 9–11.
1955 Middle Atlantic Coast Region. *Audubon Field Notes,* 9 (1): 15–17.

Potter, J. K., and F. R. Scott
1958 Middle Atlantic Coast Region. *Audubon Field Notes,* 12 (4); 340–341.

Pough, R. H.
1932 Wholesale Killing of Hawks in Pennsylvania. *Bird-Lore,* 34 (6): 429–430.
1936 Pennsylvania and the Hawk Problem. *Pennsylvania Game News,* 7 (4): 8, 23.
1939 Marsh Hawks Crossing to Cuba. *Florida Naturalist,* 12: 105.
1984 Recollections of a Pioneer. *Hawk Mountain News,* 62: 15–18.

Powell, S. E.
1957 Notes on a Migration of Broad-winged Hawks. *Maine Field Observer,* 2 (10).

Pratt, D.
1967 Observation of Broad-winged Hawk Migrations at Table Rock, N. C. *Chat,* 31 (4): 95.

Pray, R. H.
1953 Middle Pacific Coast Region. *Audubon Field Notes,* 7 (1): 33–35.

Proctor, N. S.
1978 *25 Birding Areas in Connecticut.* Pequot Press, Chester, Conn.

Proescholdt, B.
1961 Weather and a Hawk Flight in Iowa. *Iowa Bird Life,* 31 (2): 33–35.

Puckette, D.
1982a Southern Appalachians. *HMANA Newsletter,* 7 (1): 17–18.
1982b Southern Appalachians. *HMANA Newsletter,* 7 (2): 24–27.
1983a Southern Appalachians. *HMANA Newsletter,* 8 (1): 15–16.
1983b Southern Appalachians. *HMANA Newsletter,* 8 (2): 24–26.
1983c The Fall Migration of Hawks Across Virginia. *Virginia Society of Ornithology Newsletter,* 29 (1): 9.

1983d   Hawkwatching on the Heights. *Virginia Society of Ornithology Newsletter,* 29 (3): 9.

1984a   Southern Appalachians. *HMANA Newsletter,* 9 (1): 19.

1984b   Southern Appalachians. *HMANA Newsletter,* 9 (2): 27–29.

Purdue, J. R., et al.

1972   Spring Migration of Swainson's Hawk and Turkey Vulture through Veracruz, Mexico. *Wilson Bulletin,* 84 (1): 92–93.

Purrington, R. D.

1971   Central Southern Region. *American Birds,* 25 (1): 66–71.

1973   Central Southern Region. *American Birds,* 27 (1): 70–75.

1974   Central Southern Region. *American Birds,* 28 (1): 63–67.

1975   Central Southern Region. *American Birds,* 29 (1): 68–74.

1976   Central Southern Region. *American Birds,* 30 (1): 82–87.

1978   Central Southern Region. *American Birds,* 32 (2): 215–220.

1979   Central Southern Region. *American Birds,* 33 (2): 185–188.

1980   Central Southern Region. *American Birds,* 34 (2): 170–172.

1982   Central Southern Region. *American Birds,* 36 (2): 186–188.

1983   Central Southern Region. *American Birds,* 37 (2): 189–192.

Rabenold, P. P.

1983   The Communal Roost in Black and Turkey Vultures—an Information Center? In *Vulture Biology and Management.* University of California Press, Berkeley, Ca. Pp. 303-321.

Radis, R.

1976   Watch for the Hawks! *Wonderful West Virginia,* (September): 12–15.

Raffaele, H. A.

1983   *A Guide to the Birds of Puerto Rico and the Virgin Islands.* Addison-Wesley Publishing Company, Reading, Mass.

Rapp, W. F., Jr.

1954   The Status of the Western Turkey Vulture in Nebraska. *Nebraska Bird Review,* 22 (1): 3–5.

Rasmussen, E.

1974   Highlands Audubon Society Mt. Peter, N. Y.—Hawk Watch 1973. *Highlands Audubon Society Newsletter,* 4 (1): 7–10.

Raspet, A.

1960   Biophysics of Bird Flight. *Science,* 132 (3421): 191–199.

Rathert, J.

1977   Notes on Hawk Migration. *Bluebird,* 44 (4): 16.

Redmond, E. C., and R. A. Breck

1961   Montclair Hawk Lookout Sanctuary. *New Jersey Nature News,* 16: 118, 123.

Reese, J. G

1973   Bald Eagle Migration Along the Upper Mississippi River in Minnesota. *Loon,* 45 (1): 22–23.

Remsen, V. and D. A. Gaines

1973   Middle Pacific Coast Region. *American Birds,* 27 (4): 813–818.

1974   Middle Pacific Coast Region. *American Birds,* 28 (1): 98–106.

Renaud, W. E.

1973   Northern Great Plains. *American Brids,* 27 (4): 785–788.

Rew, F. M.

1971   Rough-legged Hawk Flight. *Prothonotary,* 37 (3): 42.

Rhoads, W. P.
1958 A Flight of Marsh Hawks at Henderson, Henderson County. *Kentucky Warbler*, 34 (3): 44–45.

Rice, J. N.
1969 A Peregrine Population Index on the Maryland-Virginia Coast. In *Peregrine Falcon Populations: Their Biology and Decline*. University of Wisconsin Press, Madison, Wis. Pp. 279–280.

Richardson, W. J.
1975 Autumn Hawk Migration in Ontario Studied with Radar. *Proc. North American Hawk Migration Conference 1974*. Pp. 47–60.

Ridgely, R. S.
1981 *A Guide to the Birds of Panama*. Princeton University Press, Princeton, N. J.

Robards, F., and A. Taylor
n. d. *Bald Eagles in Alaska*. U.S. Fish and Wildlife Service and U.S. Forest Service, Washington, D.C.

Robbins, C. S.
1950 Hawks Over Maryland, Fall of 1949. *Maryland Birdlife*, 6 (1): 2–11.
1956 Hawk Watch. *Atlantic Naturalist*, 11 (5): 208–217.
1975 A History of North American Hawkwatching. *Proc. North American Hawk Migration Conference*. Pp. 29–40.

Robbins, S. D.
1970 Western Great Lakes Region. *Audubon Field Notes*, 24 (1): 51–54.

Roberson, D.
1980 *Rare Birds of the West Coast of North America*. Woodcock Publications, Pacific Grove, Calif.

Roberts, P. M.
1977a Further Aids to Hawk Identification. *Bird Observer of Eastern Massachusetts*, 5 (2): 44–49.
1977b Where to Watch Hawks in Massachusetts. *Bird Observer of Eastern Massachusetts*, 5 (4): 107–111.
1978 The Spring Hawk Migration: Toward Understanding an Enigma. *Bird Observer of Eastern Massachusetts*, 6 (1): 10–22.
1984 Why Count Hawks? A Continental Perspective. *Hawk Mountain News*, 62: 46–51.

Roberts, T. S.
1932 *The Birds of Minnesota*. Volume 1. University of Minnesota Press, Minneapolis, Minn.

Robertson, W. B., Jr.
1970 Florida Region. *Audubon Field Notes*, 24 (1): 33–38.
1971 Florida Region. *American Birds*, 25: 51.
1972 Florida Region. *American Birds*, 26 (1): 50–54.

Robertson, W. B., Jr., and J. C. Ogden
1968 Florida Region. *Audubon Field Notes*, 22 (1): 25–31.
1969 Florida Region. *Audubon Field Notes*, 23 (1): 35–41.

Robertson, W. B., Jr., and D. R. Paulson
1961 Florida Region. *Audubon Field Notes*, 15 (1): 26–35.

Robinson, J. C.
1960 Hawk Migration—Huntsville (Brownsboro). *Alabama Birdlife*, 8 (4): 28.

Robinson, L. J.
1978 Hawk Migration: A New Horizon for Birders. *Bird Observer of Eastern Massachusetts*, 6 (2): 58–60.
1979 Devices for Recording Weather Data. *HMANA Newsletter*, 4 (2): 7.

1980 The Continental Summary. *HMANA Newsletter,* 5 (2): 1–4.
1981 Autumn Broadwing Flights at Wachusett Mountain. *Bird Observer of Eastern Massachusetts,* 9 (4): 175–176.

Rogers, D. T., Jr.
1982 Southeast. *HMANA Newsletter,* 7 (2): 18–19.
1983a Southeast. *HMANA Newsletter,* 8 (1): 11.
1983b Southeast. *HMANA Newsletter,* 8 (2): 27–28.
1984 Southeast. *HMANA Newsletter,* 9 (2): 29–30.

Rogers, R.
1971 Highlands Audubon Society Mt. Peter, N. Y., Hawk Watch, 1971. *California Condor,* 6 (4): 13.
1972 Hawk Watch '72. *Highlands Audubon Society Newsletter,* 2 (12): 6–7.

Rosenfield, R. N., and D. L. Evans
1980 Migration Incidence and Sequence of Age and Sex Classes of the Sharp-shinned Hawk. *Loon,* 52 (2): 66–69.

Rosengren, A.
1978 New Haven Area. In *Hawk Migration 1977 Report.* Privately published.
1979a New Haven Area. In *Hawk Migration 1978 Report.* Privately published.
1979b The New Haven County Hawk Watch. In *1979 Hawk Migration New England Hawk Watch.* Connecticut Audubon Council, Seymour, Conn.
1980 The New Haven County Hawk Watch—Fall 1980. In *1980 Hawk Migration New England Hawk Watch.* Connecticut Audubon Council, Seymour, Conn.
1981 Lighthouse Point, New Haven. In *1981 Hawk Migration New England Hawk Watch.* Connecticut Audubon Council, Seymour, Conn.
1982 Lighthouse Point and New Haven Area Hawk Watch. In *1982 Hawk Migration New England Hawk Watch.* Audubon Council of Connecticut, Sharon, Conn.
1983 Lighthouse Point Park. In *1983 Hawk Migration New England Hawk Watch.* Audubon Council of Connecticut, Sharon, Conn.

Ross, C. C
1953 Swallow-tailed Kite Over the Wissahicken. *Cassinia,* 39: 25.

Rowlett, R. A.
1977 *Vermont Hawk Migration Fall 1976.* Vermont Institute of Natural Science, Woodstock, Vt.
1978 Close-Site Study of Broadwing Migration Through the Texas Coastal Bend. *HMANA Newsletter,* 3 (2): 18–20.
1980 Migrant Broad-winged Hawks in Tobago. *J. Hawk Migration Assn. North America,* 2 (1): 54.

Rusling, W. J.
1935a "Witmer Stone Wildlife Sanctuary" Established August 1, 1935. Unpublished report.
1935b Hawk Flight August 23–November 15, 1935, Cape May Point, New Jersey. Unpublished report.
1936 The Study of the Habits of Diurnal Migrants, as Related to Weather and Land Masses during the Fall Migration on the Atlantic Coast, with Particular Reference to the Hawk Flights of the Cape Charles (Virginia) Region. Unpublished manuscript.

Russell, S. M.
1964 A Distributional Study of the Birds of British Honduras. *Ornithological Monographs No. 1.* American Ornithologists' Union.

Rymer, R.
  1985   Making A Spectacle of Themselves. *National Wildlife*, 23 (1): 4–7.
Sanders, J., and L. Yaskot
  1977   *Top Birding Spots Near Chicago*. Evanston North Shore Bird Club, Win-
         netka, Ill.
Saunders, G. B.
  1931   Report of George B. Saunders Field Representative/National Association
         of Audubon Societies/Cape May, New Jersey October 20, 1931. Un-
         published report.
Saunders, W. E.
  1909   The Sharp-shinned Hawk Migration. *Ottawa Naturalist*, 23: 156–160.
Schneider, D. G.
  1978a Western Great Lakes. *HMANA Newsletter*, 3 (1): 13.
  1978b Western Great Lakes. *HMANA Newsletter*, 3 (2): 28–29.
  1979   Western Great Lakes. *HMANA Newsletter*, 4 (1): 14.
Schneider, E. J.
  1950   A Flight of Broadwinged Hawks at Otter Creek Park. *Kentucky Warbler*,
         25 (4): 67.
Schumacher, G. H.
  1952   Berkshire Hawk Flight in March. *Bulletin Massachusetts Audubon Society*,
         36 (6): 238–239.
Schutsky, R. M.
  1982   Migration of the Osprey Along the Lower Susquehanna River. *Cassinia*,
         59: 37–42.
Scott, F. R.
  1976   Middle Atlantic Coast Region. *American Birds*, 30 (1): 47–51.
  1977   Middle Atlantic Coast Region. *American Birds*, 31 (2): 160–163.
  1978   Middle Atlantic Coast Region. *American Birds*, 32 (2): 189–193.
Scott, F. R., and D. A. Cutler
  1961   Middle Atlantic Coast Region. *Audubon Field Notes*, 15 (1): 19–23.
  1963   Middle Atlantic Coast Region. *Audubon Field Notes*, 17 (1): 18–22.
  1964   Middle Atlantic Coast Region. *Audubon Field Notes*, 18 (1): 20–24.
  1965   Middle Atlantic Coast Region. *Audubon Field Notes*, 19 (1): 21–24.
  1966   Middle Atlantic Coast Region. *Audubon Field Notes*, 20 (1): 22–27.
  1967   Middle Atlantic Coast Region. *Audubon Field Notes*, 21 (1): 15–19.
  1968a Middle Atlantic Coast Region. *Audubon Field Notes*, 22 (1): 18–22.
  1968b Middle Atlantic Coast Region. *Audubon Field Notes*, 22 (4): 510–514.
  1969   Middle Atlantic Coast Region. *Audubon Field Notes*, 23 (1): 28–32.
  1970a Middle Atlantic Coast Region. *Audubon Field Notes*, 24 (1): 26–31.
  1970b Middle Atlantic Coast Region. *Audubon Field Notes*, 24 (4): 585–588.
  1972a Middle Atlantic Coast Region. *American Birds*, 26 (1): 41–45.
  1972b Middle Atlantic Coast Region. *American Birds*, 26 (4): 745–748.
  1973a Middle Atlantic Coast Region. *American Birds*, 27 (1): 36–40.
  1973b Middle Atlantic Coast Region. *American Birds*, 27 (4): 754–757.
  1974   Middle Atlantic Coast Region. *American Birds*, 28 (1): 33–37.
  1975a Middle Atlantic Coast Region. *American Birds*, 29 (1): 34–40.
  1975b Middle Atlantic Coast Region. *American Birds*, 29 (4): 833–837.
Scott, F. R., and J. K. Potter
  1960   Middle Atlantic Coast Region. *Audubon Field Notes*, 14 (1): 20–23.
Seeber, E. L.
  1969   Hawk Flight Observations—April 27, 1969. *Prothonotary*, 35 (8): 116–118.

1970 Hawk Watch Project April 26, 1970. *Prothonotary*, 36 (5): 65–68.

Senner, S. E.
1984a The Model Hawk Law—1934 to 1972. *Hawk Mountain News*, 62: 29–36.
1984b Why Count Hawks? A Hawk Mountain Perspective. *Hawk Mountain News*, 62: 42–45.

Serr, E. M.
1976a Northern Great Plains. *American Birds*, 30 (1): 87–90.
1976b Northern Great Plains. *American Birds*, 30 (4): 855–858.
1977a Northern Great Plains. *American Birds*, 31 (2): 190–194.
1977b Northern Great Plains Region. *American Birds*, 31 (5): 1013–1016.
1978a Northern Great Plains. *American Birds*, 32 (2): 220–223.
1978b Northern Great Plains Region. *American Birds*, 32 (5): 1021–1024.
1979 Northern Great Plains Region. *American Birds*, 33 (2): 188–191.

Serrao, J.
1976 Hawks Over Greenbrook. *Members Bulletin*. Palisades Nature Assn., Alpine, N.J.
1977a The 1977 Hawk Migration. *Members' Bulletin*. Palisades Nature Assn., Alpine, N.J.
1977b *The Birds of Greenbrook Sanctuary*. Palisades Nature Assn., Alpine, N.J.

Servheen, C.
1976 Bald Eagles Soaring into Opaque Cloud. *Auk*. 93 (2): 387.

Sexton, C.
1982 South Central. *HMANA Newsletter*, 7 (2): 19–23.
1983a South Central. *HMANA Newsletter*, 8 (1): 11–14.
1983b South Central. *HMANA Newsletter*, 8 (2): 32–35.
1984a South Central. *HMANA Newsletter*, 9 (1): 24–26.
1984b South Central. *HMANA Newsletter*, 9 (2): 33–34.

Shadowen, H. E.
1968 Broad-winged Hawk Migration. *Kentucky Warbler*, 44 (2): 34.

Sharadin, R.
1972 1971 Hawk Migrations. *News Letter to Members No. 44*. Hawk Mountain Sanctuary Assn., Kempton, Pa.
1973 1972 Hawk Migration. *News Letter to Members No. 45*. Hawk Mountain Sanctuary Assn., Kempton, Pa.
1974 Migration 1973. *News Letter 46*. Hawk Mountain Sanctuary Assn., Kempton, Pa.

Sheldon, W.
1965 Hawk Migration in Michigan and the Straits of Mackinac. *Jack-Pine Warbler*, 43 (2): 79–83.

Shreve, A.
1970 Broadwings Over Kanawha County, September 1969. *Redstart*, 37 (4): 115–116.

Simons, M. M., Jr.
1977 Reverse Migration of Sharp-shinned Hawks on the West Coast of Florida. *Florida Field Naturalist*, 5 (2): 43–44.

Simpson, T. W.
1952a The Coming Season for Hawk Watching in the Carolinas. *Chat*, 16 (3): 66.
1952b A Preliminary Note on the 1952 Hawk Migration Project. *Chat*, 16 (4): 92.
1954 The Status of Migratory Hawks in the Carolinas. *Chat*, 18 (1): 15–21.

Single, E. W.
  1974 Record Numbers of Hawks at the 1973 Hook Mountain Watch. *Linnaean News-Letter,* 27 (9): 1–2.
  1975 Record Accipiter Flight at 1974 Hook Mountain Watch. *Linnaean News-Letter,* 28 (9): 1–3.
  1980 Wind Drift and the Broad-winged Hawk Flight in the New York City Region. *J. Hawk Migration Assn. North America.* 2 (1): 19–23.
Single, E., and D. W. Copeland
  1978 The Continental Summary. *HMANA Newsletter,* 3 (2): 1–3.
Single, E. W., and S. Thomas
  1975 Hook Mountain: Highway for Hawks. *Kingbird,* 25 (3): 124–131.
Skinner, J.
  1981 Sandy Hook Hawk Watch Totals and Peak Flights, Spring 1981. *Peregrine Observer,* 4 (2): 12.
Skutch, A. F.
  1945 The Migration of Swainson's and Broad-winged Hawks through Costa Rica. *Northwest Science,* 19 (4): 80–89.
  1969 Notes on the Possible Migration and Nesting of the Black Vulture in Central America. *Auk,* 86: 726–731.
Slack, R. S., and C. B. Slack
  1980 Osprey Fall Migration at the Ninigret Barrier Beach Conservation Area, Rhode Island. *Raptor Research,* 14 (2): 56–58.
  1981 Fall Migration of Peregrine Falcons Along the Rhode Island Coast. *J. Field Ornithology,* 52 (1): 60–61.
Slud, P.
  1964 *The Birds of Costa Rica. Bulletin 128.* American Museum of Natural History, New York, N.Y. Pp. 1–430.
Small, A.
  1956 Southern Pacific Coast Region. *Audubon Field Notes,* 10 (4): 363–364.
  1974 *The Birds of California.* Winchester Press, New York, N.Y.
Smart, R.
  1969 A Week-End Full of Hawks. *New Hampshire Audubon Quarterly,* 22 (4): 148.
Smith, G. A.
  1973 The 1972 Spring Migration at Derby Hill with Remarks on the Period 1963–1971. *Kingbird,* 23 (1): 13–27.
Smith, G. A., and D. G. Muir
  1978 Derby Hill Spring Hawk Migration Update. *Kingbird,* 28 (1): 5–25.
Smith, J. L.
  1982 The Golden Eagle in Eastern West Virginia. *Redstart,* 49 (3): 94–97.
Smith, M. A.
  1965 Cohoe, Alaska. *Audubon Field Notes,* 19 (1): 66–68.
  1966 Cohoe, Alaska. *Audubon Field Notes,* 20 (1): 80–81.
Smith, N. G.
  1973 Spectacular Buteo Migration Over Panama Canal Zone, October, 1972. *American Birds,* 27 (1): 3–5.
  1980 Hawk and Vulture Migrations in the Neotropics. In *Migrant Birds in the Neotropics: Ecology, Behavior, Distribution, and Conservation.* Smithsonian Institution Press, Washington, D.C. Pp. 51–65.
  1985 Dynamics of the Transisthmian Migration of Raptors between Central and South America. In *Conservation Studies on Raptors.* ICBP Technical Publication No. 5. International Council for Bird Preservation, Cambridge, England. Pp. 271–290.

Smithe, F. B.
1966   *The Birds of Tikal.* Natural History Press, Garden City, N.Y.
Snider, R. P.
1965   Southwest Region. *Audubon Field Notes,* 19 (1): 64–66.
1969   Southwest Region. *Audubon Field Notes,* 23 (1): 88–91.
Snyder, D. B.
1961   Buteo Migration in North Central Indiana. *Indiana Audubon Quarterly,* 39
        (2): 26–27.
Snyder, D. E.
1945   Eagles Along the Hudson. *Feathers,* 7 (11): 79–82.
Snyder, N. F. R., and E. V. Johnson
1985   Status Reports: Counting Condors. *Eyas,* 8 (1): 4–7.
Snyder, N. F. R., et al.
1973   Organochlorines, Heavy Metals, and the Biology of North American
        Accipiters. *BioScience,* 23 (5): 300–305.
Soucy, L. J., Jr.
1976   Raptor Report—Kittatinny Mountains. *North American Bird Bander,* 1 (3):
        108–113.
1983   First Gyrfalcon Banded in New Jersey. *Records of New Jersey Birds,* 9 (1):
        2–3.
Soulen, T. K.
1970   Western Great Lakes Region. *Audubon Field Notes,* 24 (4): 604–608.
1971   Western Great Lakes Region. *American Birds,* 25 (1): 61–64.
1972   Western Great Lakes Region. *American Birds,* 26 (1): 66–70.
Speirs, J. M.
1939   Fluctuations in Numbers of Birds in the Toronto Region. *Auk,* 56: 411–
        419.
Spiker, C. J.
1933   A Flight of Broad-winged Hawks. *Wilson Bulletin,* 45: 79.
Spofford, S. H.
1978   Southwest. *HMANA Newsletter,* 3 (2): 20–22.
1979   Southwest. *HMANA Newsletter,* 4 (2): 28–29.
1980   Southwest. *HMANA Newsletter,* 5 (1): 10–11.
Spofford, W. R.
1949   Fall Flight of Hawks at Fall Creek Falls State Park. *Migrant,* 20 (1): 16.
1969   Hawk Mountain Counts as Population Indices in Northeastern America.
        In *Peregrine Falcon Populations: Their Biology and Decline.* University of
        Wisconsin Press, Madison, Wis.
Sprunt, A., Jr.
1954   *Florida Bird Life.* Coward-McCann, New York, N.Y.
Sprunt, A., Jr., and E. B. Chamberlain
1970   *South Carolina Bird Life.* Revised Edition, with a Supplement. University
        of South Carolina Press, Columbia, S.C.
Squires, W. A.
1952   *The Birds of New Brunswick. Monograph Series No. 4.* New Brunswick
        Museum, Saint John, N.B.
1976   *The Birds of New Brunswick.* 2nd Edition. *Monograph Series No. 7.* New
        Brunswick Museum, Saint John, N.B.
Stallcup, R., D. DeSante, and R. Greenberg
1975   Middle Pacific Coast Region. *American Birds,* 29 (1): 112–119.
Stamm, A. L.
1957   Broad-winged Hawks Migrating Over Jefferson County Forest. *Kentucky
        Warbler,* 33 (2): 42–43.

1961   Broad-winged Hawks Over Louisville. *Kentucky Warbler,* 37 (1): 20.
1965   Broad-winged Hawk Migration. *Kentucky Warbler,* 42 (3): 51–52.
1969   Broad-winged Hawk Flights. *Kentucky Warbler,* 45 (1): 23.
1972   Migrating Hawks Over Southeastern Kentucky. *Kentucky Warbler,* 48 (2): 25–26.

Stearns, E. I.
1948a   A Study of the Migration of the Broad-winged Hawk through New Jersey. *Urner Field Observer,* 3 (3 & 4): 1–4.
1948b   Blawking: The Study of Hawks in Flight from a Blimp. *Urner Field Observer,* 3 (5 & 6): 2–9.
1949   The Study of Hawks in Flight from a Blimp. *Wilson Bulletin,* 61: 110.

Stedman, B. H., and S. J. Stedman
1981   Notes on the Raptor Migration at Chilhowee Mountain. *Migrant,* 52 (2): 38–40.

Steenhof, K.
1983   Activity Patterns of Bald Eagles Wintering in South Dakota. *Raptor Research,* 17 (2): 57–62.

Steenhof, K., M. N. Kochert, and M. Q. Moritsch
1984   Dispersal and Migration of Southwestern Idaho Raptors. *J. Field Ornithology,* 55 (3): 357–368.

Stephens, H. A.
1966   Observations on Eagles in Kansas. *Kansas Ornithological Society Bulletin,* 17 (3): 23–25.

Stevenson, H. M.
1955   Florida Region. *Audubon Field Notes,* 9 (1): 19–22.
1956a   Florida Region. *Audubon Field Notes,* 10 (1): 18–22.
1956b   Florida Region. *Audubon Field Notes,* 10 (4): 325–329.
1958   Florida Region. *Audubon Field Notes,* 12 (1): 21–26.
1960a   Florida Region. *Audubon Field Notes,* 14 (1) 25–29.
1960b   Florida Region. *Audubon Field Notes,* 14 (4): 379–383.
1962   Florida Region. *Audubon Field Notes,* 16 (1): 21–25.
1963   Florida Region. *Audubon Field Notes,* 17 (4): 397–399.
1967   Florida Region. *Audubon Field Notes,* 21 (1): 22–25.
1973   Florida Region. *American Birds,* 27 (1): 45–49.

Stewart, P. A.
1977   Migratory Movements and Mortality Rates of Turkey Vultures. *Bird-Banding,* 48 (2): 122–124.

Stewart, R. B.
1975   Hawk Migration—Prince Edward Point. *Blue Bill,* 22 (2): 26, 33.

Stewart, R. E., and C. S. Robbins
1958   Birds of Maryland and the District of Columbia. *North America Fauna* No. 62: 1–401.

Stirrett, G. M.
1960   *The Spring Birds of Point Pelee National Park Ontario.* Ministry of Northern Affairs and National Resources, Ottawa.

Stoddard, H. L., Sr.
1978   *Birds of Grady County, Georgia. Bulletin 21.* Tall Timbers Research Station, Tallahassee, Fla.

Stokes, S.
1980   A Buteo Sighting from Western Kentucky. *Kentucky Warbler,* 56 (4): 88.

Stone, W.
1887   A Migration of Hawks at Germantown, Pennsylvania. *Auk,* 4: 161.

1894   *The Birds of Eastern Pennsylvania and New Jersey.* Delaware Valley Ornithological Club, Philadelphia, Pa.

1909   *The Birds of New Jersey. Annual Report New Jersey State Museum* 1908: 11–347.

1922   Hawk Flights at Cape May Point, N.J. *Auk,* 39: 567–568.

1937   *Bird Studies at Old Cape May.* Volume 1. Delaware Valley Ornithological Club, Philadelphia, Pa.

Street, P. B.
1954   Birds of the Pocono Mountains, Pennsylvania. *Cassinia,* 41: 3–76.
1975   Birds of the Pocono Mountains, 1955–1975. *Cassinia,* 55: 3–16.

Stull, J. H., and J. G. Stull
1966   Field Notes Spring Migration 1966 Area No. 1. *Sandpiper,* 8 (4): 62–64.
1967   Field Observations. *Sandpiper,* 9 (4): 50.
1968   Field Observations. *Sandpiper,* 10 (4): 52–53.

Sundquist, K.
1973   Hawk Watch Duluth 1972 Summary. *Hawk Ridge Nature Reserve Newsletter,* 1: 1–9.

Sutton, C. C., Jr.
1979a   Sight Records of European Raptors at Cape May Point. *Cassinia,* 58: 19.
1979b   Kites at Cape May. *Cassinia,* 58: 20–21.
1980   The 1980 Spring Hawk Watch at Cape May. *Peregrine Observer,* 3 (2): 10–11.
1981   1981 Spring Hawk Migration at Cape May. *Peregrine Observer,* 4 (2): 14.
1984   Coastal Plain. *HMANA Newsletter,* 9 (2): 19, 20–22.

Sutton, C. C., and P. T. Sutton
1984   The Spring Hawk Migration at Cape May, New Jersey. *Cassinia,* 60: 5–18.

Sutton, G. M.
1928a   *An Introduction to the Birds of Pennsylvania.* J. Horace McFarland Company, Harrisburg, Pa.
1928b   Notes on a Collection of Hawks from Schuylkill County, Pennsylvania. *Wilson Bulletin,* 40: 84–95, 193–194.
1928c   Abundance of the Golden Eagle in Pennsylvania in 1927–28. *Auk,* 45: 375.
1929   How Can the Bird-Lover Help to Save the Hawks and Owls? *Auk,* 46: 190–195.
1931   The Status of the Goshawk in Pennsylvania. *Wilson Bulletin,* 43 (2): 108–113.
1967   *Oklahoma Birds: Their Ecology and Distribution, with Comments on the Avifauna of the Southern Great Plains.* University of Oklahoma Press, Norman, Okla.

Sutton, G. M., and O. S. Pettingill, Jr.
1942   Birds of the Gomez Farias Region, Southwestern Tamaulipas. *Auk,* 59: 1–34.

Sutton, W. D.
1956   Hawk Cliff History. *Cardinal,* 22: 7–12.

Swenk, M. H.
1939   A Slaughter of Swainson's Hawks in Merrick County. *Nebraska Bird Review,* 7 (1): 10–11.

Swiderski, J.
1975   Swallow-tailed Kite at Brasstown Bald, Ga. *Oriole,* 40 (3): 34–35.

Swindell, M.
1959   Broad-winged Hawks Caught in Crosswinds. *Alabama Birdlife,* 7 (3–4): 24.

Tabb, E. C.
1973   A Study of Wintering Broad-winged Hawks in Southwestern Florida, 1968–1973. *EBBA News*, 36 (Supple.): 11–29.
1979   Winter Recoveries in Guatemala and Southern Mexico of Broad-winged Hawks Banded in South Florida. *North American Bird Bander*, 4 (2): 60.

Tanner, J. T.
1936   Witmer Stone Wildlife Sanctuary 1936. Unpublished report.
1950   Fall Flights of Broad-winged Hawks in the Southern Appalachians. *Migrant*, 21 (4): 69–70.

Taylor, C.
1984   1984 Hawk Banding Preliminary Report. *Little Gull*, 10 (6): 36–38.

Taylor, J. W.
1971   President's Message. *News Letter to Members No. 43*. Hawk Mountain Sanctuary Assn., Kempton, Pa.

Temple, S.
1962   Red-tailed Hawks Migrate Along Lake Shore. *Cleveland Bird Calendar*, 58 (4): 40.

Tessen, D. D.
1976   Western Great Lakes Region. *American Birds*, 30 (1): 71–77.
1981   Western Great Lakes Region. *American Birds*, 35 (2): 184–187.

Teulings, R. P.
1971   Southern Atlantic Coast Region. *American Birds*, 25 (1): 40–44.
1972   Southern Atlantic Coast Region. *American Birds*, 26 (1): 45–50.
1974a   Southern Atlantic Coast Region. *American Birds*, 28 (1): 37–40.
1974b   Southern Atlantic Coast Region. *American Birds*, 28 (4): 788–790.
1975   Southern Atlantic Coast Region. *American Birds*, 29 (1): 40–43.
1976a   Southern Atlantic Coast Region. *American Birds*, 30 (1): 51–54.
1976b   Southern Atlantic Coast Region. *American Birds*, 30 (4): 826–828.
1977   Southern Atlantic Coast Region. *American Birds*, 31 (2): 163–166.

Thielen, B.
1967   Hawks and the Hawk People. *New Hampshire Audubon Quarterly*, 20 (1): 3–9.

Thiollay, J. M.
1980   Spring Hawk Migration in Eastern Mexico. *Raptor Research*, 14 (1): 13–20.

Thomas, B.
1943   The Swallow-tailed Kite at Augusta. *Oriole*, 8 (3–4): 22.

Thomas, S.
1971a   Hook Mountain, New York, Hawk Watch. *California Condor*, 6 (3): 13.
1971b   Hook Mt. Hawk Watch—Autumn, 1971. *California Condor*, 6 (5): 10–12.
1973   Annual Report Hook Mountain Hawk Watch. Mimeographed report.

Thomson, T.
1983   *Birding in Ohio*. Indiana University Press, Bloomington, Ind.

Tilly, F.
1972a   The 1971 Raccoon Ridge Autumn Hawkwatch. *New Jersey Nature News*, 27 (1): 22–28.
1972b   Raccoon Ridge Daily Hawk Counts—Autumn 1971. *Science Notes No. 7*. New Jersey State Museum, Trenton, N. J.
1972c   Raccoon Ridge Hawk Count. *New Jersey Nature News*, 27 (3): 113–114.
1973   The 1972 Autumn Hawk Count at Raccoon Ridge, Warren County, New Jersey. *Science Notes No. 13*. New Jersey State Museum, Trenton, N. J.

1979   Raccoon Ridge Autumn 1978 Hawkwatch. *New Jersey Audubon,* 5 (1): 1–2.
1980   The Hawk Migration. *Montana Magazine,* (Sept.–Oct.): 66–67.
1981a   Northwest. *HMANA Newsletter,* 6 (1): 21–22.
1981b   Northwest. *HMANA Newsletter,* 6 (2): 27.
1983a   Northwest. *HMANA Newsletter,* 8 (1): 20–21.
1983b   Northwest. *HMANA Newsletter,* 8 (2): 35.
1984   Northwest. *HMANA Newsletter,* 9 (1): 26.
1985   Eastern Mexico. *HMANA Newsletter,* 10 (1): 16–17.

Tinsley, J. C., Jr.
1971   Hawks Over Peters Mountain. *Raven,* 42 (2): 23–25.

Titus, K., and J. A. Mosher
1982   The Influence of Seasonality and Selected Weather Variables on Autumn
        Migration of Three Species of Hawks Through the Central Ap-
        palachians. *Wilson Bulletin,* 94 (2): 176–184.

Todd, H.
1970   Hawks in Center City, Philadelphia. *Cassinia,* 52: 34.

Todd, W. E. C.
1904   The Birds of Erie and Presque Isle, Erie County, Pennsylvania. *Annals
        Carnegie Museum,* 2: 481–613.
1940   *Birds of Western Pennsylvania.* University of Pittsburgh Press, Pittsburgh,
        Pa.
1963   *Birds of the Labrador Peninsula and Adjacent Areas.* University of Toronto
        Press, Toronto, Ontario, Canada.

Tomkins, I. R.
1962   More Data on the Mississippi Kite in the Lower Coastal Plain of Georgia.
        Oriole, 27 (2): 15–16.

Townsend, W.
1892   Migration of Hawks. *Forest and Stream,* 39: 311.

Tracy, J. C.
1943   Hawk Slaughter. *National Parks Magazine,* 17/74: 15–18.

Trautman, M. B., and M. A. Trautman
1968   Annotated List of the Birds of Ohio. *Ohio J. Science,* 68 (5): 257–332.

Trichka, C. J.
1981   1980 Hawk Watch. *Connecticut Warbler,* 1 (1): 4–6.
1982   Hawk Watch 1981. *Connecticut Warbler,* 2 (1): 12.
1984   Larsen Hawk Watch. *Connecticut Warbler,* 4 (1): 13.

Trowbridge, C. C.
1895   Hawk Flights in Connecticut. *Auk,* 12: 259–270.

Tufts, R. W.
1962   *The Birds of Nova Scotia.* Nova Scotia Museum, Halifax.
1973   *The Birds of Nova Scotia.* 2nd Ed. Nova Scotia Museum, Halifax.

Turnbull, W. P.
1869   *The Birds of East Pennsylvania and New Jersey.* Henry Grambo & Co.,
        Philadelphia, Pa.

Turner, H.
1934a   Notes on Some Birds of Prey Observed in Webster County in 1933.
        Nebraska Bird Review, 2 (1): 7.
1934b   The 1934 Fall Hawk Migration in Adams County. *Nebraska Bird Review,* 2
        (4): 120.
1944   Notes on the Scissor-tailed Flycatcher and Hawk Migration in Adams
        County. *Nebraska Bird Review,* 12 (2): 42.

1947   Occurrence Records of Hawks and Owls in the Bladen (Webster County) Area. *Nebraska Bird Review*, 15 (2): 44–45.

Turner, L. J.
1981   1980 Autumn Hawk Count. *Migrant*, 52: 87–90.

Tyrrell, W. B.
1934   Bird Notes from Whitefish Point, Michigan. *Auk*, 51: 21–26.
1935   Bird Notes from Honga, Maryland. *Bulletin Natural History Society of Maryland*, 5 (7): 36–40.

Ulrich, E. C.
1942   Hawk Migration in Western New York in 1942. *Prothonotary*, 8 (5): 4–5.

Van Eseltine, W. P.
1952   A Hawk Migration Project for the Carolinas. *Chat*, 16 (1): 6.

Varland, D.
1982   Raptor Banding and Migration in the Loess Bluffs of Southwest Iowa. *Iowa Bird Life*, 52 (2): 43–47.

Vickery, P. D.
1977   Northeastern Maritime Region. *American Birds*, 31 (5): 972–977.
1978a   Northeastern Maritime Region. *American Birds*, 32 (2): 174–180.
1978b   Northeastern Maritime Region. *American Birds*, 32 (5): 977–981.
1979a   Northern Maritime Region. *American Birds*, 33 (2): 154–158.
1979b   Northeastern Maritime Region. *American Birds*, 33 (5): 751–753.
1980   Northeastern Maritime Region. *American Birds*, 34 (5): 754–757.
1982   Northeastern Maritime Region. *American Birds*, 36 (2): 152–155.
1983   Northeastern Maritime Region. *American Birds*, 37 (2): 155–158.

von Lengerke, J.
1908   Migration of Hawks. *Auk*, 25: 315–316.

Wagner, R.
1973   Report of April Hawk Count. *Prothonotary*, 39 (5): 57–58.

Wallace, G. J.
1960   The 1958 Fall Migration at Whitefish Point. *Jack-Pine Warbler*, 38 (4): 140–144.

Wander, W., and S. A. Brady
1977   Hawk-Watching Hot Spot. *New Jersey Outdoors*, 4 (5): 18–19.

Ward, C. J.
1958   Hawk Flights at Jones Beach, Long Island. *Kingbird*, 8 (2): 42–43.
1960a   Hawk Flights at Jones Beach. *Linnaean News-Letter*, 14 (No. 2).
1960b   Hawk Migrations at Jones Beach. *Kingbird*, 10 (4): 157–159.
1963   Fall Hawk Migrations of Region 10. *Kingbird*, 13 (1): 22–23.

Ward, D.
1981   Expanded Hawk Migration Project—1980. *Peregrine Observer*, 4 (1): 8–9.

Ward, F. P., and R. B. Berry
1972   Autumn Migrations of Peregrine Falcons on Assateague Island, 1970–71. *J. Wildlife Management*, 36 (2): 484–492.

Warren B. H.
1890   *Report on the Birds of Pennsylvania*. Second Edition, Revised and Augmented. E. K. Meyers, Harrisburg, Pa.

Wasson, I. B. and M. Shawvan
1953   Hawk Migration Over Thatcher Woods. *Audubon Bulletin*, 88: 1–2.

Watson, F. G.
1955a   South Texas Region. *Audubon Field Notes*, 9 (1): 38–41.
1955b   South Texas Region. *Audubon Field Notes*, 9 (4): 341–344.

Weaver, A. S.
  1977   Spring Hawks (Keweenaw Peninsula, Michigan). *Birding*, 9 (2): 79–80.
Webster, F. S., Jr.
  1956a   South Texas Region. *Audubon Field Notes*, 10 (1): 37–40.
  1956b   South Texas Region. *Audubon Field Notes*, 10 (4): 345–350.
  1958a   South Texas Region. *Audubon Field Notes*, 12 (1): 41–44.
  1958b   South Texas Region. *Audubon Field Notes*, 12 (4): 365–370.
  1959a   South Texas Region. *Audubon Field Notes*, 13 (1): 45–49.
  1959b   South Texas Region. *Audubon Field Notes*, 13 (4): 383–388.
  1960a   South Texas Region. *Audubon Field Notes*, 14 (1): 52–56.
  1960b   South Texas Region. *Audubon Field Notes*, 14 (4): 401–407.
  1961a   South Texas Region. *Audubon Field Notes*, 15 (1): 56–59.
  1961b   South Texas Region. *Audubon Field Notes*, 15 (4): 424–427.
  1962a   South Texas Region. *Audubon Field Notes*, 16 (1): 52–58.
  1962b   South Texas Region. *Audubon Field Notes*, 16 (4): 428–433.
  1963   South Texas Region. *Audubon Field Notes*, 17 (1): 46–50.
  1964a   South Texas Region. *Audubon Field Notes*, 18 (1): 52–57.
  1964b   South Texas Region. *Audubon Field Notes*, 18 (4): 466–472.
  1965a   South Texas Region. *Audubon Field Notes*, 19 (1): 56–60.
  1965b   South Texas Region. *Audubon Field Notes*, 19 (4): 490–497.
  1966a   South Texas Region. *Audubon Field Notes*, 20 (1): 66–72.
  1966b   South Texas Region. *Audubon Field Notes*, 20 (4): 525–532.
  1967a   South Texas Region. *Audubon Field Notes*, 21 (1): 54–59.
  1967b   South Texas Region. *Audubon Field Notes*, 21 (4): 520–524.
  1968a   South Texas Region. *Audubon Field Notes*, 22 (1): 60–69.
  1968b   South Texas Region. *Audubon Field Notes*, 22 (4): 550–557.
  1969a   South Texas Region. *Audubon Field Notes*, 23 (1): 77–81.
  1969b   South Texas Region. *Audubon Field Notes*, 23 (4): 604–607.
  1970a   South Texas Region. *Audubon Field Notes*, 24 (1): 66–69.
  1970b   South Texas Region. *Audubon Field Notes*, 24 (4): 622–624.
  1971   South Texas Region. *American Birds*, 25 (1): 78–80.
  1972a   South Texas Region. *American Birds*, 26 (1): 84–88.
  1972b   South Texas Region. *American Birds*, 26 (4): 780–783.
  1973a   South Texas Region. *American Birds*, 27 (1): 82–85.
  1973b   South Texas Region. *American Birds*, 27 (4): 793–795.
  1974a   South Texas Region. *American Birds*, 28 (1): 76–78.
  1974b   South Texas Region. *American Birds*, 28 (4): 822–825.
  1975a   South Texas Region. *American Birds*, 29 (1): 82–85.
  1975b   South Texas Region. *American Birds*, 29 (4): 875–878.
  1976a   South Texas Region. *American Birds*, 30 (1): 95–97.
  1976b   South Texas Region. *American Birds*, 30 (4): 863–865.
  1977a   South Texas Region. *American Birds*, 31 (2): 197–199.
  1977b   South Texas Region. *American Birds*, 31 (5): 1020–1022.
  1978a   South Texas Region. *American Birds*, 32 (2): 227–230.
  1978b   South Texas Region. *American Birds*, 32 (5): 1028–1031.
  1979a   South Texas Region. *American Birds*, 33 (2): 193–195.
  1979b   South Texas Region. *American Birds*, 33 (5): 787–789.
  1980a   South Texas Region. *American Birds*, 34 (2): 179–180.
  1980b   South Texas Region. *American Birds*, 34 (5): 794–796.
  1981   South Texas Region. *American Birds*, 35 (2): 201–204.
  1983a   South Texas Region. *American Birds*, 37 (2): 199–201.
  1983b   South Texas Region. *American Birds*, 37 (5): 889–890.

Wedgwood, J. A.
1982a   Prairie Provinces Region. *American Birds*, 36 (2): 188–190.
1982b   Prairie Provinces Region. *American Birds*, 36 (5): 864–866.
Weir, R. D.
1974   Prince Edward Point—The Point Pelee of Lake Ontario. *Blue Bill*, 21 (2): 40–42.
1983a   Ontario Region. *American Birds*, 37 (2): 173–177.
1983b   Ontario Region. *American Birds*, 37 (5): 863–867.
Welch, W. A.
1975   Inflight Hawk Migration Study, *J. Hawk Migration Assn. North America*, 1 (1): 14–22.
1980   Scanning for Hawks. *HMANA Newsletter*, 5 (2): 11–12.
1981   Calculating Populations. *HMANA Newsletter*, 6 (2): 4–5.
Wellman, C.
1957   September Hawk Flight. *New Hampshire Bird News*, 10 (3): 77.
West, E. M.
1974   Spring Migration of Broad-winged Hawks in Cumberland County. *Migrant*, 45: 19.
West, F. H.
1947   Hawk Day in the Helderbergs. *Feathers*, 9 (11): 93.
Weston, F. M.
1965   *A Survey of the Birdlife of Northwestern Florida. Bulletin 5*. Tall Timbers Research Station, Tallahassee, Fla.
Wetmore, A.
1965   *The Birds of the Republic of Panama*. Part 1. Smithsonian Miscellaneous Collections. Vol. 150. Smithsonian Institution, Washington, D. C.
Wetzel, F. W.
1969   The Hawk Season. *News Letter to Members No. 41*. Hawk Mountain Sanctuary Assn., Kempton, Pa.
1970   1969 Hawk Migration. *News Letter to Members No. 42*. Hawk Mountain Sanctuary Assn., Kempton, Pa.
1971   1970 Hawk Migration. *News Letter to Members No. 43*. Hawk Mountain Sanctuary Assn., Kempton, Pa.
Wharton, C. H.
1941   A Swallow-tailed Kite from the Atlanta Area. *Oriole*, 6 (4): 50.
Whelan, D. B.
1934   A Flight of Buzzard Hawks. *Nebraska Bird Review*, 2 (4): 120.
Wilbur, S. R.
1978a   Supplemental Feeding of California Condors. In *Endangered Birds: Management Techniques for Preserving Threatened Species*. University of Wisconsin Press, Madison, Wis. Pp. 135–140.
1978b   The California Condor, 1966–76: A Look at Its Past and Future. *North American Fauna*, 72: 1–136.
1983   The Status of Vultures in the Western Hemisphere. In *Vulture Biology and Management*. University of California Press, Berkeley, Calif. Pp. 113–123.
Wilbur, S. R., and J. A. Jackson
1983   *Vulture Biology and Management*. University of California Press, Berkeley, Calif.
Wilds, C.
1983   *Finding Birds in the National Capital Area*. Smithsonian Institution Press, Washington, D.C.

Wilhelm, E.
1951   Hawks at Columbia. *Bluebird*, 18 (10): 2.
Will, T.
1974   *Vermont Hawk Migration Study Fall, 1974.* Vermont Institute of Natural Science, Woodstock, Vt.
1975   *Vermont Hawk Migration Study Spring 1975.* Vermont Institute of Natural Science, Woodstock, Vt.
1976   *Vermont Hawk Migration Fall, 1975.* Vermont Institute of Natural Science, Woodstock, Vt.
1980   The Vermont Hawk Migration Study: A Low Density Hawk Watch. *J. Hawk Migration Assn. North America*, 2 (1): 10–18.
Williams, E. A.
1962   Sight Record of White-tailed Kite in Georgia. *Oriole*, 27 (4): 52–53.
Williams, F.
1968   Southern Great Plains Region. *Audubon Field Notes*, 22 (1): 57–60.
1970   Southern Great Plains Region. *Audubon Field Notes*, 24 (1): 63–66.
1971   Southern Great Plains Region. *American Birds*, 25 (1): 74–78.
1974   Southern Great Plains Region. *American Birds*, 28 (1): 70–75.
1980   Southern Great Plains Region. *American Birds*, 34 (2): 176–179.
1981   Southern Great Plains. *American Birds*, 35 (2): 198–201.
1982   Southern Great Plains. *American Birds*, 36 (2): 192–194.
1983   Southern Great Plains. *American Birds*, 37 (2): 196–199.
1984   Southern Great Plains Region. *American Birds*, 38 (2): 218–221.
Williams, M. D.
1980   Recoveries of Some Raptors Banded in Tennessee. *Migrant*, 51 (2): 26
Williams, N. J.
1941   Migration of Swainson's Hawks in Western Iowa. *Iowa Bird Life*, 11 (2): 35.
Williams, W.
1983   Kiptopeke: A Place You Can Count On. *HMANA Newsletter*, 8 (2): 1, 4–5.
Wilshusen, J. P.
1983   *Geology of the Appalachian Trail in Pennsylvania. General Geology Report 74.* Pennsylvania Geological Survey, Harrisburg, Pa.
Wilson, G.
1925   Some Kentucky Bird Notes. *Wilson Bulletin*, 37 (1): 44.
Wingate, D. B.
1973   *A Checklist and Guide to the Birds of Bermuda.* Island Press, Ltd., Bermuda.
Winter, J. and S. A. Laymon
1979   Middle Pacific Coast Region. *American Birds*, 33 (2): 209–212.
Wolf, D. J.
1984   Early Records of Spring Hawk Migrations at Braddock Bay. *Goshawk*, 40 (4): 29–36.
Wolfarth, F. P.
1952   Hawk Flights in the Watchungs and Vicinity. *Urner Field Observer*, 5 (1): 2–4.
1975   Raccoon Ridge Fall Hawk Count. *New Jersey Audubon*, 1 (1): 10.
1980   Great Hawk Flight of March 2, 1972. *Urner Field Observer*, 17 (1): 17–18.
Wood, D. S., and G. D. Schnell
1984   *Distributions of Oklahoma Birds.* University of Oklahoma Press, Norman, Okla.
Wood, M.
1979   *Birds of Pennsylvania.* Pennsylvania State University, University Park, Pa.

Wood, N. A.
  1933   The Birds of Keweenaw Point, Michigan. *Papers Mich. Academy Science, Arts, & Letters*, 17: 713–733.
  1951   *The Birds of Michigan. Misc. Publications No. 75. Museum of Zoology, University of Michigan, Ann Arbor, Mich.*
Woodford, J.
  1962   Ontario-Western New York Region. *Audubon Field Notes*, 16 (4): 404–408.
  1963   Ontario-Western New York Region. *Audubon Field Notes*, 17 (1): 28–31.
Woodford, J., and D. E. Burton
  1961a   Ontario-Western New York Region. *Audubon Field Notes*, 15 (1): 35–39.
  1961b   Ontario-Western New York Region. *Audubon Field Notes*, 15 (4): 405–409.
Woodford, J. and J. Lunn
  1962   Ontario-Western New York Region. *Audubon Field Notes*, 16 (1): 25–31.
Worth, C. B.
  1936   Summary and Analysis of Some Records of Banded Ospreys. *Bird-Banding*, 7: 156–160.
Wray, D. L., and H. T. Davis
  1959   *Birds of North Carolina.* Revised Edition. Bynum Printing Co., Raleigh, N. C.
Wylie, B.
  1976   Migrating Osprey 'Packing a Lunch'. *Redstart*, 43 (3): 116.
Youngworth, W.
  1935   Hawks in Iowa During the Fall of 1934. *Iowa Bird Life*, 5 (1): 4.
  1959   A Broad-winged Hawk Flight. *Iowa Bird Life*, 29 (4): 105.
  1960   Swainson's Hawk in the General Sioux City Area. *Iowa Bird Life*, 30 (1): 12–13.
  1963   The Pigeon Hawk in Western Iowa. *Iowa Bird Life*, 33 (3): 72–73.
Zoch, J.
  1983   Fifth Annual Fall Hawk Watch. *Manitoba Naturalists Society Bulletin*, 7: 11.

INDEX

DONALD S. HEINTZELMAN has been Associate Curator of Natural Science at the William Penn Memorial Museum and was for some years Curator of Ornithology at the New Jersey State Museum. He is a wildlife consultant, lecturer, and writer who has published many books, including *A Guide to Eastern Hawk Watching, Autumn Hawk Flights, The Hawks and Owls of North America,* and *A Guide to Hawk Watching in North America.*